Functional Molecules from Natural Sources

Functional Molecules from Natural Sources

Edited by

Stephen K. Wrigley
Hypha Discovery Ltd., Brunel Science Park, Uxbridge, UK

Robert Thomas
University of Sussex, Brighton, UK

Colin T. Bedford
Department of Chemistry, University College London, London, UK

Neville Nicholson
Independent Scientist, UK

RSC Publishing

The proceedings from Functional Molecules from Natural Sources held at Magdalen College, Oxford, UK on 6-8 July 2009.

Special Publication No. 320

ISBN: 978-1-84755-259-4

A catalogue record for this book is available from the British Library

Published by The Royal Society of Chemistry,
Thomas Graham House, Science Park, Milton Road,
Cambridge CB4 0WF, UK

Registered Charity Number 207890

For further information see our web site at www.rsc.org

Printed and bound in Great Britain by CPI Antony Rowe, Chippenham and Eastbourne

Preface

This volume is divided into five sections and eighteen chapters and outlines the proceedings of *Functional Molecules from Natural Sources,* a conference held at the University of Oxford from 6-8 July 2009 and organised by the Royal Society of Chemistry's Biotechnology Group. Individual chapters are transcripts, for the most part, of the lectures delivered by leading authorities from international industrial, academic and government research laboratories and describe recent studies of the chemistry and biosynthesis of natural products with potentially useful biological activity. A few of the chapters were derived from some of the twenty five posters which were presented by young researchers from both industry and academia. The meeting was held in the exceptionally attractive surroundings of Magdalen College, and the College's fine facilities together with the high quality of lectures led to stimulating discussions between the twenty one lecturers and many of the other fifty two participants, who came from twenty different countries.

This is the third book describing the proceedings of a conference organised by the RSC Biotechnology Group[1] and follows those describing the proceedings of conferences held at the University of Sussex[2] in 1996 and at the University of St. Andrews[3] in 1999. A common theme of all three conferences was the search for new leads of value to the pharmaceutical and agrochemical sectors and other major industrial areas such as nutraceuticals and cosmeceuticals. Strategic approaches to this goal are based on systematic chemical development and in depth biosynthetic studies together with the increasing exploitation of genetic engineering techniques.

The evolution of modern synthetic and biosynthetic strategems dates back at least as far as the classic one pot synthesis by Robert Robinson in 1917 of the bicyclic alkaloid tropinone from succinic dialdehyde (which he suggested may be biosynthesised in plants from ornithine), methylamine and calcium acetonedicarboxylate.[4]

The diverse structures of antibiotics and other bioactive products have long motivated chemical syntheses by outstanding chemists, none more prolific than Robert Woodward, who synthesised many highly complex molecules including tetracycline, magnamycin, strychnine, cephalosporin C and chrismasterol (so-called because it was crystallised on Christmas Day) to name but a few.[5]

The availability of radioactive precursors containing ^{14}C in the late 1940s facilitated experimental studies of biosynthetic pathways, which in 1953 enabled validation of Birch's seminal polyketide hypothesis through investigation of the predicted labelling pattern of the ^{14}C-acetate derived fungal metabolite 6-methylsalicylic acid.[6] Prior to the development of modern NMR and mass spectrometric analytical procedures, the identification of radiolabelled atoms was achieved through the systematic and time-consuming chemical degradation of isotopically labelled metabolites.

One of the earliest known bioactive microbial constituents whose biosynthetic origin was investigated prior to the availability of modern spectroscopic techniques, was the fungal antibacterial metabolite alternariol methyl ether, which as anticipated proved to be a heptaketide.[7] Other labelling studies during this decade included the elegant chemical degradations of 2-^{14}C-mevalonate-labelled squalene by Cornforth and Popjak and co-workers, which clarified the nature of many enzymes involved in the formation of cholesterol.[8]

The subsequent rapid development of genetic engineering methodology in the 1990s and the integration of this technology with existing chemical procedures, has greatly enhanced our understanding of biosynthetic pathways and provided many options for their fruitful manipulation. The utility of this interdisciplinary approach has been highlighted by the landmark studies by David Hopwood's group on the mode of polyketide formation in *Streptomyces coelicor*[9] and is currently an essential adjunct to biosynthetic studies.

As in the published proceedings of preceding conferences,[2,3] the opening section entitled 'Modern and Emerging Perspectives on Natural Product Utilisation' includes a review by David Newman of investigations of natural product drug discovery at the US National Cancer Institute. A particularly noteworthy observation is that of the 162 small molecule agents approved for antitumour treatment since the late 1930s, almost half are either natural products or slightly modified natural products. The other chapter in this section is by Neville Nicholson and concerns the transmembrane receptor and pain target MrgX1 and illustrates limitations of medicinal research based on lower animals. A novel method for the selection of receptors for modulation in neurological disease derives from this research.

There are four chapters in the second section entitled 'Improved Strategies for Natural Product Exploitation' ranging from the description of antibiotics with novel modes of action, notably platensimycin and platencin by Sheo Singh, and bioactive metabolites of marine micro-organisms by Andrew Mearns Spragg, to a unique platform for the discovery of anti-malarial compounds by Michael Kemmler, followed by a review of the fast isolation of natural products by counter current chromatography by Ian Garrard.

Section 3 entitled 'Functional Molecules and their Enhancement' comprises six chapters which describe individual studies of the chemical optimisation of antibiotics and lead structures ranging from rapamycin by Guy Carter, pleuromutilins by Richard Jarvest and oxazolomycins and equisetins by Mark Moloney, to the rocket salad isothiocyanate constituent erucin by Antonietta Melchini, the discovery of new bioactive imino sugar components of medicinal plants by Robert Nash and the effects of lignans on enhancing the cytotoxicity of doxorubicin to multi-drug resistant cancer cells by Jiří Slanina.

The fourth section, 'Biosynthesis and Genetics' is primarily devoted to biosynthetic pathways and covers *Streptomyces coelicolor* antibiotic biosynthesis by Greg Challis, anticancer drugs and combinatorial biosynthesis by Jürgen Rohr, anthrax siderophore biosynthesis by Daniel Oves-Costales, quaternary structure of modular polyketide synthases by Alison Hill and Jim Staunton and the biosynthesis of polyenes in *Auxarthron umbrinum* by Benjamin Clark.

The fifth and final section 'Summaries of Other Lectures' by Steve Wrigley provides synopses of those lectures for which individual manuscripts were not available. These were: *Biosynthetic Engineering of Nonribosomal Lipopeptides* by Jason Micklefield (Univ. Manchester); *Development of Lantibiotics for Treatment of Nosocomial Infections* by Mike Dawson (Novacta); *Acetogenic Anthraquinones and Alkaloids – Online Structural Elucidation, Biosynthesis, Bioactivities and Total Synthesis* by Gerhard Bringmann (Univ. Würzburg); *Chemical Diversity by Other Means: the Biosynthesis of Polyketide Drugs* by Peter Leadlay (Univ. Cambridge); *New Paradigms in Fungal Polyketide Biosynthesis* by Tom Simpson (Univ. Bristol); *Switching on Defence Genes with Plant Secondary Metabolites* by John Pickett (Rothamsted Research) and *Natural Products as Leads for New and Innovative Crop Protection Chemicals* by John Clough (Syngenta). One other conference lecture is not covered in this section because the content was new and there was no published literature to cite: *Finding New Antibacterials: Opening a Window on the Black Box of Natural Product Drug Discovery* by Tony Buss (MerLion Pharmaceuticals).

Steve Wrigley also wrote a full summary of the conference for the *Journal of Antibiotics*,[10] the editor of which had agreed to run a full page advertisement of the conference in its April and May issues.

The Organising Committee is pleased to acknowledge the receipt of generous sponsorship from: Advanced Chemistry Development UK Ltd., Bracknell; Unilever R&D, Colworth Science Park, Bedford; and the RSC's Chemistry Biology Interface Forum (CBIF).

In addition, we are grateful to Richard Kelly, Deputy Editor of the RSC journals *Organic and Biomolecular Chemistry* and *Natural Product Reports*, for advertising the conference. The Committee also thanks the RSC's Education Section for provision of Student Bursaries and the staff of the RSC Publications Section for assistance in the preparation of this book.

Robert Thomas

References

1 RSC Biotechnology Group Conference Organising Committee: Professor Bob Thomas (Chair, University of Sussex), Dr. Colin Bedford (University College London), Dr. Martin Hayes (Astra Zeneca, Sweden), Mr. Neville Nicholson (Consultant), Dr. Martine Verhoeyen (Unilever R&D) and Dr. Stephen Wrigley (Hypha Discovery Ltd.).

2 S. Wrigley, M. Hayes, R. Thomas and E. Chrystal, eds., *Phytochemical Diversity: a Source of New Industrial Products,* Royal Society of Chemistry, Cambridge, 1997.

3 S.K. Wrigley, M.A. Hayes, R. Thomas, E.J.C. Chrystal and N.H. Nicholson, eds. *Biodiversity: New Leads for the Pharmaceutical and Agrochemical Industries,* Royal Society of Chemistry, Cambridge, 2000.

4 R. Robinson, *J. Chem. Soc.*, 1917, **111**, 762.

5 A.R. Todd and J.W. Cornforth, *Robert Burns Woodward, Biographical Memoirs of Fellows of the Royal Society,* 1981, **27,** 628.

6 A.J. Birch and F.W. Donovan, *Aust. J. Chem.*, 1953, **6**, 360.

7 R. Thomas, *Proc. Chem. Soc.*, 1959, 88.

8 J.W. Cornforth, R.H. Cornforth, M.G. Horning, A. Pelter and G. Popjak, in *CIBA Foundation Symposium: Biosynthesis of Terpenes and Sterols*, eds. G.E.W. Wolstenholme and M. O'Connor, Little Brown, Boston, 1959, p. 119.

9 D.A. Hopwood, *Chem. Rev.*, 1997, **27**, 2465.

10 S.K. Wrigley, *J. Antibiotics,* 2009, **62**, 649.

Contents

1 Modern and Emerging Perspectives on Natural Product Utilisation

NATURAL PRODUCTS, DERIVATIVES AND MIMICS AS ANTITUMOUR AGENTS

David J. Newman and Gordon M. Cragg

Natural Products Branch, Developmental Therapeutics Program, NCI-Frederick, P. O. Box B, Frederick, MD, 21702, USA

(Note: The opinions expressed in this article are the opinions of the authors, not necessarily those of the US Government)

1 INTRODUCTION

From early in the 1940s, the quest for agents that may ameliorate the scourge of the manifold diseases clustered under the term cancer, has involved all aspects of chemistry and pharmacology and throughout all these years, compounds from natural sources, microbes, plants, and latterly, marine invertebrates, have played an extremely important part.

The roles played by these natural products have changed with time and the increase of scientific knowledge. Thus, their initial role was as the major source of drugs used for direct treatment. This was followed by use as active scaffolds upon which chemists would practice their skill and in current times, natural products and derivatives are acting as modulators of specific cellular pathways in the tumour cell.

Currently, the 164 (as of June 2009) small molecule compounds available to the physician (depending upon their individual country) as antitumour agents can be categorized as follows: N (natural products: 25; 17%), ND (natural product-derived, usually semi-synthetic modifications: 50; 31%), S (totally synthetic drugs: 42; 26%), S/NM (synthetic drugs/natural product mimics: 16; 10%), S* (synthetic drugs inspired by natural products: 20; 12%) and S*/NM (synthetic drugs inspired by natural products/natural product mimics: 7; 4%), using the nomenclature of Newman *et al.*[1,2] Recently another author, using a somewhat different series of definitions where only "direct and slightly modified natural products are counted as such", has shown the influence of natural product scaffolds using a different schematic and readers should consult Bailly's article for further information.[3]

As a further refinement, the influence of microbial secondary metabolites on approved agents in cancer treatment can be seen in Table 1. Of the 47 commercially available compounds listed (including everolimus, which was approved for anticancer use after being approved for immunosuppression), 26 (or 55%) are either directly from microbes or are derivatives of microbial secondary metabolites. Trabectedin **1** is probably from microbial commensals rather than from the marine tunicate, *Ecteinascidia turbinata*. The remaining 21 agents (45%) are nominally synthetic but the major proportion are either modified nucleosides whose chemical antecedents can be traced back to Bergmann's discoveries of marine-sourced nucleosides containing arabinose,[4-6] or are isosteres of ATP.

Table 1
[a] Codes from references [1, 2]

Generic name	Year Introduced	Source[a]	Microbe
carzinophilin	1954	N	yes
sarkomycin	1954	N	yes
mitomycin C	1956	N	yes
chromomycin A3	1961	N	yes
mithramycin	1961	N	yes
actinomycin D	1964	N	yes
bleomycin	1966	N	yes
doxorubicin	1966	N	yes
daunomycin	1967	N	yes
neocarzinostatin	1976	N	yes
aclarubicin	1981	N	yes
peplomycin	1981	N	yes
pentostatin	1992	N	yes
trabectedin	2007	N	yes
streptozocin	Pre-1977	N	yes
epirubicin HCl	1984	ND	yes
pirarubicin	1988	ND	yes
idarubicin hydrochloride	1990	ND	yes
zinostatin stimalamer	1994	ND	yes
valrubicin	1999	ND	yes
gemtuzumab ozogamicin	2000	ND	yes
amrubicin HCl	2002	ND	yes
hexyl aminolevulinate	2004	ND	yes
ixabepilone	2007	ND	yes
temirolimus	2007	ND	yes
everolimus	2009	ND	yes
imatinib mesilate	2001	S/NM	
gefitinib	2002	S/NM	
bortezomib	2003	S/NM	
erlotinib hydrochloride	2004	S/NM	
dasatinib	2006	S/NM	
sunitinib maleate	2006	S/NM	
lapatinib ditosylate	2007	S/NM	
nilotinib hydrochloride	2007	S/NM	
cytosine arabinoside	1969	S*	
floxuridine	1971	S*	
enocitabine	1983	S*	
doxifluridine	1987	S*	
fludarabine phosphate	1991	S*	
gemcitabine HCl	1995	S*	
capecitabine	1998	S*	
clofarabine	2005	S*	
nelarabine	2005	S*	
decitabine	2006	S*	
azacytidine	Pre-1977	S*	
vorinostat	2006	S*/NM	

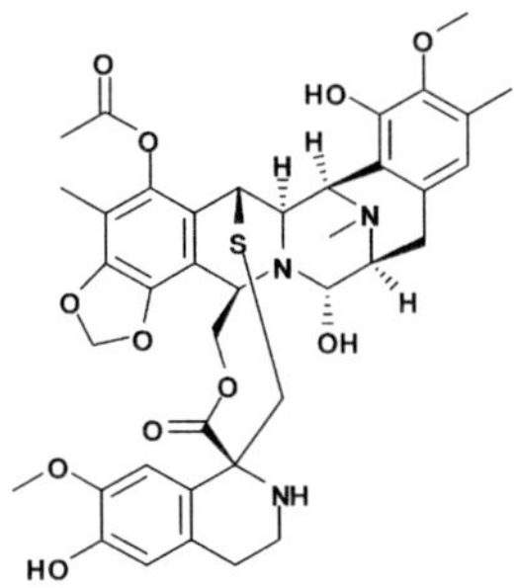

1. Trabectedin (Et743)

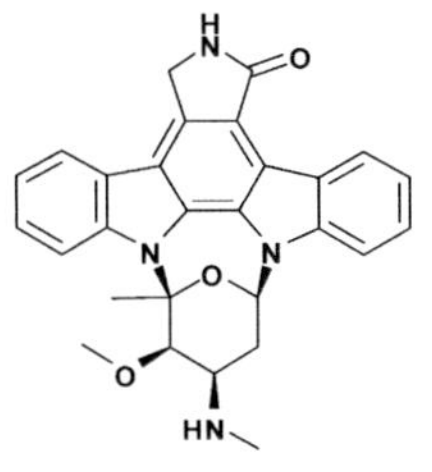

2. Staurosporin

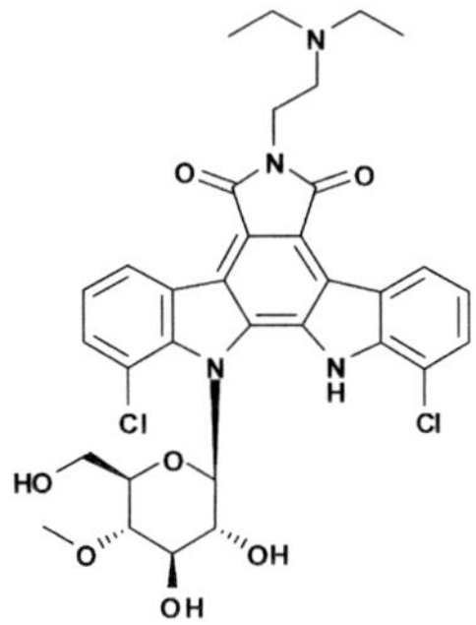

3. Becatecarin

Since the full history of such sources with their manifold digressions into chemical and pharmacological space would be a massive undertaking, we have elected to highlight the influence of certain natural product classes that have derivatives in or approaching clinical trials in order to demonstrate how even today, in the midst of massive efforts related to combinatorial chemical techniques coupled to rational drug design, natural products from many ecological niches are still a major source of novel scaffolds upon which to base potential antitumour agents. For a discussion of the earlier history and derivation of the major classes of compounds derived from nature that are in current use, the reader is directed to the 2005 compendium on Anticancer Agents from Natural Products[7] and the articles therein.

Due to space constraints generally we will discuss only the agents that are currently in Phase II and Phase III clinical trials in general, though where a particular class of compounds spans the range from preclinical to Phase III, we will include most of the molecules in order to demonstrate the breadth of possibilities.

2 NON-MARINE MICROBIAL AGENTS CURRENTLY IN CLINICAL TRIALS

Currently there are at least 17 agents derived from identified microbial sources in various clinical trials against a wide variety of human tumours in both paediatric and adult patients. These encompass structures based on staurosporins, anthracyclines, and bicyclic thiobridged compounds, with mechanisms of action (MOA) including inhibitors of topoisomerase I , II and histone deacetylase (HDAC).

2.1 Staurosporin Based Molecules

Staurosporin 2 is a pan-protein kinase C (PKC) inhibitor and the role of the base structure is evident if one looks at the number of derivatives that are in clinical trials at this moment. Becatecarin **3,** a rebeccamycin **4** analogue with a sugar moiety that has been extensively modified, is claimed not to be a PGP substrate, though it is transported by ABCG2 and can induce ABCG2 expression in cancer cells.[8] It is in trials for leukaemia and other tumour types both with and without other cytotoxins. A recent report shows efficacy in a Phase II trial against advanced biliary cancers with an overall response rate (partial response in 5% and stable disease in 35%) of 40%, which has encouraged extension to a Phase III trial in due course.[9] A current listing of trials with this agent can be found *via* the "clinicaltrials.gov" web site and data in early 2009 can also be found in the "Gateways to Clinical Trials" report by Tomillero and Moral.[10]

A reasonably close relative to staurosporin is the indolocarbazole K-252a **5,** derivatives of which have been placed into clinical trials in neurodegenerative diseases as well as in cancer. For cancers, the simple derivative lestaurtinib **6** is currently in five trials, with four being at the Phase II level (two in myelofibrosis and one each in acute myelogenous leukaemia and polycythemia vera), and one at Phase I (paediatric neuroblastoma). The same basic structure is also in Phase II/III clinical trials as midostaurin **7** and the ring-opened version of staurosporin, known as enzastaurin **8,** is in multiple Phase II trials with some recent positive effects[11] and has reportedly just entered Phase III trials in lymphoma. Enzastaurin also has potential in the treatment of human transitional cell carcinoma in a preclinical model with gemcitabine.[12] Although not formally an antitumour compound, the modified maleimide, sotrastaurin **9** is also a PKC inhibitor and is in Phase II trials for psoriasis, which is a skin disease that antitumour agents may well help in controlling.[13] This compound also has activity as an immunomodulator[14] and is in Phase II clinical trials for this indication as well. Whether it will follow in the footsteps of the rapamycin analogues (see below) has yet to be determined.

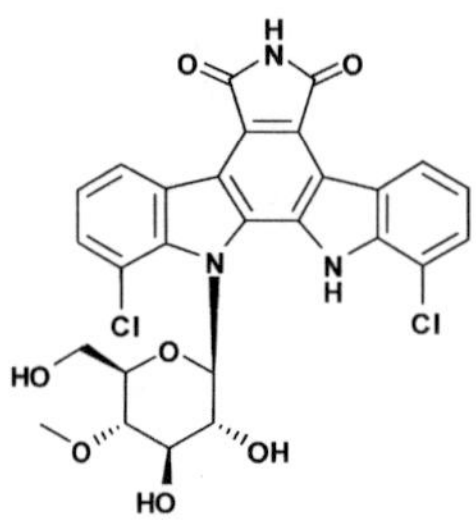

4. Rebeccamycin

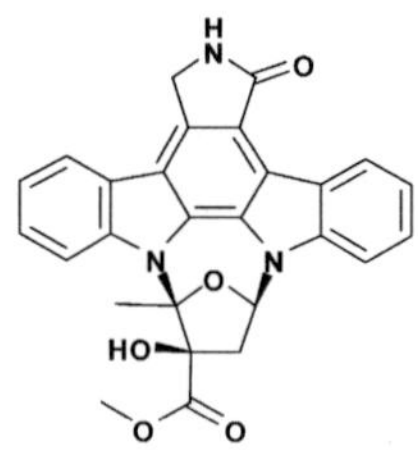

5. K252a

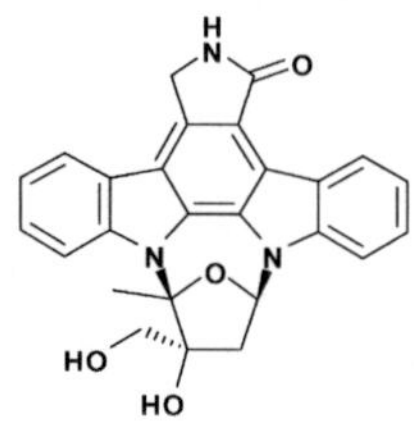

6. Lestaurtinib (CEP-701)

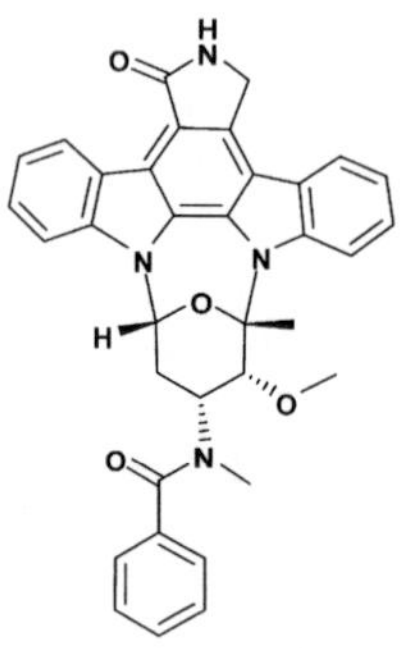

7. Midostaurin

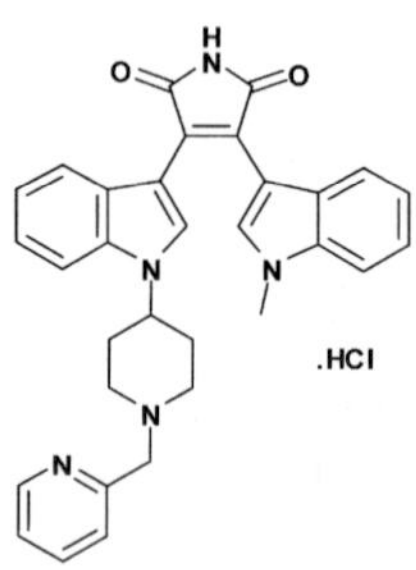

8. Enzastaurin

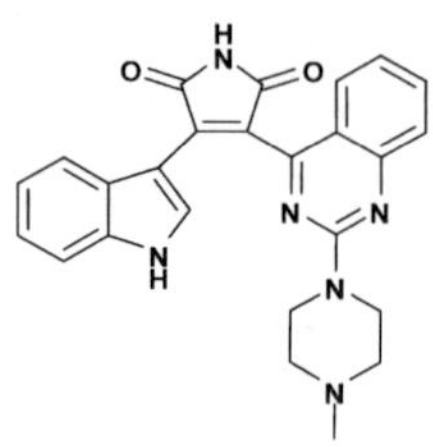

9. Sotrastaurin

2.2 Anthracycline Based Molecules

The anthracyclines, isolated from bacteria of the order *Actinomycetales*, are probably the most utilized microbial chemical class of antitumour agents, both directly from nature and as semi-synthetic modifications, with two of the most useful being daunorubicin **10** and its natural derivative doxorubicin (adriamycin) **11**. Currently, doxorubicin is a major component of the treatment regimen for breast cancer. Although there have been many similar molecules isolated and described in the literature, it is doxorubicin and its more modern derivatives such as epirubicin **12**, pirirubicin **13**, idarubicin **14** and, more recently, valrubicin **15** and amrubicin **16** that have been approved for cancer treatment. The 2005 review by Arcamone[15] should be consulted for details of their history.

Just to demonstrate that the base structure is still in contention as a drug in the 21st century, there are several liposomal, PEGylated or proprietary formulations of doxorubicin and daunorubicin being investigated in phase I/II/III clinical trials, including DOXO-EMCH **17**, NK-911 (a pegylated doxorubicin), SP-1049C (doxorubicin in a Biotransport™ carrier), CPX-351 (liposomal mixture of daunorubicin and cytarabine) and Sarcodoxome (liposomal doxorubicin). The targeted version of doxorubicin (TAP-doxorubicin **18**) where a four residue peptide aids in transportation and acts as an inhibitor until cleaved off by intracellular peptidases, is in phase II trials in Africa and Europe under the code number DTS-201 for the treatment of breast cancer and hormone-refractory prostate cancer respectively.

In addition to doxorubicin-based drug candidates, there are other modified anthracyclines at various stages of development. Thus in phase I/II trials, there is CNDO-101 **19** in which a quinone moiety has been chemically modified and the cardiac toxicity is reported to be absent as a result.[16] There are three more chemical modifications of the

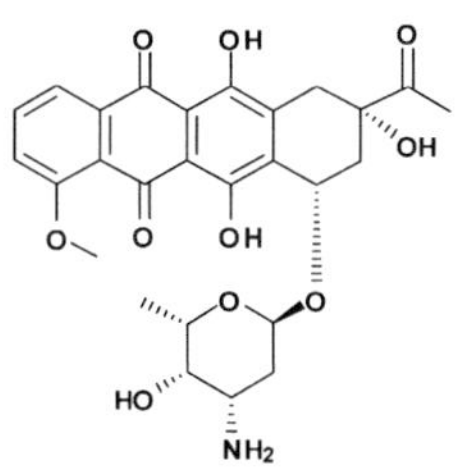

10. Daunorubicin

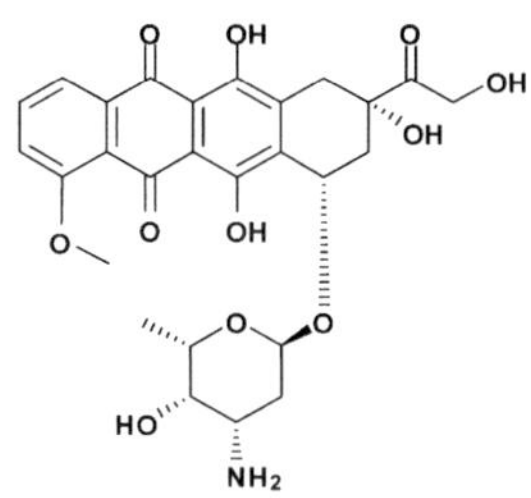

11. Doxorubicin (Adriamycin)

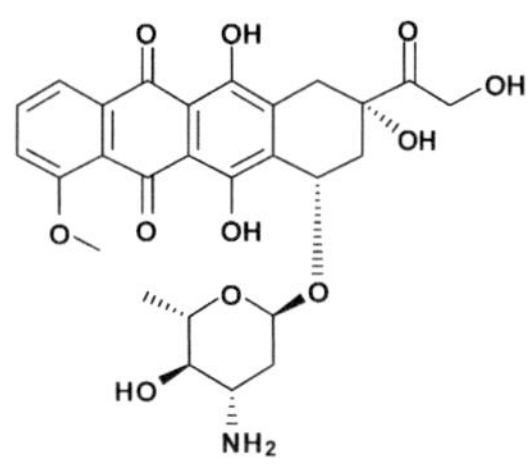

12. Epirubicin

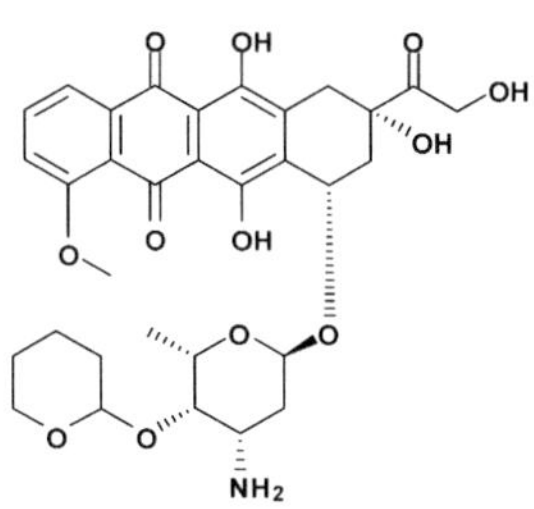

13. Pirarubicin

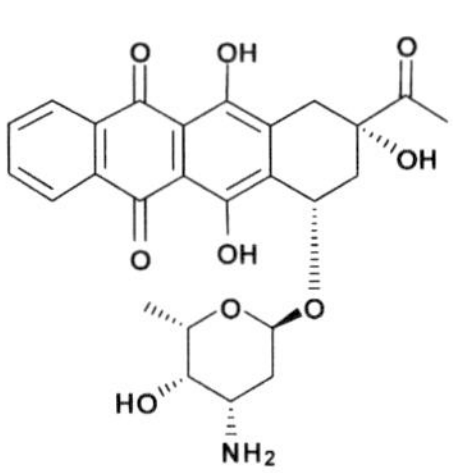

14. Idarubicin

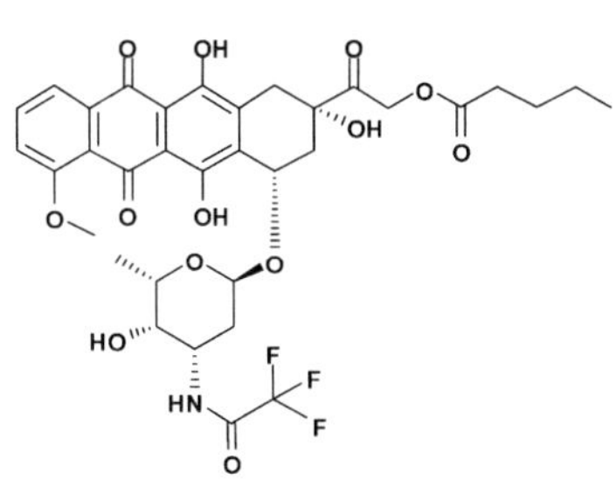

15. Valrubicin

doxorubicin skeleton in phase II trials: annamycin **20**, a liposomal variant of which is also in phase I/II trials for leukaemia; sabarubicin **21**, in which the major structural modification is an extra sugar moiety; berubicin hydrochloride **22**, which has a benzyl ether attached to the sugar moiety, reported to permit crossing of the blood-brainbarrier, was removed from Phase II clinical trials against glioma as a result of a business decision, but the structural features are still important.

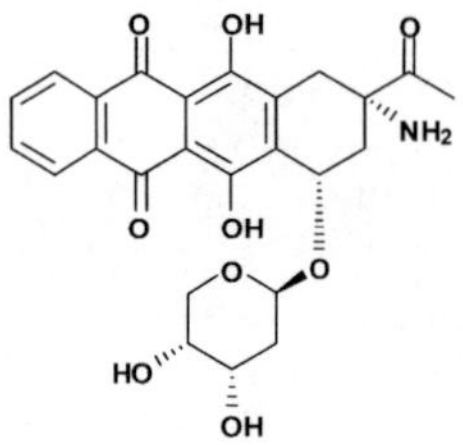

16. Amrubicin

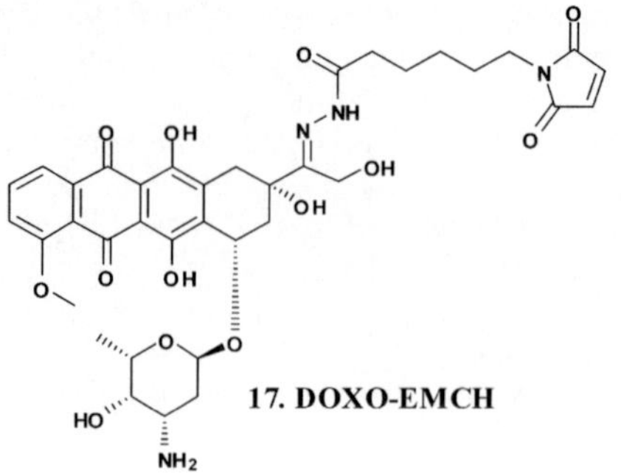

17. DOXO-EMCH

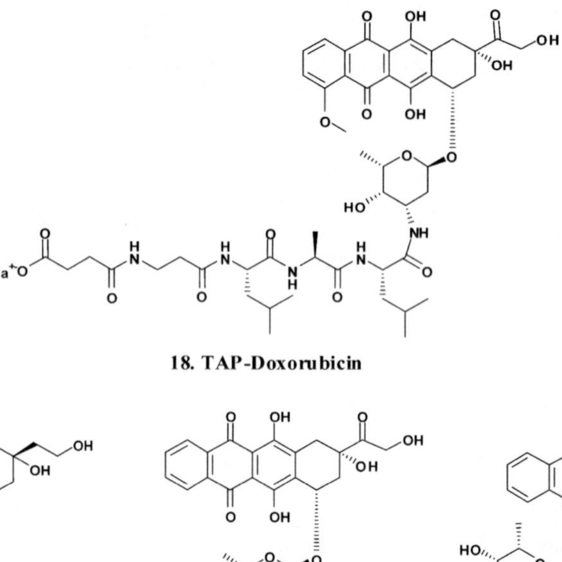

18. TAP-Doxorubicin

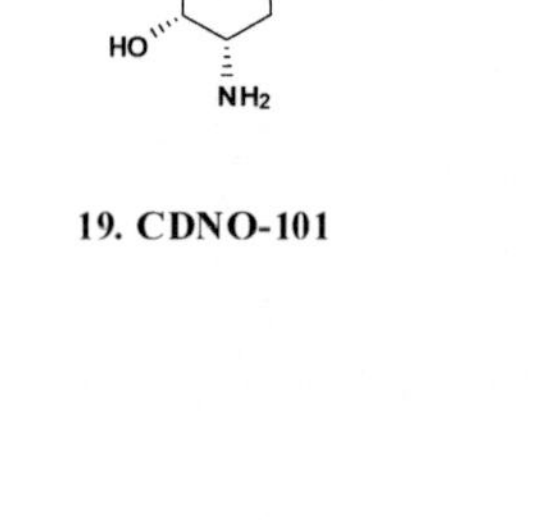

19. CDNO-101

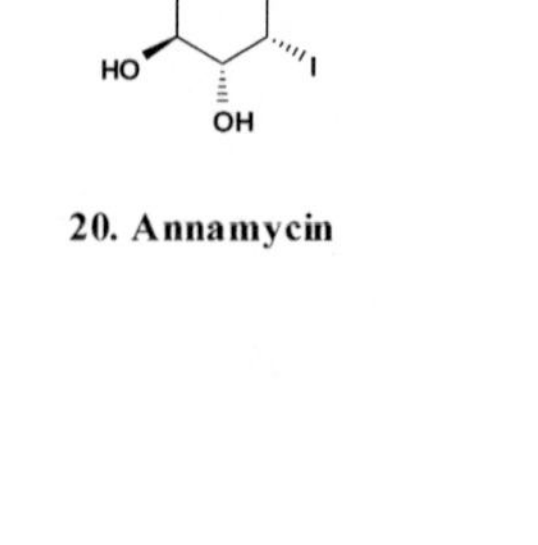

20. Annamycin

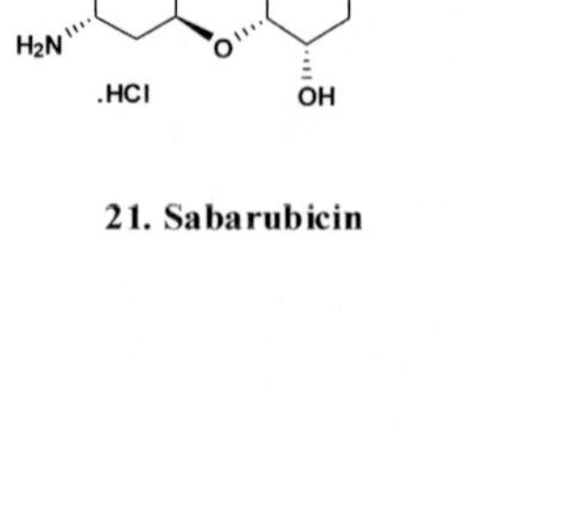

21. Sabarubicin

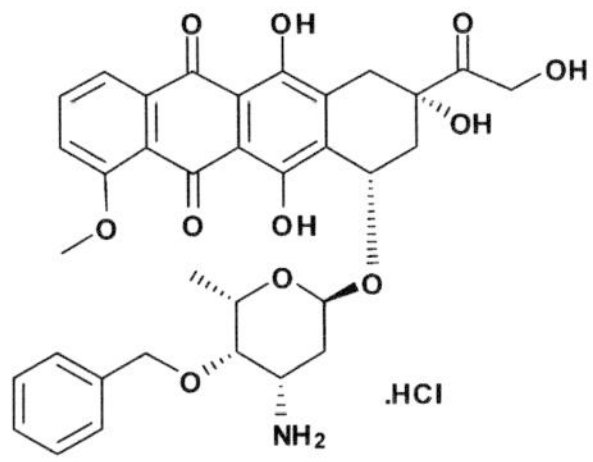

22. Berubicin hydrochloride

2.3 Enediynes

One of the most important recently approved microbial compounds (as the warhead for a monoclonal antibody [Mab]-linked delivery system) is the enediyne, calicheamicin γ_1^I **23**. The reason for its importance is that although Calicheamicin γ_1^I has *in vitro* cytotoxic activity at the sub-picomolar level, it was not developed further as a single compound as it was just too toxic to pursue in spite of its exquisite activity. In addition to its very potent activity, it was also the first non-protein bound structure identified of a new chemical class, the enediynes; though the neocarzinostatin chromophore **24** was identified in 1985.[17] These agents currently number 13 with the two basic structural types (the so-called 9- and 10-membered endiynes), differing in the number of carbon atoms in the endiyne system. Two of the 13, sporolides A and B **(25, 26)** and cyanosporasides A and B **(27, 28)** are rearrangement products of putative enediynes. These agents have been well reported by various groups, the Wyeth (originally Lederle) discoverers of the calicheamicins from *Micromonospora echinospora* ssp *calichensis* covering the ten-membered category,[18] while the chemistry and biosynthesis of the molecules is reviewed by a number of investigators.[19,20]

Following binding to DNA and subsequent activation, these agents undergo an unprecedented rearrangement which causes cleavage of both strands of DNA leading to the death of the cell. As mentioned earlier, Wyeth gained FDA approval for gemtuzumab ozogamicin (Mylotarg®), in 2000. This monoclonal antibody-warhead construct for use against chronic myologenous leukaemia, is possibly the most potent antitumour agent yet approved for clinical use. Due to the success of this type of construct, calicheamicin has

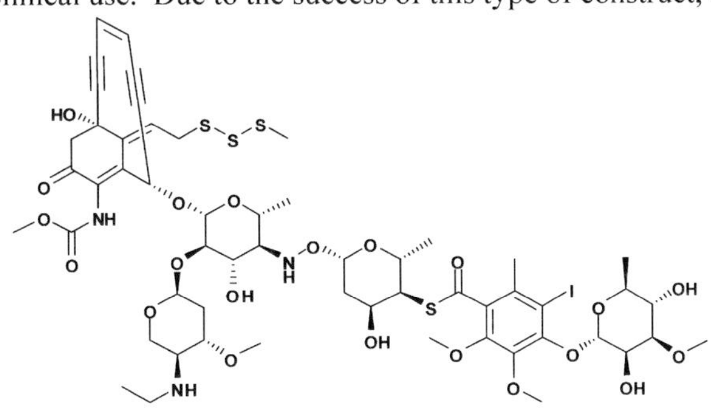

23. Calicheamicin gamma1-I

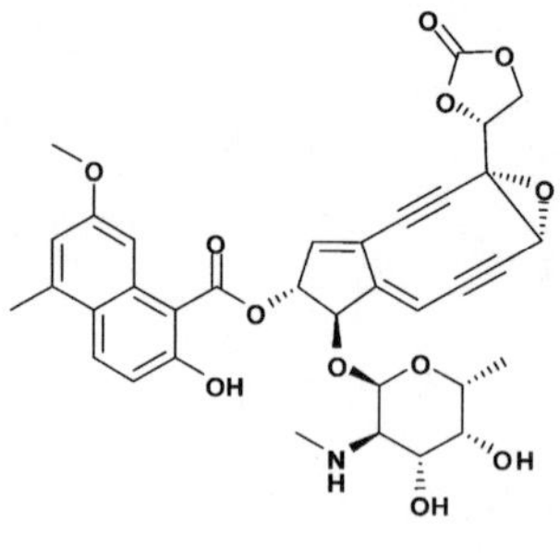

24. Neocarzinostatin Chromophore

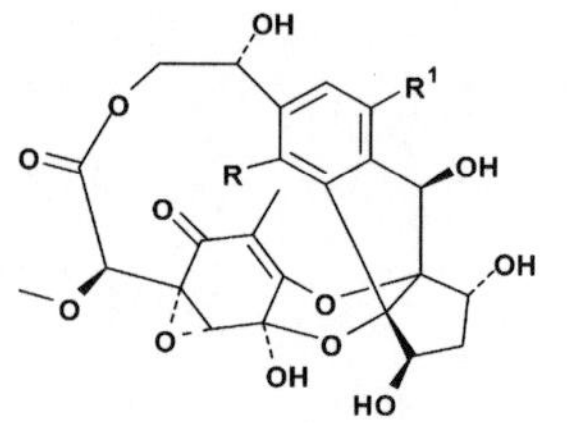

25. Sporolide A; R = Cl, R^1 = H
26. Sporolide B; R = H, R^1 = Cl

been linked to a variety of other monoclonal antibodies which are in various clinical trials. CMD-193, where the antibody is directed against an anti-Lewis Y antigen[21] has now been withdrawn from Phase I trials but CMC-544 or inotuzumab ozogamicin, which is a conjugate of calicheamicin and an anti-CD22 monoclonal antibody[22] is in five Phase I/II/III trials against a variety of lymphomas. If further information is desired on such agents, the recent review article by Castillo *et al*[23] gives details of the current impact of monoclonal antibodies, with or without microbial-sourced attachments, on cancer treatments.

In addition to the isolation, development and biosynthetic work mentioned above on the overall class, a significant number of papers covering aspects of synthesis have been published in the last few years. These include work that permitted reassessment of the original structures by direct comparison of the possible isomers obtained by synthesis with the data from the natural product.

Perhaps the best current example in this respect is the synthetic effort around uncialamycin. The original structure was determined on less than 500 micrograms of material by Davies *et al.*[24] following isolation from an unidentified streptomycete extracted from the lichen *Cladonia uncialis*, that appeared to be related to *Streptomyces cyanogenus*. Due to the lack of material, the discoverers were unable positively to determine the stereochemistry at position 26 in the natural product. Nicolaou's group rapidly synthesized both of the enantiomers and demonstrated that the 26 (*R*) enantiomer **29** was the natural product and the 26 (*S*) or *epi*-enantiomer **30** the unnatural one. Further biological investigations have shown that both epimers were very potent antibiotics (confirming the original work of Davies *et al*), and also that they were potent cytotoxins in the NCI 60 cell line panel, with 5 to 10 fold less activity for the 26-*epi* enantiomer.[25] This work demonstrated that even molecules as complex as these are amenable to current synthetic methods.

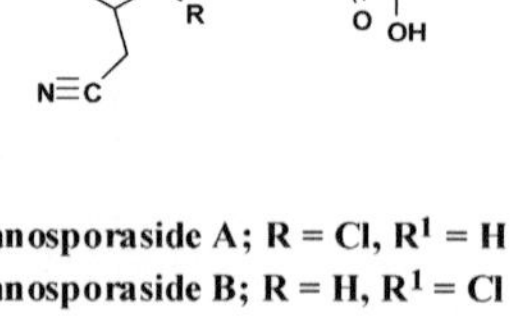

27. Cyanosporaside A; R = Cl, R^1 = H
28. Cyanosporaside B; R = H, R^1 = Cl

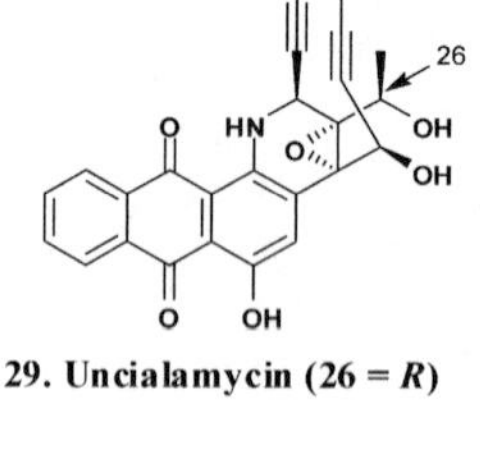

29. Uncialamycin (26 = *R*)

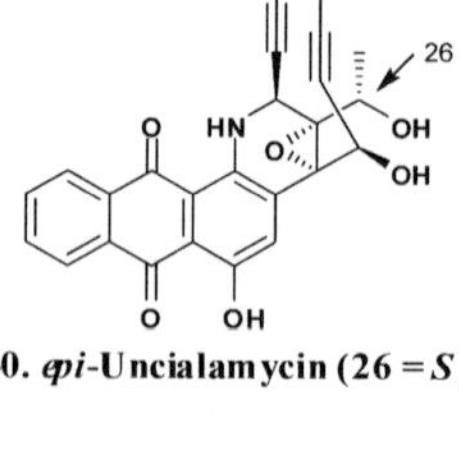

30. *epi*-Uncialamycin (26 = *S*)

2.4 Rapamycins and Epothilones

Rapamycin **30** was originally reported in 1975 as a potential antifungal agent under the code number AY-22,989 from an Easter Island isolate identified as a *Streptomyces hygroscopicus*.[26-28] Unfortunately, the antifungal activity was not adequate for further development, but in 1984 its potential as a possible antitumour agent was reported by workers at Ayerst Canada using syngeneic murine tumours together with a hint of oral activity.[29]

The antitumour activity was not developed at that time but the rapamycin skeleton has now spawned a plethora of molecules with a variety of different pharmacologic activities including cancer. The first modifications were at one site (the carbon atom at C^{43}) and led to four more clinical drugs in addition to rapamycin. The rapamycin base molecule was approved as sirolimus **31** in 1999, initially as an immune-suppressive agent and now the same molecule is in Phase I/II trials against various cancers.

Everolimus **32**, the second variant, was launched in 2004 as an immunosuppressive agent and it is currently in further Phase II/III trials for various cancers in the EU, Japan and the USA having recently been approved for renal cancer in the EU and the USA under the name Afinitor®. The third variation temsirolimus (CCI-779) **33** was approved as a treatment for renal carcinoma in the USA in 2007 under the name Toricel® and is in a number of Phase I/II/III trials against various carcinomas in the USA, with a number under the auspices of the National Cancer Institute. The fourth, zotarolimus **34** was launched in the USA in 2005 for treatment of restenosis as part of a drug-eluting stent and is still in a number of trials comparing its activity against various different drug stents. The fifth, biolimus A9 **35** was launched in 2008 where the rapamycin derivative was linked to the stent *via* a biodegradable polymer.[30]

Although there are five clinical agents as shown above, the skeleton is still under refinement with the following agents at various stages of development. The closest compound to possible approval at the present time is deforolimus (A23573) **36** which is in Phase III clinical trials against sarcoma. There are two prodrugs of rapamycin currently in

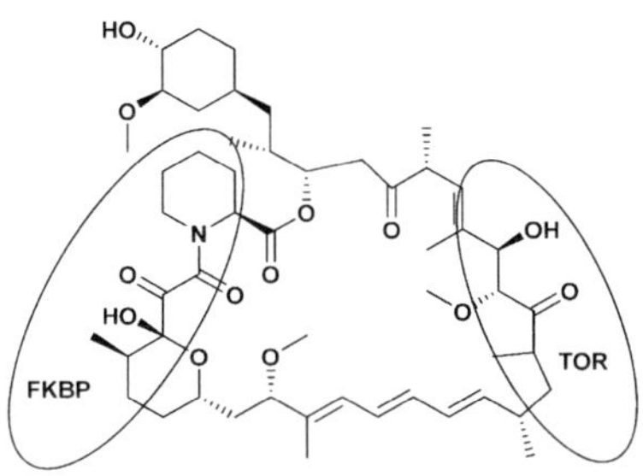

31. Rapamycin (Sirolimus)

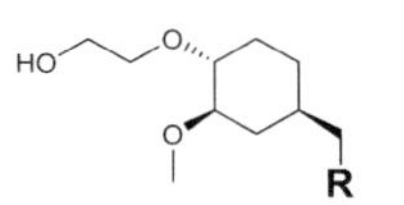

32. Everolimus,
R = Macrolide Ring

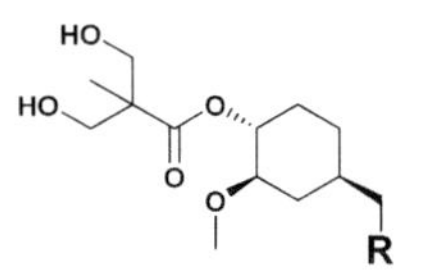

33. Temsirolimus,
R = Macrolide Ring

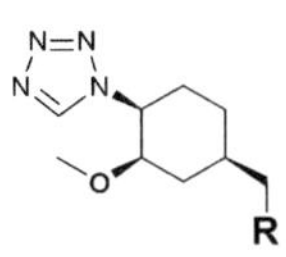

34. Zotarolimus,
R = Macrolide Ring

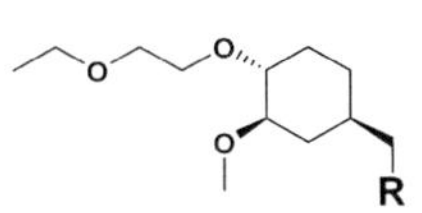

35. Biolimus A9,
R = Macrolide Ring

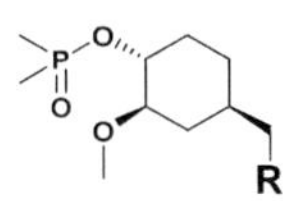

36. Deforolimus,
R = Macrolide Ring

phase I trials, Abraxis' ABI-009 (which is a nanoparticle encapsulated formulation of rapamycin) in cancer and Isotechnika's TAFA-93 (structure not yet published), which is an immunosuppressive. It should be noted that all of these (where structures are known) are either the base molecule or have been modified at only the one site, the C^{43} alcoholic hydroxyl group that avoids both the FKBP-12 and the Target of Rapamycin (TOR) binding sites. Modifications in other areas were thought to negate the basic biological activity of this molecule.[31,32]

However, a rapamycin derivative ILS-920 **37** with a modified macrolide ring structure is currently in Phase 0 (first in man) clinical trials and is headed for Phase I trials shortly. ILS-920 has a modification in the triene portion of the molecule, designed to disrupt mTOR binding and appears to have a different target as it is a nonimmunosuppressive neurotrophic rapamycin analogue and has demonstrated over a 200 fold higher binding affinity for FKBP52 over FKBP12. It has been reported to promote neuronal survival and outgrowth *in vitro* and to bind to the β1 subunit of L-type calcium channels (CACNB1).[33,34] Inhibition of FKBP 52 is reported to affect tubulin interactions in cells[35] so there is an interesting possibility that this agent may also have antitumour activity, though no reports of such activity have yet been published. Finally, the patent literature reveals other variations that are in preclinical studies in a variety of pharmacologic areas.

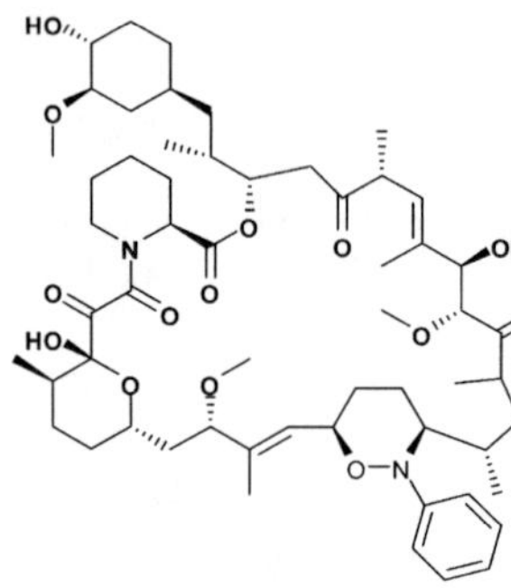

37. ILS-920

Reichenbach and Hoefle identified the myxobacterial products epothilones A and B (**38, 39**) as tubulin stabilizers (a similar mechanism to that of paclitaxel) in the middle to late 1980s,[36,37] which precipitated an avalanche of chemical, biochemical and even genomic modifications of the base structure in order to further explore the utility of the skeleton. This culminated in the approval in October 2007 of the semisynthetic epothilone, 16-aza-epothilone B, known generically as ixabepilone **40** (Ixempra®) by the FDA for treatment of breast cancer.

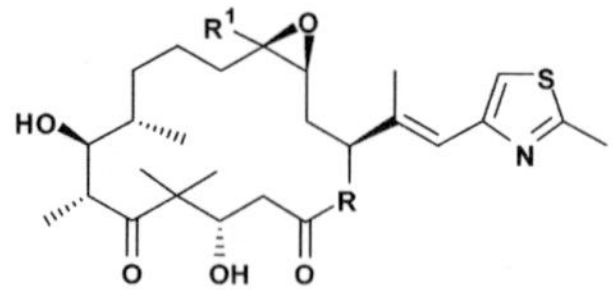

38. Epothilone A; R = O, R^1 = H
39. Epothilone B; R = O, R^1 = CH_3
40. 16-Aza-epothilone B; R = NH, R^1 = CH_3

At the present time, there are five other compounds listed in the Prous Integrity® database based upon the epothilone skeleton and in active development as anticancer agents. The natural product epothilone B (patupilone; **39**) is in Phase III cancer trials sponsored by a collaboration of Novartis with the original discoverers. A totally synthetic derivative, though very close to the base skeleton, Sagopilone (ZK-EPO) **41** is in Phase II cancer trials under Bayer-Schering.[38] Much fuller details of this and the opportunities for synthesis of other agents are given in three recent reviews[39-41] and one book chapter[42] which should be consulted for the finer details. There are two agents derived from work originating in Danishefsky's laboratory at Memorial Sloan-Kettering being developed by Kosan, now part of BristolMyers Squibb. The first is (*E*)-9,10-didehydroepothilone D or dehydelone **42** currently in Phase II trials and chemists at Kosan have demonstrated modifications of the structure whilst retaining the activity.[43] The second is isoxazolefludelone[44] **43,** which is currently in preclinical evaluation[45] and a discussion of the chemistry leading to the compound and other derivatives produced *en route* was given by the group in late 2008.[46] Isoxazolefludelone was scheduled for an IND in late 2008, but no details of such a filing are currently available though further preclinical data was presented in abstract form in 2009.[47]

Finally, there is a very interesting folate receptor-targeted molecule synthesized as a result of a collaboration between Endocyte and BristolMyers Squibb (BMS-753493) **44,** where folic acid has been linked to an aza-modified epothilone.[48] The compound is now in Phase I/II trials under the name epofolate against folate-receptor-positive tumours.

Now that the genetic sequence of the original producing organism has been reported, we can expect not just combinatorial biosyntheses giving modifications of the base skeleton, but that a number of different secondary metabolites from the same microbe may well be reported in the literature in the near future.[49-51]

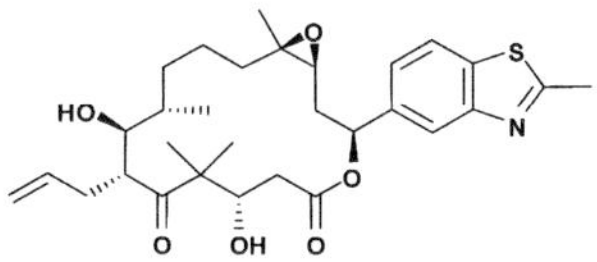

41. Sagopilone

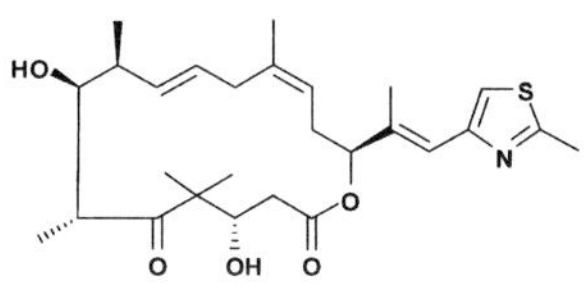

42. Dehydelone

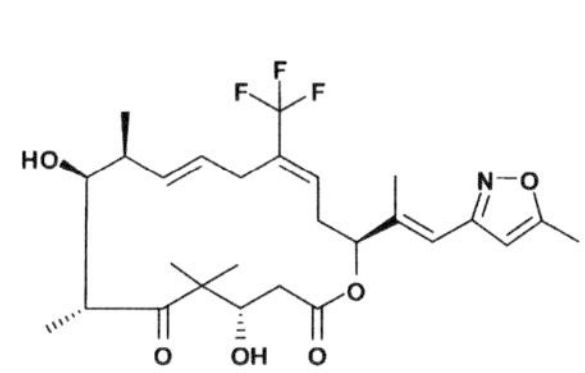

43. Isoxazolefludelone

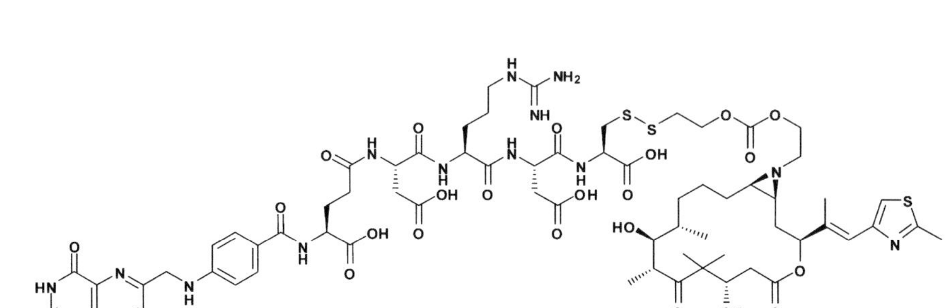

44. Epofolate

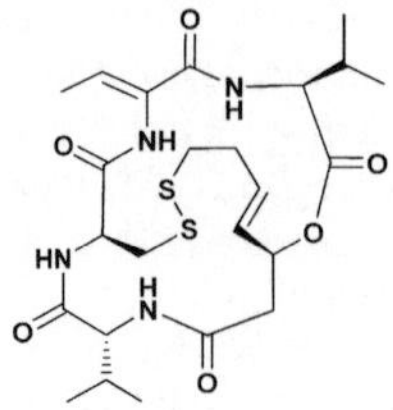

45. FK-228

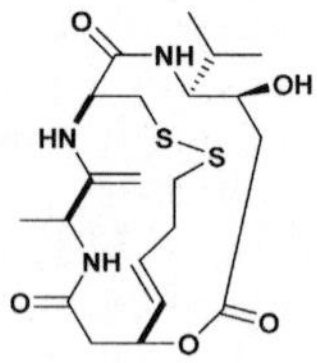

46. Spiruchostatin A

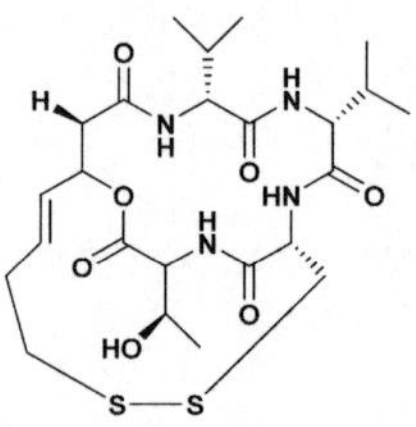

47. FR901375

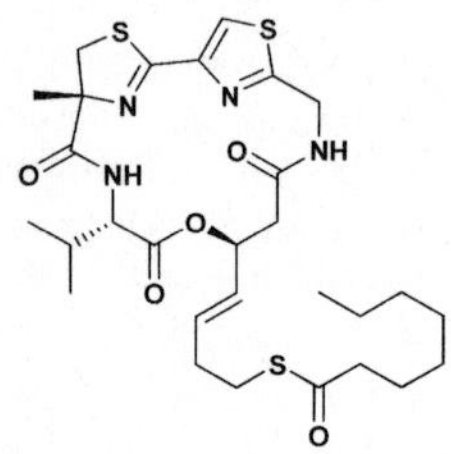

48. Largazole

2.5 Selected Histone Deacetylase Inhibitors (HDACs)

FK228 **45,** now known as romidepsin, an example of a microbial metabolite from the Gram negative plant pathogen *Chromobacterium violaceum* No. 968, was identified as an HDAC inhibitor after it entered Phase I clinical trials as a cytotoxic agent. It has a very unusual bicyclic structure with a di-sulphide bridge and is one of four naturally occurring sulphur-containing HDAC inhibitors from natural sources. The others are spiruchostatin A **46**, FR901375 **47,** both with the di-sulphide bridge and the marine cyanobacterial product, largazole **47** (placed here rather than in the marine microbial section below because of the common mechanism), which has a thioester side chain. FK228 has been in 39 clinical trials (with 10 now actively recruiting patients) and in early September 2009, an FDA review committee recommended approval for treatment of cutaneous T-cell leukaemia.

The use of other agents in conjunction with HDAC inhibitors has been discussed frequently in the clinical literature with a recent paper by Frew *et al* covering some of the possibilities.[52] For chemists, the very recent review by Newkirk *et al* that covers natural product derived HDAC inhibitors, their syntheses and potential should be consulted for in depth coverage from that perspective.[53]

3. MARINE SOURCES

It has now become obvious that a significant proportion of pharmacologically active agents isolated from marine invertebrates in all pharmacologic areas are produced by single-celled microbes[54,55] in, on or around the invertebrate from which the active agents were isolated. Thus, where the actual source of compounds discussed in this section has been identified, it will be noted, even if the source was only identified years later.

Although a plethora of compounds have been reported with cytotoxic activities in the last 30 or so years, only a relative few have been definitely proven to be microbial in origin. For example, the bryostatins (see 3.1, below), are probably produced by an as yet uncultured microbe that is vertically transmitted during reproduction. There are, however, a significant number where there is sufficient circumstantial evidence to invoke a microbe or microbes in their biosynthesis.[56] There are also a significant number of compounds isolated from blue-green algae (which are prokaryotes) and from bacteria and fungi directly isolated from sediments (both shallow and abyssal) or from the invertebrate, where there is no question as to the actual producer as they have been fermented to produce the compounds of interest. Materials isolated from direct microbial involvement are discussed in a later section.

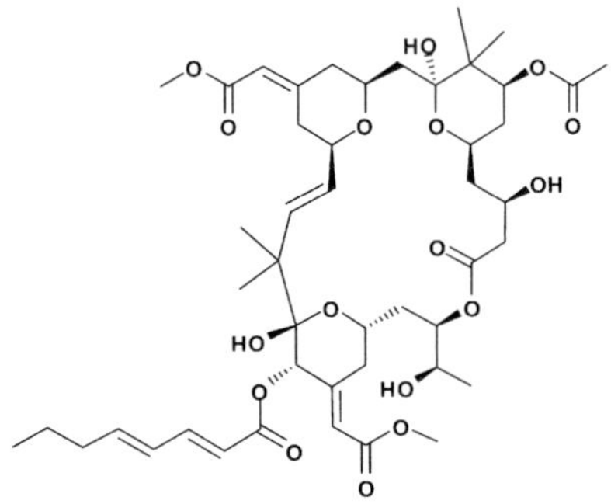

49. Bryostatin 1

3.1 Bryostatins, Dolastatins and Analogues

Perhaps the two quintessential examples of marine cytotoxins from the earlier work would be bryostatin 1 **49** and dolastatin 10 **50**. In both cases, heroic efforts were performed in the 1970s to early 1980s to obtain enough material for initial clinical trials from the nominally-producing invertebrate *Bugula neritina* in the case of the former, and the nudibranch *Dolabella auricularia* in the latter case. Details of the initial work on both materials has been presented in many reviews and original papers and the early to relatively late history of both can be found in the 2005 reviews by Newman for the bryostatins[57] and Flahive and Srirangam for the dolastatins.[58]

Later work has demonstrated that in the case of the bryostatins (of which there are at least twenty naturally occurring variations[57]) the probable source is an, as yet uncultured, microbe from which the gene cluster that would produce the base ring structure of the bryostatins has been isolated and cloned, but not yet expressed.[59] Currently there are no active trials of bryostatin as an anti-tumour agent. Contemporaneous with the clinical and biochemical work, medicinal chemists were extremely busy producing modifications of the basic structure and some very recent publications describe rapid methods of obtaining bryostatins,[60-63] using them to identify specific PKC targets[64] and demonstrating that bryostatin-1 may have potential in amelioration of cognitive diseases.[65,66]

In contrast to bryostatin, the producing organism for dolastatin 10 (and other similar peptides) was subsequently directly isolated and identified as a cyanophyte of the genus *Symploca*[67] with later work showing the vast capabilities of cyanophytes.[68] Dolastatin 10 entered clinical trials in the early 1980s and over the next twenty or so years, it was followed by a variety of similar synthetic molecules derived from the base structure. At the time of writing, though, there is some (published) controversy over the current status of dolastatin 10 as inspection of the NCI Clinical Trials database shows no ongoing clinical trials. Of the derivatives, only auristatin PE (soblidotin) **51** completed two Phase II clinical trials, but it appears to have ceased development as a single agent at this moment. There are some very interesting modifications made by medicinal chemists in order to deliver this close relative to dolastatin-10 using specific monoclonal antibodies targeted at specific epitopes.[69,70].

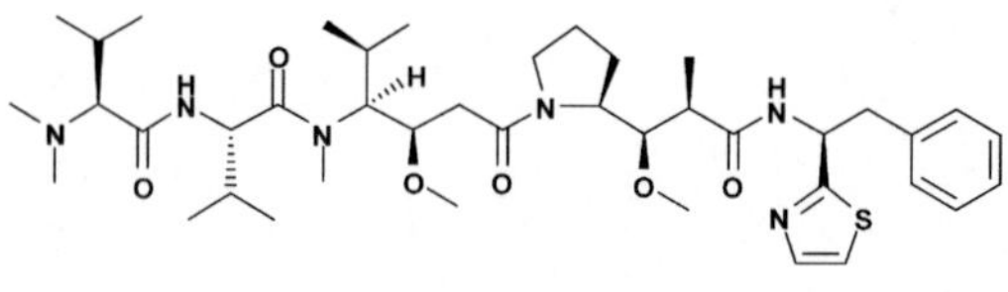

50. Dolastatin 10

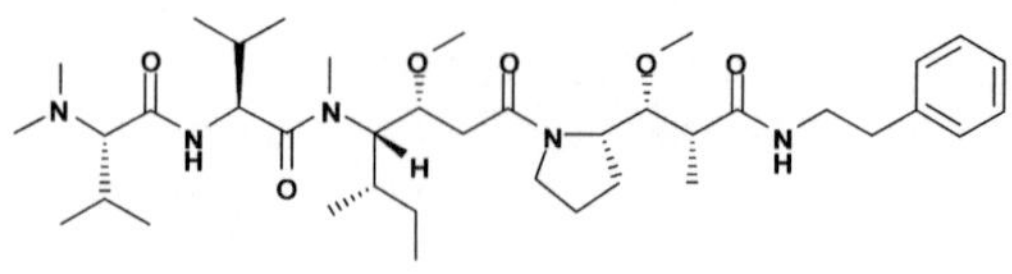

51. Auristatin PE (Soblidotin)

A number of combinations of this base molecule with varying monoclonal antibodies are in preclinical to Phase II/ III clinical trials predominately against haematologic cancers. Thus Seattle Genetics has SGN-35 which is a monoclonal antibody-drug conjugate consisting of the chimeric anti-CD30 antibody cAC10 (SNG-30) linked through the Val-Cit linker to 8 molecules of monomethylauristatin E (MMAE)[71] in Phase II trials and it is currently heading for Phase III and preregistration in the EU. CuraGen has CR011, in which the antibody-drug conjugate consists of monomethylauristatin E (MMAE) linked to the anti-CG56972 fully human monoclonal antibody through a Val-Cit dipeptide linker, in Phase II trials for metastatic breast cancer and melanoma.[72,73] Progenics has PSMA-ADC, a dimeric-specific PSMA antibody also conjugated to an auristatin E derivative, in Phase I against prostate cancer.[74] In all of these examples, although the exact linkages between the auristatin molecule and the antibody are subtly different, all are licensed from Seattle Genetics. Hopefully these dolastatin-inspired adducts will propel this class of molecules to higher phases of clinical trials as the base molecule, auristatin PE, oscillated between Phases I and II as it changed hands amongst a variety of small and large pharmaceutical companies.

3.2 Et-743 (Trabectedin or Yondelis®)

The first (and currently only) compound "direct from the sea" to be approved for the treatment of cancer is Et743 **1**, which was approved by the EMEA in September 2007 for the treatment of sarcoma and was launched in Sweden, Germany and the UK late in 2007. Currently there are 24 clinical trials listed in the NCI clinical trials list involving this agent and more in the Integrity® database. Et-743 is one of a series originally reported by two groups in back-to-back papers in 1990.[75,76] The molecule was recognized as a derivatised member of the saframycin class antibiotics and thus might have a microbial component in its production by the marine invertebrate.

The compound was licenced by the discoverers to the Spanish pharmaceutical company PharmaMar,and followed a chequered development path involving large-scale collections of the source tunicate *Ecteinascidia turbinata* and in-sea and on-land aquaculture,all of which produced enough harvested materials for early clinical trials. The compound, though synthesized early in its study by Corey's group, was ultimately produced for later clinical trials by semi-synthesis starting with cyanosafracin B **52**. This intermediate compound was produced by fermentation of a marine-derived *Pseudomonas fluorescens* and fuller details of this agent can be found in the 2005 review by Henriquez *et al.*[77]

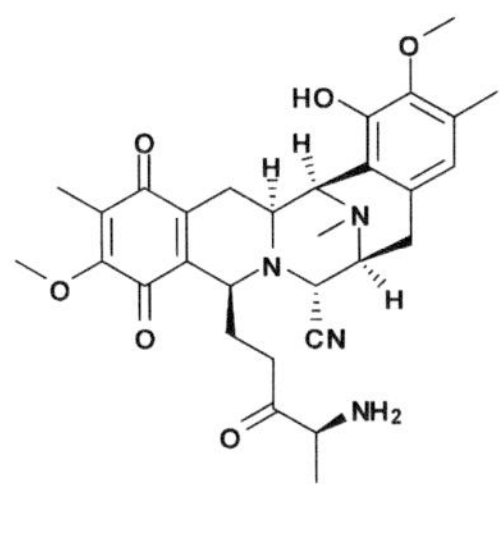

52. Cyanosafracin B

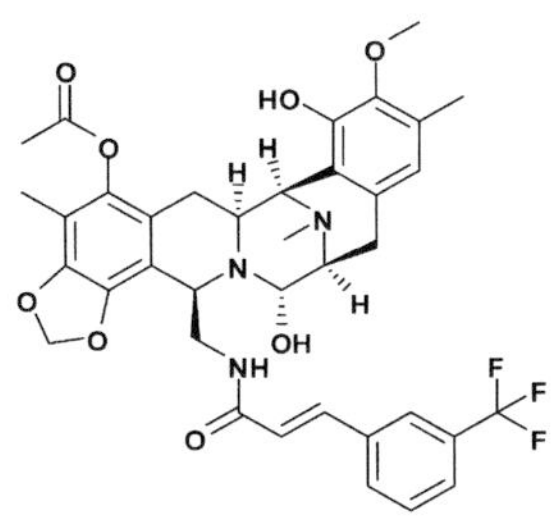

53. Zalypsis(R); PM-10450

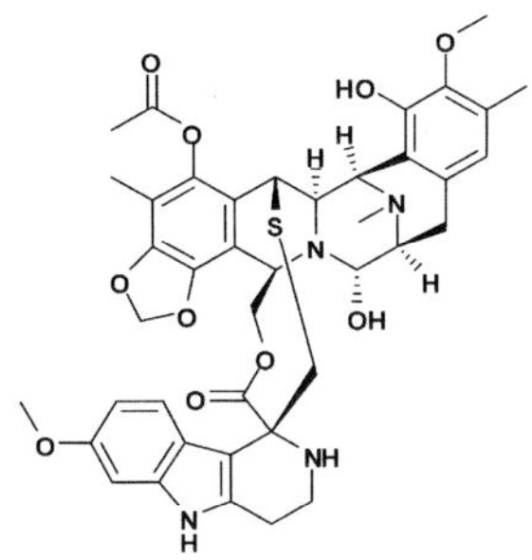

54. PM-01183

In addition to this review, Velasco *et al* have also reported the identification of the producing gene cluster for saframycin, thus opening up possibilities for combinatorial biosyntheses in the future.[78] A report in 2003 identified potential bacteria from mediterranean *E. turbinata*[79] and a recent examination of the microbial assemblages related to *Ecteinascidia turbinata* in both the caribbean and mediterranean seas has identified five potentially persistent bacteria, with one, *Candidatus Endoecteinascidia frumentensis* occurring in *E. turbinata* in both geographic locations. Though not absolute proof, this is indicative of microbial involvement in the biosynthesis and will need further investigation[80] analogous to the work carried out with the putative bryostatin producer (*vide supra*).

In addition to Et-743, there are two other closely related alkaloids of this general class in clinical trials, the first is zalypsis® **53** or PM-10450, which is a modification of a metabolite known as jorumycin originally isolated from a nudibranch.[81,82] Zalypsis is in Phase II trials for cervical cancer and PM-01183 **54**, a subtle modification of the Et-743 base molecule, is currently in Phase I trials against solid tumours.[83]

3.3 Aplidine

Aplidine, formally dehydrodidemnin B **55**, was first reported in a patent application in 1989 with an UK patent issued in 1990, and then referred to in a paper[84] from Rinehart's group in 1996 on the structure-activity relationships amongst the didemnins. The initial work on aplidine, its entry into Phase I and II trials and the preferred method of synthesis was described in detail through late 2004 by PharmaMar scientists.[77] At that time the dose limiting toxicity was muscle pain that was responsive to either dose limitation or addition of carnitine. Interestingly, in the presence of carnitine, the maximum tolerated dose could be increased by 40% to $7mg.M^{-2}$. A number of Phase I/II clinical trials are currently underway in Europe and the USA in acute lymphoblastic leukaemia (ALL), lymphoma, myeloma and melanoma, both with and without other cytotoxins or signal transduction inhibitors. Some recent reports from Phase II studies are now entering the literature in other than abstract format.[85, 86]

The precise MOA of this agent is not yet known, but it appears to prevent VEGF secretion and blocks the corresponding VEGF-VEGF-Receptor-1 (also known as *flt-1*) autocrine loop in leukaemic cells.[87] A series of *ex vivo* studies with leukaemic blasts from paediatric and adult patients confirmed that cells underwent massive apoptosis at levels of 5nM, below the blood levels achievable in man.[88] Using these and other results, workers at PharmaMar were able to develop a pharmacogenomic model that led to a molecular

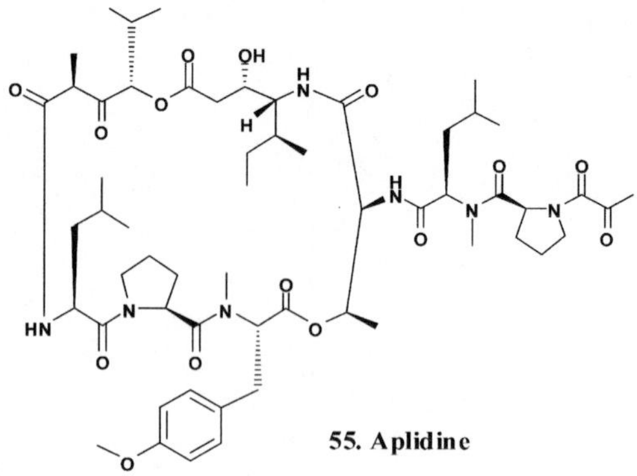

55. Aplidine

fingerprint for sensitivity to this agent using the "Oncochip" array. Full details were given in a 2006 paper.[89]

Three more recent papers have further developed the molecular characteristics of aplidine. One demonstrates that membrane cholesterol levels may influence both cellular binding and Rac1/JNK pathway activation; lowering of the cholesterol levels eliminates the activation of Rac1 and hence induces apoptosis.[90] In the second, published in 2008 an extension of the Rac1/c-Jun pathway involvement was identified,[91] and in the third, investigators demonstrated that the level of $p27^{kip1}$ was inversely proportional to the sensitivity to aplidine.[92]

In addition to the array data mentioned earlier, a recent paper gives an excellent meta-analysis of the pharmacokinetics of aplidine treatment (as a single agent) across nearly 300 patients.[93] It will be interesting to analyse how the predetermination of genomic targeting for prospective patients coupled to the levels that can be achieved in clinical settings, will aid in the more rapid development of agents such as these.

What is very interesting both chemically and pharmacologically, is that the removal of two hydrogen atoms, *i.e.* conversion of the lactyl side chain (in didemnin B) to a pyruvyl side chain (in aplidine), appears to significantly alter the toxicity profile. This is the only formal change in aplidine when compared to didemnin B, though the comments on dosage regimens used for the didemnin B studies by Vera and Joullie[94] should be taken into account when such toxicity comparisons are made in the future. The resemblance to didemnin B is emphasized by the work of Cardenas *et al* who reported in 2003[95] that in DMSO solution, aplidine, like didemnin B, does not exhibit a formal β-turn in its side chain in approximately 20% of its solution conformers, thus suggesting that the presence of such a turn is not required for biological activity. Further work by the same group using constrained didemnin and aplidine derivatives imply that at least part of the biological activities may be due to interaction with protein-prolyl isomerases. This is analogous to interactions reported for FK506, cyclosporin and rapamycin with similar enzymes and all of the didemnin class of molecules (didemnins, aplidine and tamandarines) do exhibit immunomodulatory activities.[96] Further synthetic work on tamandarin derivatives has recently been reported, thus extending the structure-activity relationships in this overall series.[97]

3.4 Kahalalide F

The cyclic depsipeptide kahalalide F **56** was isolated from the sacoglossan mollusk, *Elysia rufescens,* following grazing by the mollusk on a green macroalga, *Bryopsis* sp.

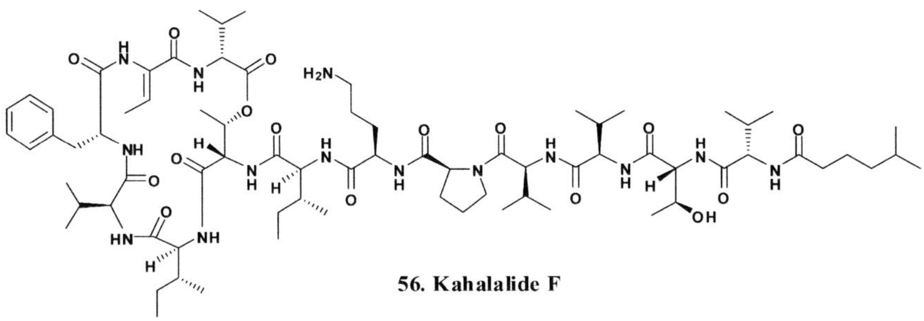

56. Kahalalide F

Subsequently it was discovered that the depsipeptide also occurs in the alga, but on a wet weight basis, the mollusk concentrated the depsipeptides significantly.[98] The compound was licensed to PharmaMar by the University of Hawaii in the 1990s and entered preclinical testing. It was synthesized in a very efficient manner using solid phase peptide techniques by a group in the chemistry department at the University of Barcelona.[99]

The base compound entered Phase I clinical trials in Europe in December 2000 for the treatment of androgen-independent prostate cancer. It progressed to a variety of Phase II clinical trials but, according to the NCI clinical trials web site, it is currently in Phase I trials against advanced solid tumours with and without a second agent under the code number PM-02734. Recently, however, Pharmamar reported that kahalalide F is in Phase II trials for psoriasis and that they have licensed the compound to Medimetriks for indications other than oncology and neurology. Earlier information (up to 2005) on the preclinical and clinical development is available in the comprehensive review by Henriquez *et al.*[77]

There are a variety of mechanisms attributed to this compound. It was known to target lysosomes,[100] thus suggesting selectivity for tumour cells with high lysosomal activity such as prostate tumours. In addition, Suarez *et al*[101] demonstrated that kahalalide F induced cell death via "oncosis" (the progression of cellular processes leading to necrotic cell death) possibly initiated by lysosomal membrane depolarization in both prostate and breast cancer cell lines. Then in 2005 Sewell *et al*[102] reported that HepG2 cells demonstrated significant alterations in their membrane permeability with cell swelling and blebbing implying specific interactions with membranes and/or proteins at 300nM. It was also reported that it induces a necrosis-like cell death that involves inhibition of Akt signalling and depletion of ErbB3.[103] This latter work was extended in a report in 2006 that demonstrated that ErbB3 may well be a marker for progress against suitable tumour types in patients and also implied that an ErbB3 kinase inhibitor may well increase efficacy.[89]

Recently, Piggott and Karuso in Australia reported results from utilizing a very elegant reverse chemical proteomics study with a T7 cDNA phage display that identified human RPS25 as a binding partner for kahalalide F. They then showed that the compound bound to phage-displayed RPS25 in a dose-dependent manner with a conservative dissociation constant of approximately 50μM.[104]

Work is still being reported by both the Hamann group at the University of Mississippi and by PharmaMar and collaborators on other synthetic derivatives of kahalalide F as methods of extending the various active structures derived from this agent.[105,106] In 2005 a PCT International Application was filed by Hill *et al* claiming production of kahalalide F and other derivatives from a *Vibrio* microbe isolated from *Bryopsis* alga and also from the sea slug, *Elysia rufescens*, implying that the invertebrate obtains the producing microbe from the alga and then maintains the microbe(s) as

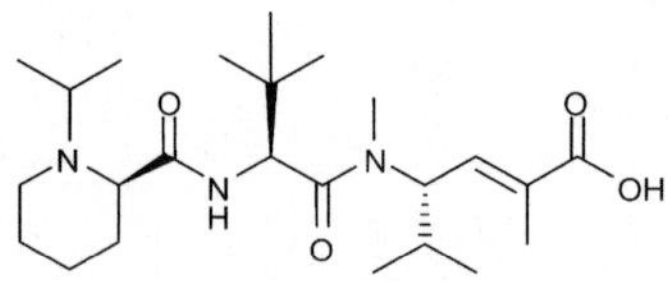

57. E-7974

symbionts.[107] Thus there is a potential renewable source of these agents by use of fermentation.

Hawaii is not the only area with these peptides, since Ashour *et al* have reported new cyclic peptides of this class from the Indian Ocean mollusk, *Elysia grandifolia* with a higher cytoxicity *in vitro* than reported for kahalalide F.[108] A significant number of further natural variations on these peptides has recently been reported from Hawaiian invertebrates.[109]

3.5 E-7974

Compound **57** is a synthetic variation from Eisai modelled upon the tripeptides of the hemiasterlin class and arose due to the use of a pipecolic acid derivative as the first aminoacid in the tripeptide This substituent is not claimed in the patent literature and can be considered as homoproline. The literature is sparse with only issued patents[110] and abstracts available at the present time for this Phase I compound. The latest clinical abstracts are from the 2007 ASCO Meeting[111,112] and the 2008 Gastrointestinal Cancers Symposium.[113] It will be interesting to see how this compound, which is reported to be a poor MDR substrate will proceed.

3.6 Eribulin (E7389)

Compound **58**, modelled from the naturally occurring antitubulin compound halichondrin B **59,** arose from a synthetic *tour-de-force* utilizing the synthetic method for halichondrin B first reported by Kishi's group in 1992,[114] and the realization by Kishi that the active part of the molecule resided in the macrolide ring (approx MW of 600) and not in the tail (the remaining 400 MW). Chemists at the then Eisai Research Institute in Massachusetts working very closely with the Kishi group at Harvard, synthesized over 200 molecules and in conjunction with Developmental Therapeutics Program (DTP) at NCI, chose the modified truncated macrocyclic ketone (E-7389) as the candidate compound. The molecule, like its parent, is a tubulin interactive agent with very potent activity at the nanomolar level *in vitro*[115] and binding at or close to the *vinca* site from recent modelling studies.[116]

Fuller details of the synthetic and base biological information were published by the leaders of the studies in 2005[117] and in 2007 a short article covering the latest clinical citations was published by Wang.[118] Recently, fuller details of two Phase I studies,

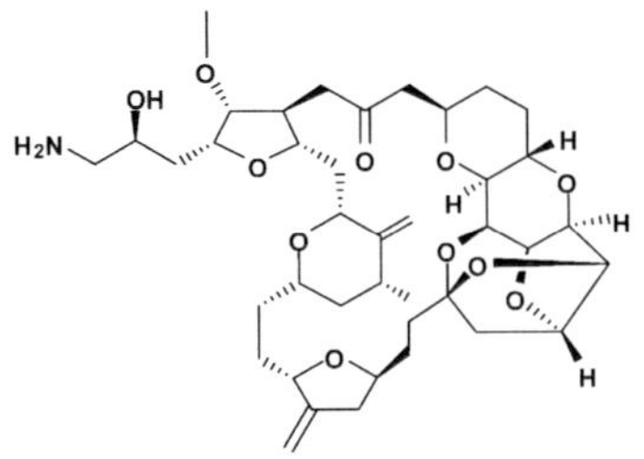

58. Eribulin; E7389

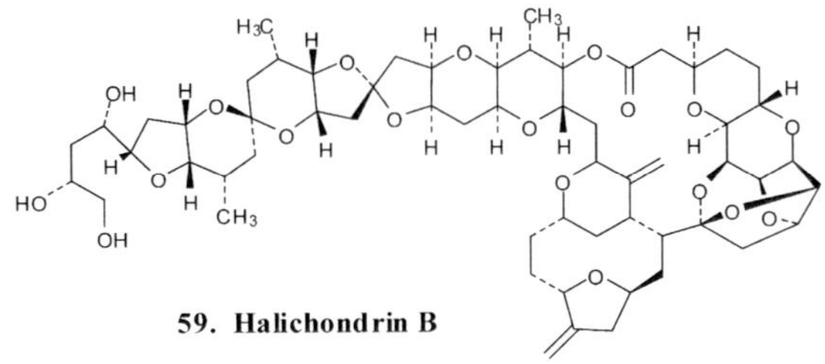

59. Halichondrin B

together with a commentary, have been published[119-121] and a report on a Phase II study with heavily pretreated patients has also appeared in the scientific literature.[122] These articles should be read by the interested reader for fuller details of the evolution of this compound.

Inspection of the recent Eisai press releases (September 2009) show that an MAA has been filed in Switzerland for treatment of breast cancer with similar filings in the EU/USA following in early 2010. A search of the NCI clinical trials web site shows 13 listed trials at the Phase I/II level in the Americas and a Phase III trial which has just finished in Europe. Thus, this compound has gone from Phase I to an MAA in roughly 7 years, as it went into Phase I clinical trials under NCI/Eisai auspices in 2002, having been approved for clinical trials by the then Decision Network Committee of the NCI in the middle of 2001.

4. MARINE MICROBIAL SECONDARY METABOLITES

Over the last twenty or so years, there have been many comments and presentations implying that a number of the agents found in marine invertebrates had "microbe(s) in their background". This has subsequently been shown to be probable by using genomic information in the cases of bryostatin and ET743 and by direct isolation of microbes from the *Bryopsis* alga in the case of Kahalalide F.[107,123] However, it was the thesis of Fenical and Jensen that there were deep sea free-living microbes that could be cultivated and novel agents produced by modifications of methods used for other microbial flora. That their thesis was correct is shown by extensive publications over the last fifteen or so years, with a number of excellent reviews giving their methods and results.[124-127]

4.1 Salinosporamide A

Salinoporamide A **60** was first reported[128] from a marine streptomycete of a new genus and species named initially as *Salinospora tropica*. The genus was subsequently changed to *Salinispora* but the compound retained its original name. The structure was reminiscent of the terrestrial bacterial product lactacystin and its rearrangement product, omuralide **61**, a known proteasome inhibitor and that activity was reported for the new compound. Salinosporamide A had an unusual chlorine substitution and within a year or so of the publication, two academic groups had synthesized the molecule,[129,130] and this was followed by a synthetic paper from Nereus Pharmaceutical scientists (who developed the compound).[131] Subsequently, many groups have reported improved syntheses with an excellent review covering most of those through 2006.[132] Since that time, there have been

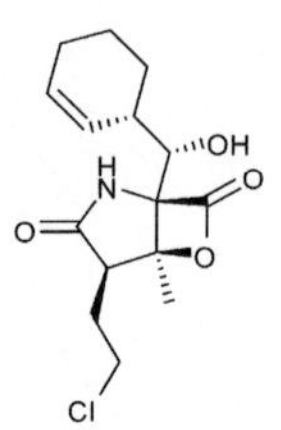

60. Salinosporamide A

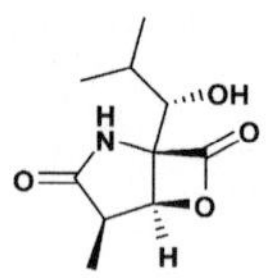

61. Omuralide

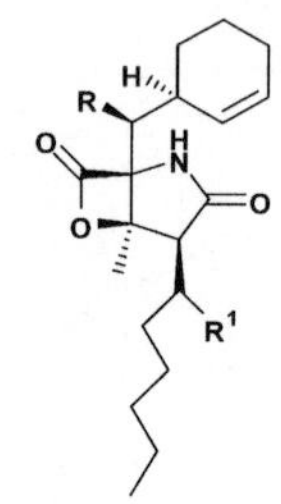

62. Cinnabaramide A; R = OH, R[1] = H
63. Cinnabaramide B; R = R[1] = OH
64. Cinnabaramide C; R = R[1] = H

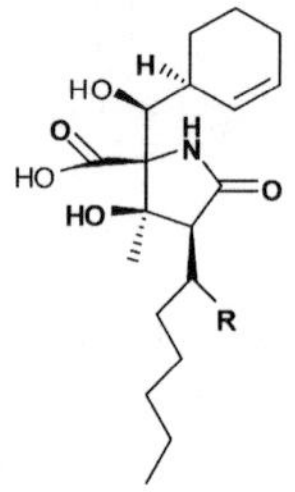

65. Cinnabaramide D; R = OH
66. Cinnabaramide E; R = H

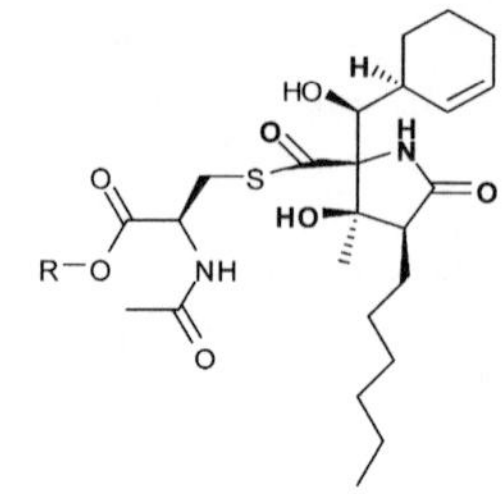

67. Cinnabaramide F; R = H
68. Cinnabaramide G; R = CH_3

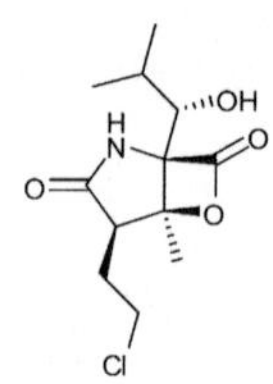

69. Antiprotealide

a number of papers covering an enantiospecific synthesis,[133] a chemoenzymatic technique[134] and a full description of modifications that improve the potency of this class of compounds by Nereus scientists,[135] plus a recent paper by the Omura group demonstrating the use of multiple techniques in a stereospecific synthesis.[134]

In addition to providing synthetic methods, Nereus scientists in concert with the now defunct fermentation group at IRL in New Zealand, were able to produce the necessary current good manufacturing practices (cGMP) product for clinical trials by fermentation in a saline environment, the first time that this task had been successfully performed on any scale with a marine-sourced microbe. During these runs, a significant number of other salinosporamide derivatives were also isolated and other secondary metabolites were explored.[127,136-140] Concomitant with this production work, the mechanism of salinosporamide as an inhibitor was worked out using excellent X-ray crystallographic studies[141] together with studies on the stability of the β-lactone ring.[142]

The compound entered Phase I clinical trials in May of 2006 initially against solid tumours and leukaemias and in April of 2007, another Phase I trial against multiple myeloma was initiated. Currently (September 2009), the NCI Clinical Trials web site shows four Phase I trials that are ongoing, including one in conjunction with the HDAC inhibitor vorinostat and the Nereus web site implies that Phase II trials are being planned. Some results from Phase I trials have been reported demonstrating a clinical response in patients with solid tumours or lymphomas.[143-145] A thorough discussion on the background of the discovery and development was recently published by the Scripps and Nereus scientists involved and should be consulted for background information.[146]

In the last year, some interesting papers have been published on the isolation of similar molecules from a terrestrial microbial source, the cinnabaramides A-G **62-68**,[147] and on the isolation and identification of other salinosporamides from fermentation experiments, with varying proteasome activities.[137] Although the producing organism is a marine streptomycete, salinosporamide production is dependent upon sodium not halogen levels,[136] and can occur in low chloride levels[140] or even with substitution of potassium for sodium,[148] and these are close to conditions normally used for terrestrial fermentations, giving further credence to the cinnabaramide story.

5. GENOMIC INFLUENCE ON MICROBIAL METABOLITES

Although the production of secondary metabolites in microbial systems is obviously controlled genomically, only in the last few years has it become clear that researchers have only observed the tip of the iceberg. For over 40 years, many companies, both large and small fermented millions of soil isolates in order to find novel agents from microbes and a constant theme was that "we keep discovering the same compounds". For this and other reasons, almost all companies ceased their fermentation-based discovery programs from roughly the middle of the 1980s to the early 2000s. However, with the advent of two interrelated techniques, the search for new microbial agents has entered a new phase. The first was automated DNA sequencing and the second was the discovery of the polymerase chain reaction technique. When these were coupled with the advances in computer analyses of DNA and RNA sequence data, the immense potential of the microbial universe from the perspective of secondary metabolites began to be realized. One must also mention the advent of combinatorial biosynthesis, which is the production of unnatural natural products by biosynthetic genes and parts of genes that can be mixed and matched to produce novel structures. Unnatural starting substrates may also be used in this technique. Most important for the discovery of the putative microbial producers, is the concept of the metagenome and its informatic analysis, whereby gross DNA samples can be extracted from soils, seas and invertebrate organisms and then probed for genes by PCR allowing amplification, sequencing and informatic analyses.

The first use of these techniques was in the work from Handelsman on soil antibiotics,[149,150] with perhaps the most adventurous being the work of Venter's group on the marine metagenome.[151] A current example of where this has led, is the recent paper from Piel and colleagues,[56] which demonstrates the importance of *trans* acyltransferases in the biosyntheses of a number of potentially important microbial secondary metabolites from as yet uncultured organisms.

Since the interplay of these discoveries and their current status would fill another book, let alone a single chapter, we are just going to give a few examples relevant to drug discovery. Perhaps the most important from a drug discovery and development perspective was the recognition by the Omura group in 2001 that the avermectin producer, *S. avermitilis*, contained at least twenty-five identifiable secondary metabolite gene clusters covering roughly 6% of the total DNA.[152] They revised the figure to over thirty when a more thorough analysis was performed in 2003.[153] Contemporaneous with this work, the Hopwood group reported the sequence of the quintessential streptomycete, *S. coelicolor* in 2002,[154] again demonstrating that more than 20 secondary metabolite clusters could be identified.

This has led is to the interplay of microbiology, chemistry, biosynthesis and genomics, so that it is now possible in some cases to use the information from the sequence of a microbe to identify the potential producing clusters (cryptic clusters) that have never been expressed under the normal fermentation parameters used to date. A prime recent example of this is the work on the salinosporamide producer, *S. tropica*. In 2007, Udwary *et al* reported[155] the sequence of this organism, identifying 17 potential clusters including the salinosporamide locus and over the last few years Fenical and co-workers at Scripps Institution of Oceanography have been identifying some of these agents by using particular fermentation conditions. Now that the positions of the potential producing clusters have been identified, a combination of cloning, expression and fermentation is unlocking these metabolites. As mentioned earlier, the sporolides **25, 26** and cyanosporolides **27, 28** have been isolated from this organism and their clusters have been identified.

The genomic information obtained permits one to produce unknown molecules by performing the mixing and matching referred to earlier. Thus, using information from the sequence and a knowledge of the mechanism[156] of the unique chlorinase in *S. tropica*, Eustaquio and Moore produced the fluoro analogue of salinosporamide A by mutasynthesis.[157] A further example is provided by the production of the unnatural (though see later) antiproteolide **69**, a novel proteasome inhibitor.[158] Very recently however, further investigation of the large scale fermentation products from *S. tropica* by Nereus scientists has shown that this molecule is in fact a natural product.[159] The current status of the work has been presented and reviewed with a discussion of the production of bioengineered novel metabolites as part of the *Salinispora* genome project by Nett and Moore[160,161]

These are just some of the very many examples of the potential for this combination of resources and for up to date reviews of such experimental designs, the reviews by Watanabe,[162] Brady *et al*[163] and Wenzel and Müller[164] should be consulted. For further reading on the prospects for the world of marine microbes, the papers from the Fenical and Jensen groups are recommended, particularly the recent ones demonstrating the multiplicity of non photosynthetic microbes[124,165] and this potential is extended into the cyanobacteria by the very recent review by Kalaitzis *et al.*[166]

6. PLANT SECONDARY PLANT METABOLITES

Although plant metabolites have a very long history in antitumour studies, a significant number of those currently in preclinical and clinical trials are variations on taxanes, camptothecins and other well known base structures. These have been covered quite well in a multiplicity of reviews and books over the last few years, with current examples being the articles by Hostettmann and Marston,[167] Potterat and Hamburger,[168] and Cragg and Newman.[169]

What is interesting, however, is that a number of older agents have now entered preclinical and clinical trials as "warheads" linked to various delivery systems. Thus, two derivatives of maytansine, DM1 and DM4 are in clinical trials. These compounds are in fact derivatized bacterial metabolites that are (probably) on the path to maytansine as this agent is now known to be predominately bacterial in origin,[170] The formal structures of DM1 and DM4 with various conjugates are given in the recent article by Ikeda *et al*,[171] which should be read in conjunction with the commentary by Polson and Sliwkowski[172] in order to put the results in perspective.

Currently, huN901-DM1 (BB-10901), which targets the CD56 epitope is in three Phase I and II trials under the auspices of Immunogen. The same warhead but now linked to the Roche antibody trastuzumab, is in one Phase III trial against breast cancer and in 10 other Phase I and II trials for a variety of tumour types under the name TDM-1. A significant number of other antibody combinations are in preclinical studies. A slight variation in the warhead structure gives the compound known as DM4 and there are two constructs (BT-062 and HuC242-DM4) in Phase I clinical trials at the current time, with

Biotest and Sanofi-Aventis respectively. Recently, two other DM4 constructs, AVE-9633 and hu242-DM4 were discontinued, the former in Phase I for toxicity reasons and the latter by Immunogen for business reasons.

A discussion of the techniques involved and some of the results with other such antibody-warhead constructs has recently been published by Senter of Seattle Genetics and should be consulted by interested readers.[173] This should be read in concert with the discussion by Stephan *et al* of the problems involved in assaying such conjugates from a pharmacokinetics aspect in order to appreciate the problems involved in assessing the value of such conjugates.[174]

6.1 Vinca alkaloids and Folate Delivery Systems

As mentioned in an earlier section, the use of folate linked conjugates to deliver warheads to tumours has been employed with modified epothilones. However, the much older vinblastine skeleton has also been used as a warhead, in the form of a desacetylvinblastine hydrazide folate conjugate **70**[175] and, under the code name EC-145, it is in Phase II clinical trials against platinum resistant ovarian cancer under the sponsorship of Endocyte.

6.2 Rocaglates (Silvestrol and *epi*-Silvestrol)

The cyclopenta[*b*]benzofuran compounds known generically as rocaglates are toxic to insects and tumour cells and are protein synthesis inhibitors, which induce arrest at the G_2/M stage of the cell cycle.[176] With continuing work on these interesting molecules, the Kinghorn group, initially at the University of Illinois at Chicago (UIC) and now at the Ohio State University (OSU), have shown that the most active of the molecules is silvestrol **71** and its epimer (*epi*-silvestrol) **72** is slightly less active. The clinical group at OSU has shown silvestrol to have a selective activity against B cells derived from patients with chronic lymphocytic leukaemia (CLL) relative to normal peripheral blood monocytes. Subsequent studies both *ex vivo* and *in vivo* using the *Tcl-1* transgenic murine model confirmed the T-cell sparing activity.[177] Further reports from two groups, the UIC/OSU researchers and their associates[178] and Pelletier's group at McGill,[179] have shown that the molecule has a series of interesting activities including impairment of the ribosome recruitment step of translation initiation by affecting the composition of the eukaryotic initiation factor (eIF) 4F complex.[179]

Silvestrol is in the NCI's preclinical pipeline and work is progressing on studies related to pharmacology and pharmacokinetics, which expands upon work at UIC and OSU, with the aim of developing the agent as a potential drug lead.

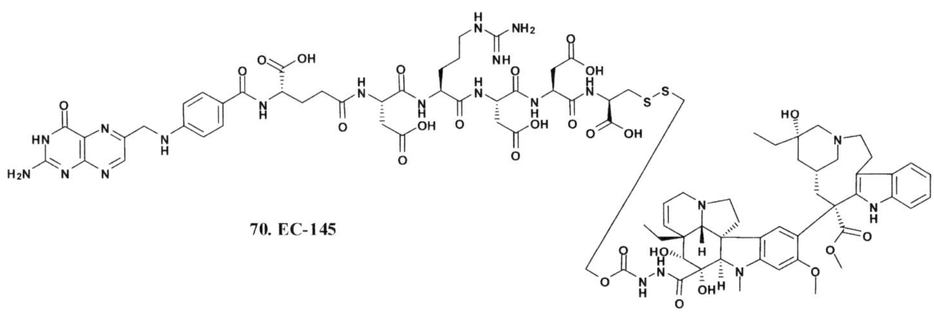

70. EC-145

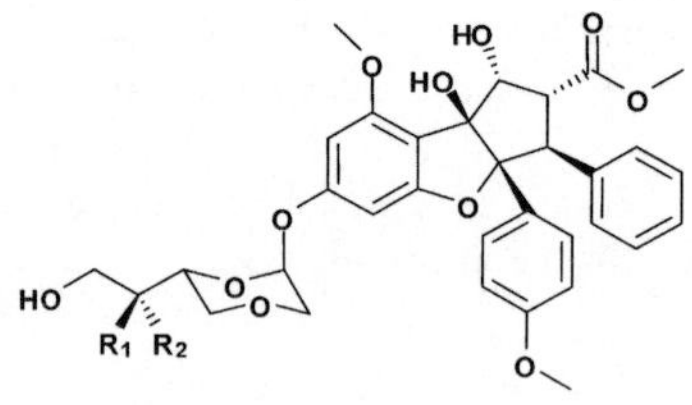

71. Silvestrol; R_1 = H, R_2 = OH
72. *epi*-Silvestrol; R_1 = OH, R_2 = H

6.3 Plant Metabolites and the Fungal Connection

Until a few years ago, questioning whether plants were the actual source of some of the important antitumour-active materials isolated from them by chemists, would probably have led to a series of derisory comments from most scientists, but from more recent work, this question is now reminiscent of similar comments and questions made by marine natural products chemists 20 or so years ago with respect to sponge-derived molecules.

Easier access to work performed in the People's Republic of China in the last 5 years, has revealed a significant number of reports that the four major classes of apparently plant-derived molecules, the taxanes, camptothecins, podophyllotoxins and even the vinca alkaloids can be produced by endophytic fungi found in plants. These fungi have been isolated, purified and fermented on scales of up to 10 litres and found to produce the same molecules that were originally isolated from the plants themselves. Strobel at Montana State had demonstrated over 15 years ago that a new genus of endophytic fungus isolated from *Taxus* trees that produced or contained paclitaxel would make very small amounts of the compound when fermented in the laboratory. This finding he later extended to identify the sources as *Taxomyces*[180] and many *Pestalotiopsis* species.[181] Following on from these reports came the realization that this was not an isolated phenomenon as, over the years, camptothecin,[182,183] podophyllotoxin,[184,185] vinblastine,[186] and vincristine[187,188] have been reported to be produced by fermentation of endophytic fungi isolated from the original source plants.

That very specific control mechanisms are in play in these fungi was demonstrated by the work of Keller's group on the common soil fungus, *Aspergillus nidulans*. The expression of the 40 plus cryptic secondary metabolite clusters that her group has identified from sequence studies was controlled by the presence of a control protein *Lae*A.[189] Over the last two years, a number of related investigations of soil fungi have been reported in five papers. One was a primary report, demonstrating the discovery of the emericellamide **73** biosynthetic pathway,[162] and four were review papers covering aspects of activation of fungal secondary metabolite genes.[190-193] The identification of the genes and gene products controlling metabolite production by these microbes could provide an entry into greatly increased production of key bioactive natural products and result in the production of a large number of previously unknown active materials that are not necessarily antitumour agents and will require the skills of synthetic and medicinal chemists in the near future.

In addition, these discoveries bring up the intriguing possibility that there exists some form of signalling analogous to bacterial quorum sensing between these fungal endophytes and the plant itself for the production of toxic secondary metabolites when attacked by a predator. Such a system would make more "economic sense" from the perspective of

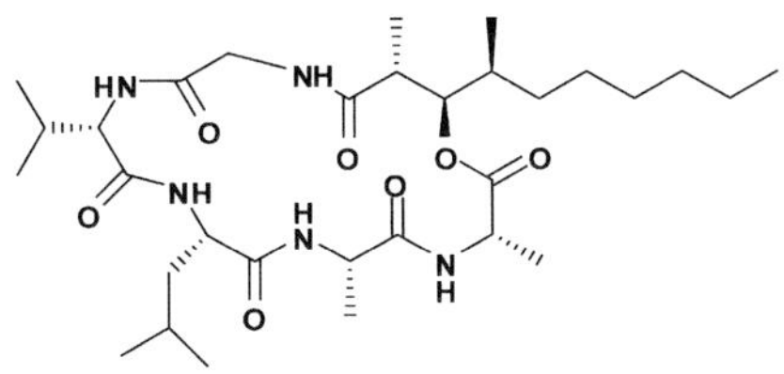

73. Emericellamide A

"energy balances" rather than postulating production centrally, followed by transport to the region under attack and then removal of a material that is toxic to the host as well as the predator. However this suggestion is simply an intriguing hypothesis from arguing by analogy.

7. IN CONCLUSION

The natural product world, with examples from all three domains of life, is a prolific producer of secondary metabolites with significant activity as both drugs and leads thereto, thus proving the validity of the chapter's title. But over the last few years, investigation of the genomic sequences of representatives of these organisms, in particular the marine and terrestrial microbial communities, including examples that were thought to be "very well investigated in prior years", has shown that the number of metabolites so far isolated and identified represent a small proportion of the biosynthetic gene clusters (cryptic clusters) that are now beginning to yield their secrets using both modified fermentation conditions and activation of specific clusters by expression in other hosts.[194]

It is interesting to speculate on the source of the plant secondary metabolites that are in use as either antitumour drugs or as leads to newer structures. Recent work on the biosynthetic machinery of the vinca alakoids in hairy root cultures of *Catharanthus roseus* does not appear to invoke endophytic microbes, but the cultures, though outwardly sterile, have not been checked for the presence of endophytes (Personal Communication, Sarah O'Connor, MIT).

Only time will tell whether some of the metabolites now isolated from plants have a "microbe in their background", as has been found for a large number of marine invertebrate-sourced metabolites. Horinouchi recently published a paper on the combinatorial biosynthesis of plant medicinal polyketides in microorganisms which makes very interesting reading.[195] This paper should be read in conjunction with his earlier work with *Streptomyces* reporting the discovery of the type III PKS known as RppA[196] and coumarate/cinnamate:CoA ligase enzymes which were once considered to be plant specific.[197]

Thus we can finish with the thought that we may run out of funding to investigate these metabolite sources, but we will not run out of sources of metabolites to investigate!

References

1 D.J. Newman and G.M. Cragg, *J. Nat. Prod.*, 2007, **70**, 461.
2 D.J. Newman, G.M. Cragg and K.M. Snader, *J. Nat. Prod.*, 2003, **66**, 1022.
3 C. Bailly, *Biochem. Pharmacol.*, 2009, **77**, 1447.

4 W. Bergmann and R.J. Feeney, *J. Am. Chem. Soc.*, 1950, **72**, 2809.
5 W. Bergmann and R.J. Feeney, *J. Org. Chem.*, 1951, **16**, 981.
6 W. Bergmann and D.C. Burke, *J. Org. Chem.*, 1955, **20**, 1501.
7 G.M. Cragg, D.G.I. Kingston and D.J.Newman, eds., *Anticancer Agents from Natural Products*, Taylor and Francis, Boca Raton, FL, 2005, pp. 1-577.
8 R.W. Robey, T. Obrzut, S. Shukla, O. Polgar, S. Macalou, J.C. Bahr, A. Di Pietro, S.V. Ambudkar and S.E. Bates, *Cancer Chemother. Pharmacol.*, 2009, **64**, 575.
9 A. Dowlati, J. Posey, R.K. Ramanathan, L. Rath, P. Fu, A. Chak, S. Krishnamurthi, J. Brell, S. Ingalls, C.L. Hoppel, P. Ivy and S.C. Remick, *Cancer Chemother. Pharmacol.*, 2009, **65**, 73.
10 A. Tomillero and M.A. Moral, *Methods Find Exp. Clin. Pharmacol.*, 2009, **31**, 263.
11 Y. Oh, R.S. Herbst, H. Burris, A. Cleverly, L. Musib, M. Lahn and G. Bepler, *J. Clin. Oncol.*, 2008, **26**, 1135.
12 W. Jian, H. Yamashita, J.M. Levitt, S.P. Lerner and G. Sonpavde, *Mol. Cancer Ther.*, 2009, **8**, 1772.
13 H. Skvara, M. Dawid, E. Kleyn, B. Wolff, J.G. Meingassner, H. Knight, T. Dumortier, T. Kopp, N. Fallahi, G. Stary, C. Burkhart, O. Grenet, J. Wagner, Y. Hijazi, R.E. Morris, C. McGeown, C. Rordorf, C.E. Griffiths, G. Stingl and T. Jung, *J. Clin. Invest.*, 2008, **118**, 3151.
14 S. Domhan, M. Zeier and A. Abdollahi, *Nephrol. Dial. Transplant.*, 2009, **24**, 1097.
15 F.M. Arcamone, 'Anthracyclines' in *Anticancer Agents from Natural Products*, ed. G. M. Cragg, D.G.I. Kingston and D.J. Newman, Taylor and Francis, Boca Raton, FL, 2005, pp. 299-320.
16 R.D. Olson, M.B. Headley, A. Hodzic, G.M. Walsh and D.G. Wingett, *Int. Immunopharmacol.*, 2007, **7**, 734.
17 K. Edo, M. Mizugaki, Y. Koide, H. Seto, K. Furihata, N. Otake and N. Ishida, *Tet. Letts.*, 1985, **26**, 331.
18 P.R. Hamann, J. Upeslacis and D.B. Borders, 'Enediynes' in *Anticancer Agents from Natural Products*, ed. G.M. Cragg, D.G.I. Kingston and D.J. Newman, Taylor and Francis, Boca Raton, Fl, 2005, pp. 451-474.
19 S.G. Van Lanen and B. Shen, *Curr. Topics Med. Chem.*, 2008, **8**, 448.
20 C. Hertweck, *Angew. Chem. Int. Ed.*, 2009, **48**, 4688.
21. A.M. Scott, R.A. Herbertson, F.T. Lee, B. Chappell, N. Micallef, S.T. Lee, T. Saunder, W. Hopkins, F.E. Smyth and N.C. Tebutt, *J. Clin. Oncol.*, 2008, **26**, Abst. 3025.
22 J.F. DiJoseph, M.M. Dougher, L.B. Kalyandrug, D.C. Armellino, E.R. Boghaert, P. R. Hamann, J.K. Moran and N.KDamle, *Clin. Cancer Res.*, 2006, **12**, 242.
23 J. Castillo, E. Winer and P. Quesenberry, *Exp. Hematol.*, 2008, **36**, 755.
24 J. Davies, H. Wang, T.T.K. Warabi, X.H. Huang and R.J. Andersen, *Org. Lett.*, 2005, **7**, 5233.
25 K.C. Nicolaou, J.S. Chen, H. Zhang and A. Montero, *Angew. Chem. Int. Ed.*, 2008, **47**, 185.
26 S.N. Sehgal, H. Baker and C. Vezina, *J. Antibiot.*, 1975, **28**, 727.
27 C. Vezina, A. Kudelski and S.N. Sehgal, *J. Antibiot.*, 1975, **28**, 721.
28 H. Baker, A. Sidorowicz, S.N. Sehgal and C. Vezina, *J. Antibiot.*, 1978, **31**, 539.
29 C.P. Eng, S.N. Sehgal and C. Vezina, *J.Antibiot.*, 1984, **37**, 1231.
30 S. Windecker, P.W. Serruys, S. Wandel, P. Buszman, S. Trznadel, A. Linke, K. Lenk, T. Ischinger, V. Klauss, F. Eberli, R. Corti, W. Wijns, M.-C. Morice, C. di Mario, S. Davies, R.-J. van Geuns, P. Eerdmans, G.-A. van Es, B. Meier and P. Jüni, *Lancet* 2008, **372**, 1163.

31 F.E. Koehn, *Curr. Opin. Biotech.*, 2006, **17**, 631.
32 C.K. Tsang, H. Qi, L.F. Liu and X.F.S. Zheng, *Drug Discov. Today*, 2007, **12**, 112.
33 B. Ruan, K. Pong, F. Jow, M. Bowlby, R.A. Crozier, D. Liu, S. Liang, Y. Chen, M. L. Mercado, X. Feng, F. Bennett, D. von Schack, L. McDonald, M.M. Zaleska, A. Wood, P.H. Reinhart, R.L. Magolda, J. Skotnicki, M.N. Pangalos, F.E. Koehn, G. T. Carter, M. Abou-Gharbia and E.I. Graziani, *Proc. Natl. Acad. Sci. USA*, 2008, **105**, 33.
34 E.I. Graziani, *Nat. Prod. Rep.*, 2009, **26**, 602.
35 B. Chambraud, H. Belabes, V. Fontaine-Lenoir, A. Fellous and E.E. Baulieu, *Faseb. J.*, 2007, **21**, 2787.
36 G. Hoefle and H. Reichenbach, 'Epothilone, a myxobacterial metabolite with promising antitumor activity' in *Anticancer Agents from Natural Products*, ed. G.M. Cragg, D.G.I. Kingston and D.J. Newman, Taylor and Francis, Boca Raton, FL, 2005, pp. 413-450.
37 H. Reichenbach and G. Hoefle, *Drugs R&D*, 2008, **9**, 1.
38 U. Klar, B. Buchmann, W. Schwede, W. Skuballa, J. Hoffmann and R.B. Lichtner, *Angew. Chem. Int. Ed.*, 2006, **45**, 7942.
39 K.-H. Altmann and J. Gertsch, *Nat. Prod. Rep.*, 2007, **24**, 327.
40 K.H. Altmann, B. Pfeiffer, S. Arseniyadis, B.A. Pratt and K.C. Nicolaou, *ChemMedChem*, 2007, **2**, 396.
41 F. Feyen, F. Cachoux, J. Gertsch, M. Wartmann and K.H. Altmann, *Acc. Chem. Res.*, 2008, **41**, 21.
42 K.H. Altmann and K. Memmert, 'Epothilones as lead structures for new anticancer drugs - pharmacology, fermentation, and structure - activity relationships' in *Progress in Drug Research*, ed. F. Petersen and R. Amstutz, Birkhauser Verlag, Basel, 2008, 65, pp. 275-334.
43 B.R. Hearn, D. Zhang, Y. Li and D.C. Myles, *Org. Lett.*, 2006, **8**, 3057.
44 Y.S. Cho, K.D. Wu, M.A.S. Moore, T.C. Chou and S.J. Danishefsky, *Drugs Fut.*, 2005, **30**, 737.
45 T.C. Chou, X.G. Zhang, H. Dong, Z. Zhong, L. Feng, Y. Li, M. Sherrill, P. Timmermanns, R.G. Johnson Jr and S.J. Danishefsky, *Proc. Am. Assoc. Cancer Res.*, 2008, **49**, Abst 1402.
46 T.C. Chou, X. Zhang, Z. Zhong, Y. Li, L. Feng, S. Eng, D. C. Myles, R. G. Johnson Jr, N. Wu, Y.I. Yin, M.W. Wilson and S.J. Danishefsky, *Proc. Nat. Acad. Sci. USA*, 2008, **105**, 13157.
47 T.C. Chou, X.G. Zhang, H. Dong and S.J. Danishefsky, *Proc. Am. Assoc. Cancer Res.*, 2009, **50**, Abst. 5565.
48 C.P. Leamon, *Curr. Opin. Investig. Drugs*, 2008, **9**, 1277.
49 B. Frank and R. Müller, *Pharmaz. Zeit.*, 2007, **152**, 64.
50 S. Schneiker, O. Perlova, O. Kaiser, K. Gerth, A. Alici, M.O. Altmeyer, D. Bartels, T. Bekel, S. Beyer, E. Bode, H.B. Bode, C.J. Bolten, J.V. Choudhuri, S. Doss, Y.A. Elnakady, B. Frank, L. Gaigalat, A. Goesmann, C. Groeger, F. Gross, L. Jelsbak, L. Jelsbak, J. Kalinowski, C. Kegler, T. Knauber, S. Konietzny, M. Kopp, L. Krause, D. Krug, B. Linke, T. Mahmud, R. Martinez-Arias, A.C. McHardy, M. Merai, F. Meyer, S. Mormann, J. Mũoz-Dorado, J. Perez, S. Pradella, S. Rachid, G. Raddatz, F. Rosenau, C. Rückert, F. Sasse, M. Scharfe, S.C. Schuster, G. Suen, A. Treuner-Lange, G.J. Velicer, F.J. Vorhölter, K.J. Weissman, R.D. Welch, S.C. Wenzel, D.E. Whitworth, S. Wilhelm, C. Wittmann, H. Blöcker, A. Pühler and R. Müller, *Nat. Biotech.*, 2007, **25**, 1281.
51 S.C. Wenzel and R. Müller, *Nat. Prod. Rep.* , 2007, **24**, 1211.

52 A.J. Frew, R.W. Johnstone and J.E. Bolden, *Cancer Lett.*, 2009, **280**, 125.

53 T.L. Newkirk, A.A. Bowers and R.M. Williams, *Nat. Prod. Rep.*, 2009, **26**, 1293.

54 M.W. Taylor, R.T. Hill, P.J.R. W. Thacker and U. Hentschel, *ISME J.*, 2007, **1**, 187.

55 P.G. Williams, *Trends Biotech.*, 2009, **27**, 45.

56 T.A. Nguyen, K. Ishida, H. Jenke-Kodama, E. Dittmann, C. Gurgui, T. Hochmuth, S. Taudien, M. Platzer, C. Hertweck and J. Piel, *Nat. Biotech.*, 2008, **26**, 225.

57 D.J. Newman, 'The Bryostatins' in *Anticancer Agents from Natural Products*, ed. G.M. Cragg, D.G.I. Kingston and D.J. Newman, Taylor and Francis, Boca Raton, FL, 2005, pp. 137-150.

58 E. Flahive and J. Srirangam, 'The Dolastatins: Novel Antitumor Agents from *Dolabella auricularia*' in *Anticancer Agents from Natural Products*, ed. G.M. Cragg, D.G.I. Kingston and D.J. Newman, Taylor and Francis, Boca Raton, FL, 2005, pp. 191-214.

59 N.B. Lopanik, J.A. Shields, T.J. Buchholz, C.M. Rath, J. Hothersall, M.G. Haygood, K. Hakansson, C.M. Thomas and D.H. Sherman, *Chem. Biol.*, 2008, **15**, 1175.

60 A. Charette, *Nature*, 2008, **456**, 451.

61 B.M. Trost and G. Dong, *Nature*, 2008, **456**, 485.

62 P.A. Wender, B.A. DeChristopher and A.J. Schrier, *J. Am. Chem. Soc.*, 2008, **130**, 6658.

63 P.A. Wender, V.A. Verma, T.J. Paxton and T.H. Pillow, *Acc. Chem. Res.*, 2008, **41**, 40.

64 G.E. Keck, M.B. Kraft, A.P. Truong, W. Li, C.C. Sanchez, N. Kedei, N.E. Lewin and P.M. Blumberg, *J. Am. Chem. Soc.*, 2008, **130**, 6660.

65 M.K. Sun, J. Hongpaisan, T.J. Nelson and D.L. Alkon, *Proc. Nat. Acad. Sci. USA*, 2008, **105**, 13620.

66 D. Wang, D.S. Darwish, B.G. Schreurs and D.L. Alkon, *Behav. Pharmacol.*, 2008, **19**, 245.

67 H. Luesch, R.E. Moore, V.J. Paul, S.L. Mooberry and T.H. Corbett, *J. Nat. Prod.*, 2001, **64**, 907.

68 A.V. Ramaswamy, P.M. Flatt, D.J. Edwards, T.L. Simmons, B. Han and W.H. Gerwick, 'The secondary metabolites and biosynthetic gene clusters of marine cyanobacteria. Applications in biotechnology' in *Frontiers in Marine Biotechnology*, ed. P. Proksch and W.E.G. Mueller, Horizon Bioscience, Wymondham, UK, 2006, pp. 175-224.

69\. C.L. Law, K.A. Gordon, B.E. Toki, A.K. Yamane, M.A. Hering, C.G. Cerveny, J. M. Petroziello, M.C. Ryan, L. Smith, R. Simon, G. Sauter, E. Oflazoglu, S.O. Doronina, D.L. Meyer, J.A. Francisco, P. Carter, P.D. Senter, J.A. Copland, C.G. Wood and A.F. Wahl, *Cancer Res.*, 2006, **66**, 2328.

70 M.S.K. Sutherland, R.J. Sanderson, K.A. Gordon, J. Andreyka, C.G. Cerveny, C. Yu, T.S. Lewis, D.L. Meyer, R.F. Zabinski, S.O. Doronina, P.D. Senter, C.L. Law and A.F. Wahl, *J. Biol. Chem.*, 2006, **281**, 10540.

71 E. Oflazoglu, K.M. Kissler, E.L. Sievers, I.S. Grewal and H-P. Gerber, *Brit. J. Haematol.*, 2008, **142**, 69.

72 V.A. Pollack, E. Alvarez, K.F. Tse, M.Y. Torgov, S. Xie, S.G. Shenoy, J.R. MacDougall, S. Arrol, Z. Zhong, R.W. Gerwein, W.F. Hahne, P.D. Senter, M.E. Jeffers, H.S. Lichtenstein and LaRochelle, *Cancer Chemother. Pharmacol.*, 2007, **60**, 423.

73 P. Hwu, M. Sznol, A. Pavlick, H. Kluger, K.B. Kim, P. Boasberg, D. Sanders, R. Simantov, E. Crowley and O. Hamid, *J. Clin. Oncol.*, 2009, **27**, Abst 9032.

74 D. Ma, C.E. Hopf, A.D. Malewicz, G.P. Donovan, P.D. Senter, W.F. Goeckeler, P. J. Maddon and W.C. Olson, *Clin. Cancer Res.*, 2006, **12**, 2591.
75 K. Rinehart, T.G. Holt, N.L. Fregeau, J.G. Stroh, P.A. Kiefer, F. Sun, L.H. Li and D.G. Martin, *J. Org. Chem.*, 1990, **55**, 4512.
76 A.E. Wright, D.A. Forleo, G.P. Gunawardana, S.P. Gunasekera, F.E. Koehn and O.J. McConnell, *J. Org. Chem.*, 1990, **55**, 4508.
77 R. Henriquez, G. Faircloth and C. Cuevas, 'Ecteinascidin 743 (Et-743; Yondelis^{TM}), Aplidin, and Kahalalide F' in *Anticancer Agents from Natural Products*, ed. G.M. Cragg, D.G.I. Kingston and D.J. Newman, Taylor & Francis, Boca Raton, FL, 2005, pp. 215-240.
78 A. Velasco, P. Acebo, A. Gomez, C. Schleissner, P. Rodriguez, T. Aparicio, S. Conde, R. Munoz, F. de la Calle, J.L. Garcia and J.M. Sanchez-Puelles, *Mol. Microbiol.*, 2005, **56**, 144.
79 C. Moss, D.H. Green, B. Perez, A. Velasco, R. Henriquez and J.D. McKenzie, *Mar. Biol.*, 2003, **143**, 99.
80 A.E. Perez-Matos, W. Rosado and N.S. Govind, *Anton. van Leeuwen.*, 2007, **92**, 155.
81 A. Fontana, P. Cavaliere, S. Wahidulla, C.G. Naik and G. Cimino, *Tetrahedron*, 2000, **56**, 7305.
82 J.W. Lane, A. Estevez, K. Mortara, O. Callan, J. R. Spencer and R.M. Williams, *Bioorg. Med. Chem. Lett.*, 2006, **16**, 3180.
83 V. Moneo, J.F. Martinez-Leal, M.J. Guillen, P. Martinez, G. Tarazona, P.M. Aviles, C. Cuevas, L.F. Garcia-Fernandez and C.M. Galmarini, *Proc Am Assoc Cancer Res., (AACR)* 2009, **50**, Abst 4525.
84 R. Sakai, K.L. Rinehart, V. Kishore, B. Kundu, G. Faircloth, J.B. Gloer, J. R. Carney, M. Manikoshi, F. Sun, R. G. Hughes Jr, D. Garcia-Gravalos, T. Garcia de Quesada, G. R. Wilson and R. M. Heid, *J. Med. Chem.*, 1996, **39**, 2819.
85 C.S. Mitsiades, E.M. Ocio, A. Pandiella, P. Maiso, C. Gajate, M. Garayoa, D. Vilanova, J.C. Montero, N. Mitsiades, C.J. McMullan, N.C. Munshi, T. Hideshima, D. Chauhan, P. Aviles, G. Otero, G. Faircloth, M.V. Mateos, P.G. Richardson, F. Mollinedo, J.F. San-Miguel and K.C. Anderson, *Cancer Res.*, 2008, **68**, 5216.
86 T. Eisen, J. Thomas, W.H. Miller Jr, M. Gore, P. Wolter, P. Kavan, J.A. Martín and P. Lardelli, *Melanoma Res.*, 2009 **19**, 185.
87 M. Broggini, S.V. Marchini, E. Galliera, P. Borsotti, G. Taraboletti, E. Erba, M. Sironi, J. Jimeno, G.T. Faircloth, R. Giavazzi and M. D'Incalci, *Leukemia*, 2003, **17**, 52.
88 E. Erba, M. Serafini, G. Gaipa, G. Tognon, S. Marchini, N. Celli, D. Rotilio, M. Broggini, J. Jimeno, G.T. Faircloth, A. Biondi and M. D'Incalci, *Brit. J. Cancer*, 2003, **89**, 763.
89 J. Jimeno, M. Aracil and J.C. Tercero, *J. Trans. Med.*, 2006, **4:3**.
90 Y. Suarez, L. Gonzalez-Santiago, N. Zarich, A. Davalos, J.F. Aranda, M.A. Alonso, M.A. Lasuncion, J.M. Rojas and A. Munoz, *Mol. Pharmacol.*, 2006, **70**, 1654.
91 M.J. Muñoz-Alonso, L. González-Santiago, N. Zarich, T. Martínez, E. Alvarez, J.M. Rojas and A. Muñoz, *J. Pharmacol. Exp. Ther.*, 2008, **324**, 1093.
92 V. Moneo, B.G. Serelde, J.F.M. Leal, C. Blanco-Aparicio, R. Diaz-Uriarte, M. Aracil, J.C. Tercero, J. Jimeno and A. Carnero, *Mol. Cancer Ther.*, 2007, **6**, 1310.
93 R. Nalda-Molina, B. Valenzuela, A. Ramon-Lopez, B. Miguel-Lillo, A. Soto-Matos and J.J. Perez-Ruixo, *Cancer Chemother. Pharmacol.*, 2009, **64**, 97.
94 M. Vera and M.M. Joullie, *Med. Res. Revs.*, 2002, **22**, 102.
95 F. Cardenas, J.M. Caba, M. Feliz, P. Lloyd-Williams and E. Giralt, *J. Org. Chem.*, 2003, **68**, 9554.

96 M. Gutierrez-Rodriguez, M. Martin-Martinez, M.T. Garcia-Lopez, R. Herranz, F. Cuevas, C. Polanco, I. Rodriguez-Campos, I. Manzanares, F. Cardenas, M. Feliz, P. Lloyd-Williams and E. Giralt, *J. Med. Chem.*, 2004, **47**, 5700.
97 J. Adrio, C. Cuevas, I. Manzanares and M.M. Joullie, *J. Org. Chem.*, 2007, **72**, 5129.
98 M.T. Hamann, C.S. Otto, P.J. Scheuer and D.C. Dunbar, *J. Org. Chem.*, 1996, **61**, 6594.
99 A. Lopez-Macia, J.C. Jimenez, M. Royo, E. Giralt and F. Alberico, *J. Amer. Chem. Soc.*, 2001, **123**, 11398.
100 M. Garcia-Rocha, P. Bonay and J. Avila, *Cancer Lett.*, 1996, **99**, 43.
101 Y. Suarez, L. Gonzalez, A. Cuadrado, M. Berciano, M. Lafarga and A. Munoz, *Mol. Cancer Ther.*, 2003, **2**, 863.
102 J.M. Sewell, I. Mayer, S.P. Langdon, J.F. Smyth, D.I. Jodrell and S.M. Guichard, *Eur. J. Cancer*, 2005, **41**, 1637.
103 M.L. Janmaat, J.A. Rodriguez, J. Jimeno, F.A.E. Kruyt and G. Giaccone, *Mol. Pharmacol.*, 2005, **68**, 502.
104 A.M. Piggott and P. Karuso, *ChemBioChem*, 2008, **9** 524.
105 A.G. Shilabin, N. Kasanah, D.E. Wedge and M.T. Hamann, *J. Med. Chem.*, 2007, **50** 4340.
106 J. C. Jiménez, A. López-Macià, C. Gracia, S. Varón, M. Carrascal, J.M. Caba, M. Royo, A.M. Francesch, C. Cuevas, E. Giralt and F. Albericio, *J. Med. Chem.*, 2008, **51**, 4920.
107 R.T. Hill, M.T. Hamann, J. Enticknap and K.V. Rao, WO 2005042720 A2, *Chem. Abs*tr, **142**, 444570
108 M. Ashour, R.A. Edrada, R. Ebel, V. Wray, W. Waetjen, K. Padmakumar, W. E. G. Mueller, W.H. Lin and P. Proksch, *J. Nat. Prod.*, 2006, **69**, 1547.
109 K.V. Rao, M.K. Na, J. C. Cook, J. Peng, R. Matsumoto and M.T. Hamann, *J. Nat. Prod.*, 2008, **71** 772.
110 J.J. Kowalczyk, G. Kuznetsov, H. Yang, B.M. Seletsky, M. Spyvee and S. Schiller, *US. Pat.*,7,192,972.
111 S. Madajewicz, N.J. Zojwalla, A.G. Lucarelli, P. Hentschel, J. Giardelli, E. L. Schuck, I. Krivelevich, B.Y. Wong, N. J. Vogelzang and S. Sharma, *J. Clin. Oncol.*, 2007, **25**, 2550 (Abs).
112 N.J. Zojwalla, C.H. Takimoto, A.G. Lucarelli, R. Clark, A.C. Mita, M.M. Mita, O. Romero, E.L. Schuck, I. Krivelevich and C.J. Sweeney, *J. Clin. Oncol.,* 2007, **25**, Abst 2543
113 C.S. Rocha Lima, E.H. Rubin, L.E. Raez, E. Poplin, J. McIntyre, A.M. Flores, J.R. Merchan, A. G. Lucarelli, E. Schuck and N. J. Zojwalla, *Gastrointest Cancers Symp*,, 2008, Abst 326.
114 T.D. Aicher, K.R. Buszek, F.G. Fang, C.J. Forsyth, S.H. Jung, Y. Kishi, M.C. Matelich, P.M. Scola, D.M. Spero and S.K. Yoon, *J. Amer. Chem. Soc.*, 1992, **114**, 3162.
115 M.A. Jordan, K. Kamath, T. Manna, T. Okouneva, H. P. Miller, C. Davis, B. A. Littlefield and L. Wilson, *Mol. Cancer Ther.*, 2005, **4**, 1086.
116 D.A. Dabydeen, J.C. Burnett, R. Bai, P. Verdier-Pinard, S.J.H. Hickford, G.R. Pettit, J.W. Blunt, M.H.G. Munro, R. Gussio and E. Hamel, *Mol. Pharmacol.*, 2006, **70**, 1866.
117 M.J.Yu, Y. Kishi and B.A. Littlefield, 'Discovery of E7389, a Fully Synthetic Macrocyclic Ketone Analog of Halichondrin B' in *Anticancer Agents from Natural Products*, ed. G.M. Cragg, D.G.I. Kingston and D.J. Newman, Taylor and Francis, Boca Raton, FL, 2005, pp. 241-265.

118 Y. Wang, *Drugs Fut.*, 2007, **32**, 681.
119 J. Jimeno, *Clin. Cancer Res.*, 2009, **15**, 3903.
120 S. Goel, A.C. Mita, M.M. Mita, E. . Rowinsky, Q.C. Chu, N. Wong, C. Desjardins, F. Fang, M. Jansen, D.E. Shuster, S. Mani and C.H. Takimoto, *Clin. Cancer Res.*, 2009, **15**, 4207.
121 A.R. Tan, E.H. Rubin, D.C. Walton, D.E. Shuster, N. Wong, F. Fang, S. Asworth and L.S. Rosen, *Clin. Cancer Res.*, 2009, **15**, 4213.
122 L.T. Vahdat, B. Pruitt, C.J. Fabian, R.R. Rivera, D.A. Smith, E. Tan-Chiu, W.J.A. R. Tan, N.A. Dacosta, E. Chuang, J. Smith, J. O'Shaughnessy, D.E. Shuster, N.L. Meneses, K. Chandrawansa, F. Fang, P.E. Cole, S. Ashworth and J.L. Blum, *J. Clin. Oncol.*, 2009, **27**, 2954.
123 D.J. Newman and R.T. Hill, *J. Ind. Microbiol. Biotechnol.*, 2006, **33**, 539.
124 W. Fenical and P.R. Jensen, *Nat. Chem. Biol.*, 2006, **2**, 666.
125 P.R. Jensen and W. Fenical, 'New natural-product diversity from marine actinomycetes' in *Natural Products*, ed. L. Zhang and A.L. Demain, Humana Press, Ltd, Totowa, NJ, 2005, pp. 315-328.
126 P.R. Jensen, T.J. Mincer, P.G. Williams and W. Fenical, *Anton. van Leeuwen.*, 2005, **87**, 43.
127 P.R. Jensen, P.G. Williams, D.C. Oh, L. Zeigler and W. Fenical, *App. Environ. Microbiol.*, 2007, **73**, 1146.
128 R.H. Feling, G.O. Buchanan, T.J. Mincer, C.A. Kauffman, P.R. Jensen and W. Fenical, *Angew. Chem. Int. Ed.*, 2003, **42**, 355.
129 L.R. Reddy, P. Saravanan and E.J. Corey, *J. Am. Chem. Soc.*, 2004, **126**, 6230.
130 A. Endo and S.J. Danishefsky, *J. Amer. Chem. Soc.*, 2005, **127**, 8298.
131 V.R. Macherla, S.S. Mitchell, R.R. Manam, K.A. Reed, T.H. Chao, B. Nicholson, G. Deyanat-Yazdi, B. Mai, P.R. Jensen, W.F. Fenical, S.T.C. Neuteboom, K.S. Lam, M.A. Palladino and B.C.M. Potts, *J. Med. Chem.*, 2005, **48**, 3684.
132 M. Shibasaki, M. Kanai and N. Fukuda, *Chem. Asian J.*, 2007, **2**, 20.
133 T. Ling, V.R. Macherla, R.R. Manam, K.A. McArthur and B.C.M. Potts, *Org. Lett.*, 2007, **9**, 2289.
134 T. Fukuda, K. Sugiyama, S. Arima, Y. Harigaya, T. Nagamitsu and S. Omura, *Org. Lett.*, 2008, **10**, 4239.
135 R.R. Manam, K.A. McArthur, T.H. Chao, J. Weiss, J.A. Ali, V. J. Palombella, M. Groll, G.K. Lloyd, M.A. Palladino, S.T.C. Neuteboom, V.R. Macherla and B.C. M. Potts, *J. Med. Chem.*, 2008, **51**, 6711.
136 K.S. Lam, G. Tsueng, K.A. McArthur, S.S. Mitchell, B.C. M. Potts and J. Xu, *J. Antibiot.*, 2007, **60**, 13.
137 K.A. Reed, R.R. Manam, S.S. Mitchell, J. Xu, S. Teisan, T.H. Chao, G. Deyanat-Yazdi, S.T.C. Neuteboom, K.S. Lam and B.C.M. Potts, *J. Nat. Prod.*, 2007, **70**, 269.
138 G. Tsueng and K.S. Lam, *J. Antibiot.*, 2007, **60**, 469.
139 G. Tsueng, K.A. McArthur, B.C.M. Potts and K.S. Lam, *Appl. Microbiol. Biotech.*, 2007, **75**, 999.
140 G. Tsueng and K.S. Lam, *Appl. Microbiol. Biotech.*, 2008, **78**, 821.
141 M. Groll, R. Huber and B.C.M. Potts, *J. Am. Chem. Soc.*, 2006, **128**, 5136.
142 N. Denora, B.C.M. Potts and V.J. Stella, *J. Pharm. Sci.*, 2007, **96**, 2037.
143 R. Kurzrock, P. Hamlin, A. Younes, D. Hong, M. Gordon, M.A. Spear, M.A. Palladino, G.K. Lloyd, A.M. Longenecker, S.T.C. Neuteboom, G.F. Cropp, A. Hannah, C. Aghajanian and A.L. Hannah, *Blood*, 2007, **110**, Abst. 4504.

144 P.A. Hamlin, C. Aghajanian, A. Younes, D.S. Hong, M.A. Palladino, A.M. Longenecker, G.K. Lloyd, A.L. Hannah, M.A. Spear and R. Kurzrock, *J. Clin. Oncol.*, 2009, **27**, Abst 3516.
145 C.C. Hofmeister, P. Richardson, T. Zimmerman, M.A. Spear, M.A. Palladino, A.M. Longenecker, G.F. Cropp, G.K. Lloyd, A.L. Hannah and K. Anderson, *J. Clin. Oncol.*, 2009, **27**, Abst 8505.
146 W. Fenical, P.R. Jensen, M.A. Palladino, K.S. Lam, G.K. Lloyd and B.C.M. Potts, *Bioorg. Med. Chem.*, 2009, **17**, 2175.
147 M. Stadler, J. Bitzer, A. Mayer-Bartschmid, H. Mueller, J. Benet-Buchholz, F. Gantner, H.V. Tichy, P. Reinemer and K.B. Bacon, *J. Nat. Prod.*, 2007, **70**, 246.
148 G. Tsueng, S. Teisan and K.S. Lam, *Appl. Microbiol. Biotech.*, 2008, **78**, 827.
149 J. Handelsman, M.R. Rondon, S.F. Brady, J. Clardy and R.M. Goodman, *Chem. Biol.*, 1998, **5**, R245.
150 M.R. Rondon, P.R. August, A.D. Bettermann, S.F. Brady, T.H. Grossman, M.R. Liles, K.A. Loiacono, B.A. Lynch, I.A. MacNeil, C. Minor, C.L. Tiong, M. Gilman, M.S. Osburne, J. Clardy, J. Handelsman and R.M. Goodman, *Appl. Environ. Micro.*, 2000, **66**, 2541.
151 D.B. Rusch, A.L. Halpern, G. Sutton, K.B. Heidelberg, S. Williamson, S. Yooseph, D. Wu, J.A. Eisen, J.M. Hoffman, K. Remington, K. Beeson, B. Tran, H. Smith, H. Baden-Tillson, C. Stewart, J. Thorpe, J. Freeman, C. Andrews-Pfannkoch, J.E. Venter, K. Li, S. Kravitz, J.F. Heidelberg, T. Utterback, Y.H. Rogers, L.I. Falcon, V. Souza, G. Bonilla-Rosso, L.E. Eguiarte, D.M. Karl, S. Sathyendranath, T. Platt, E. Bermingham, V. Gallardo, G. Tamayo-Castillo, M.R. Ferrari, R.L. Strausberg, K. Nealson, R. Friedman, M. Frazier and J.C. Venter, *PLoS Biology*, 2007, **5**, 398.
152 S. Ōmura, H. Ikeda, J. Ishikawa, A. Hanamoto, C. Takahashi, M. Shinose, Y. Takahashi, H. Horikawa, H. Nakazawa, T. Osonoe, H. Kikuchi, T. Shiba, Y. Sakaki and M. Hattori, *Proc. Nat. Acad. Sci. USA*, 2001, **98**, 12215.
153 H. Ikeda, J. Ishikawa, A. Hanamoto, M. Shinose, H. Kikuchi, T. Shiba, Y. Sakaki, M. Hattori and S. Omura, *Nat. Biotech.*, 2003, **21**, 526
154 S.D. Bentley, K.F. Chater, A.M. Cerdeno-Tarraga, G.L. Challis, N.R. Thomson, K. D. James, D.E. Harris, M.A. Quail, H. Kieser, D. Harper, A. Bateman, S. Brown, G. Chandra, C.W. Chen, M. Collins, A. Cronin, A. Fraser, A. Goble, J. Hidalgo, T. Hornsby, S. Howarth, C.H. Huang, T. Kieser, L. Larke, L. Murphy, K. Oliver, S. O'Neil, E. Rabbinowitsch, M.A. Rajandream, K. Rutherford, S. Rutter, K. Seeger, D. Saunders, S. Sharp, R. Squares, S. Squares, K. Taylor, T. Warren, A. Wietzorrek, J. Woodward, B.G. Barrell, J. Parkhill and D.A. Hopwood, *Nature*, 2002, **417**, 141
155 DW Udwary, L. Zeigler, R.N. Asolkar, V. Singan, A. Lapidus, W. Fenical, P.R. Jensen and B.S. Moore, *Proc. Nat. Acad. Sci. USA* 2007, **104**, 10376.
156 A.S. Eustaquio, F. Pojer, J.P. Noel and B.S. Moore, *Nat. Chem. Biol.*, 2008, **4**, 69.
157 A.S. Eustaquio and B.S. Moore, *Angew. Chem. Int. Ed.*, 2008, **47**, 3936.
158 R.P. McGlinchey, M. Nett, A.S. Eustaquio, R.N. Asolkar, W. Fenical and B.S. Moore, *J. Am. Chem. Soc.*, 2008, **130**, 7822.
159 R.R. Manam, V.R. Macherla, G. Tsueng, C.W. Dring, J. Weiss, S.T.C. Neuteboom, K.S. Lam and B.C.M. Potts, *J. Nat. Prod.*, 2009, **72**, 295.
160 M. Nett and B.S. Moore, *Pure Appl. Chem.*, 2009, **81**, 1075.
161 M. Nett, H. Ikeda and B.S. Moore, *Nat. Prod. Rep.*, 2009.
162 Y.M. Chiang, E. Szewczyk, T. Nayak, A.D. Davidson, J.F. Sanchez, H.C. Lo, W.Y. Ho, H. Simityan, E. Kuo, A. Praseuth, K. Watanabe, B.R. Oakley and C.C.C. Wang, *Chem. Biol.*, 2008, **15**, 527.

163 S.F. Brady, T.L. Simmons, J.H. Kim and E.W. Schmidt, *Nat. Prod. Rep.*, 2009, DOI 10.1039/b817078a.
164 S.C. Wenzel and R. Muller, *Nat. Prod. Rep.*, 2009, DOI 10.1039/b817073h.
165 E.A. Gontang, W. Fenical and P.R. Jensen, *Appl. Environ. Micro.*, 2007, **73**, 3272.
166 J.A. Kalaitzis, F.M. Lauro and B.A. Neilan, *Nat. Prod. Rep.*, 2009, DOI, 10.1039/b817074f.
167 K. Hostettmann and A. Marston, *Chimia*, 2007, **61**, 322.
168 O. Potterat and M. Hamburger, 'Drug Discovery and development with plant-derived products' in *Progress in Drug Reearch*, ed. F. Petersen and R. Amstutz, Burkhauser Verlag, Basel, 2008, **65**, pp. 47-118.
169 G.M. Cragg and D.J. Newman, *Phytochem. Rev.*, 2009, **8**, 313.
170 T.W. Yu and H.G. Floss, 'Ansamitocins (Maytansinoids)' in *Anticancer Agents from Natural Products*, ed. D.G.I. Kingston, G.M. Cragg and D.J. Newman, Taylor and Francis, Boca Raton, FL, 2005, pp. 321-337.
171 H. Ikeda, T. Hideshima, M. Fulciniti, R.J. Lutz, H. Yasui, Y. Okawa, T. Kizitepe, S. Vallet, S. Pozzi, L. Santo, B. Daiken, S. Aigner, F. Osterroth, N.C. Munshi, P. Richardson and K.C. Anderson, *Clin. Cancer Res.*, 2009, **15**, 4028.
172 A.G. Polson and M.X. Sliwkowski, *Clin. Cancer Res.*, 2009, **15**, 3906.
173 P.D. Senter, *Curr. Opin. Chem. Biol.*, 2009, **13**, 235.
174. J.P. Stephan, P. Chan, C. Lee, C. Nelson, J.M. Elliou, C. Bechtel, H. Raab, D. Xie, J. Akutagawa, J. Baudys, O. Saad, S. Prabhu, W.L.T. Wong, R. Vandlen, F. Jacobson and A. Ebens, *Bioconjugate Chem.*, 2008, **19**, 1673.
175 P.S. Low and S.A. Kularatne, *Curr. Opin. Chem. Biol.*, 2009, **13**, 256.
176 S. K. Lee, B. Cui, R.R. Mehta, A.D. Kinghorn and J.M. Pezzuto, *Chem-Biol. Interact.*, 1998, **115**, 215.
177 D.M. Lucas, R.B. Edwards, G. Lozanski, D.A. West, J.D. Shin, M.A. Vargo, M.E. Davis, D.M. Rozewski, A.J. Johnson, B.N. Su, V.M. Goettl, N.A. Heerema, T.S. Lin, A. Lehman, X. Zhang, D. Jarjoura, D.J. Newman, J.C. Byrd, A.D. Kinghorn and M.R. Grever, *Blood*, 2009, **113**, 4656.
178 A.D. Kinghorn, Y.W. Chin and S.M. Swanson, *Curr. Opin. Drug Discov. Devel.*, 2009, **12**, 189.
179 R. Cencic, M. Carrier, G. Galicia-Vazquez, M.E. Bordeleau, R. Sukarieh, A. Bourdeau, B. Brem, J.G. Teodoro, H. Greger, M.L. Tremblay, J.A. Porco, Jr. and J. Pelletier, *PLoS ONE*, 2009, **4**, e5223.
180 A. Stierle, G. Strobel and D. Stierle, *Science*, 1993, **260**, 214.
181 J.Y. Li, R.S. Sidhu, A. Bollon and G.A. Strobel, *Mycolog. Res.*, 1998, **102**, 461.
182 S.C. Puri, V. Verma, T. Amna, G.N. Qazi and M. Spiteller, *J. Nat. Prod.*, 2005, **68**.
183 T. Amna, S.C. Puri, V. Verma, J.P. Sharma, R.K. Khajuria, J. Musarrat, M. Spiteller and G.N. Qazi, *Can. J. Microbiol.*, 2006, **52**, 189.
184 A.L. Eyberger, R. Dondapati and J.R. Porter, *J. Nat. Prod.*, 2006, **69**, 1121.
185 S.C. Puri, A. Nazir, R. Chawla, R. Arora, S. Riyaz-ul-Hasan, T. Amna, B. Ahmed, V. Verma, S. Singh, R. Sagar, A. Sharma, R. Kumar, R.K. Sharma and G.N. Qazi, *J. Biotech.*, 2006, **122**, 494.
186 B. Guo, H. Li and L. Zhang, *Yunnan Daxue Xuebao, Ziran Kexueban (J. Yunnan Univ.)* 1998, **20**, 214; (*Chem. Abs.*, **129**,172900).
187 L.Q. Zhang, B. Guo, H. Li, S. Zeng, H. Shao, S. Gu and R. Wei, *Zhong Cao Yao (Chinese Tradit. Herb. Drugs)*, 2000, **31**, 805.
188 X. Yang, L. Zhang, B. Guo and S. Guo, *Zhong Cao Yao (Chinese Tradit. Herb. Drugs)*, 2004, **35**, 79.

189 J. W. Bok, D. Hoffmeister, L.A. Maggio-Hall, R. Murillo, J.D. Glasner and N. P. Keller, *Chem. Biol.*, 2006, **13**, 31.
190 D. Hoffmeister and N.P. Keller, *Nat. Prod. Rep.*, 2007, **24**, 393.
191 A.A. Brakhage, J. Schuemann, S. Bergmann, K. Scherlach, V. Schroeckh and C. Hertweck, 'Activation of fungal silent gene clusters: A new avenue to drug discovery' in *Progress in Drug Research*, ed. F. Petersen and R. Amstutz, Birkhauser Verlag, Basel, 2008, **66**, pp. 3-12.
192 E.M. Fox and B.J. Howlett, *Curr. Opin. Microbiol.*, 2008, **11**, 481.
193 M. Stadler and N.P. Keller, *Mycolog. Res.*, 2008, **112**, 127.
194 F. Lefevre, P. Robe, C. Jarrin, A. Ginolhac, C. Zago, D. Auriol, T.M. Vogel, P. Simonet and R. Nalin, *Res. Microbiol.*, 2008, **159**, 153.
195 S. Horinouchi, *Curr. Opin. Chem. Biol.*, 2009, **13**, 197.
196 N. Funa, Y. Ohnishi, I. Fujii, Shibuya, M.,Y. Ebizuka and S. Horinouchi, *Nature*, 1999, **400**, 897.
197 M. Kaneko, Y. Ohnishi and S. Horinouchi, *J. Bacteriol.*, 2003, **185**, 20.

FROM NATURAL PRODUCTS TO MEDICINAL CHEMISTRY

N.H. Nicholson

Independent Scientist, Helions Bumpstead, Suffolk, CB9 7AT, England
E-mail: nxnicholson@gmail.com

1 INTRODUCTION

Our programme of work on the pain target MrgX1 illustrates the limitations of medicinal research based on lower animals, but here I propose that natural product mechanisms and technologies can be applied to our work and to other medicinal chemistry programmes to enhance drug candidates and to direct the selection of receptors for modulation in neurological disease.

1.1 The Classical Development of Medicines from Natural Products

Natural products have made an enormous contribution to medicinal chemistry,[1] and Figure 1 portrays some of the natural products which are landmarks in modern western medicine. These have followed the classical pattern of natural product discovery and development. The active natural product is isolated from a mixture which has a desired medicinal effect, such as the ability to inhibit bacterial growth and is then modified to make it more potent or to enhance its pharmacological profile.

Thus, the stereochemistry of the hydroxyl in compactin was inverted and a methyl added to give the successful cholesterol lowering agent lovastatin. The 2'-hydroxyl was removed from coformycin to give the anti-cancer drug pentostatin and many thousands of acyl side chains have been added to the penicillin nucleus in order to tailor its activity to particular bacteria, to counter bacterial resistance and to achieve a good pharmacological profile.

Some compounds have needed no improvement and others have presented a serious challenge to the medicinal chemist. The naturally occurring beta-lactamase inhibitor, clavulanic acid could not be bettered despite programmes of modification in two major pharmaceutical companies so it was combined unaltered with amoxicillin to overcome bacterial resistance to the penicillin. The natural product artemisinin is used as an antimalarial and it is difficult for the medicinal chemist to see how to modify this compound because the peroxide moiety looks too delicate to survive *in vivo*. Morphine and the antibiotic tetracycline are still used unaltered although their side effects have encouraged a search for modern replacements.

It is in antibiotics that natural products have had the greatest impact; to such an extent that the quinolones is the only major class of antibiotic that is not derived from natural

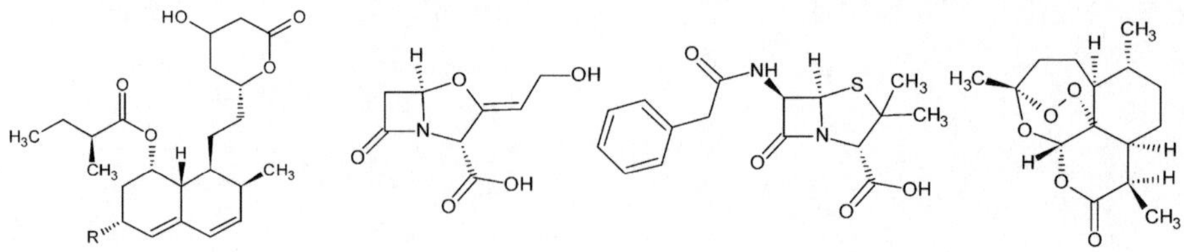

Lovastatin R = CH_3, OH = (*R*)
Compactin R = H, OH = (*S*)
Penicillium citrinum
HMG CoA reductase

Clavulanic acid
Streptomyces clavuligerus
–lactamase inhibitor

Benzyl penicillin
Penicillium notatum
Cell wall inhibitor

Artemisinin
Artmesia annua
Antimalarial

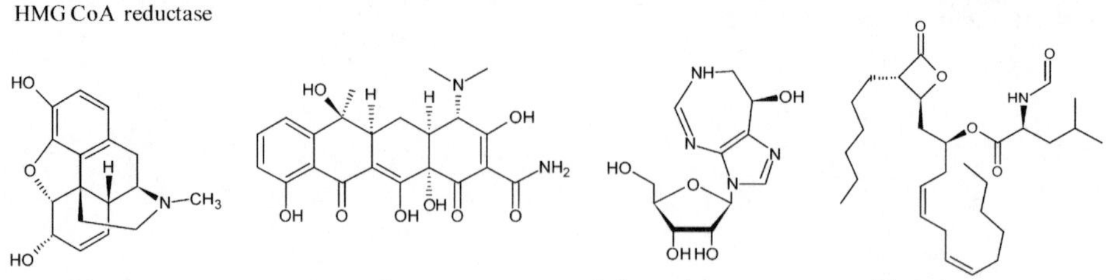

Morphine
Papaver somniferum
Neuropathic Pain

Tetracycline
Streptomyces spp
Ribosomal inhibitor

Coformycin
Streptomyces antibioticus
Adenosine deaminase inhibitor

Lipstatin
Streptomyces toxitricini
Pancreatic lipase inhibitor

Figure 1 *Landmark natural products in modern western medicine*

products. Much of the progress in anticancer drugs is now coming from natural products[2-7] and in both these cases, the activity of the natural product lead is relatively easy to enhance because it is concerned with selective destruction and can usually be tested against the target cells *in vitro*.

However, the classical development of natural products is relatively less common in neurosciences. The activity of herbal remedies often disappears during the separation of the constituent compounds[8] or the purified components have disappointingly low activity at any individual receptor. As a consequence, many of the modern lead compounds in neurosciences have been derived from high throughput screening (HTS) of libraries of synthetic chemicals against cloned human receptors.

Large pharmaceutical companies have justified the move away from natural products, citing risk, hit rate, cost and time of development as reasons for the greater efficiency of high throughput screening in generating lead compounds for new targets. A typical example of the HTS route is provided by our development of an agonist for the newly discovered MrgX1 receptor which is thought to be involved in neuropathic pain.

1.2 Discovery of a Potent and Selective MrgX1 Agonist

Neuropathic pain is the result of nerve injury caused either by mechanical damage such as loss of a limb or crushed nerves, or by diseases such as multiple sclerosis, HIV or cancer. Treatment is customarily with a mixture of drugs from several classes such as opioids, tricyclic antidepressants, anticonvulsants, NSAIDs and analgesics.[9] This gives some pain relief in about half of patients but compliance is poor due to often intolerable side effects and improved medication for the condition is urgently required.

MrgX1 is a 7-transmembrane receptor whose existence was deduced from an analysis of the human genome. It is suspected of being a pain target because it is predominantly expressed on small afferent nerves in the dorsal root ganglion. Thus, it is situated on the

Compd	Structure	MrgX1 pEC_{50}	5HT2A pKi	5HT2B pKi	5HT2C pKi
1	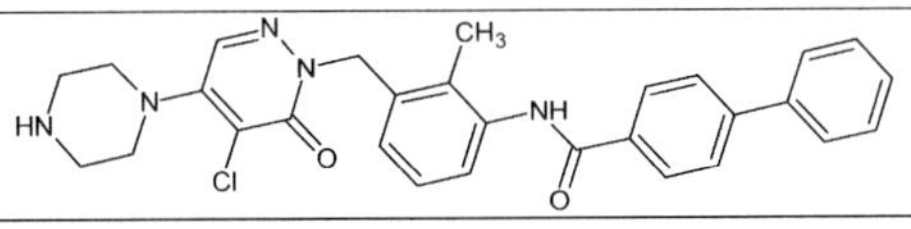	6.9	7.3	7.0	6.5
2	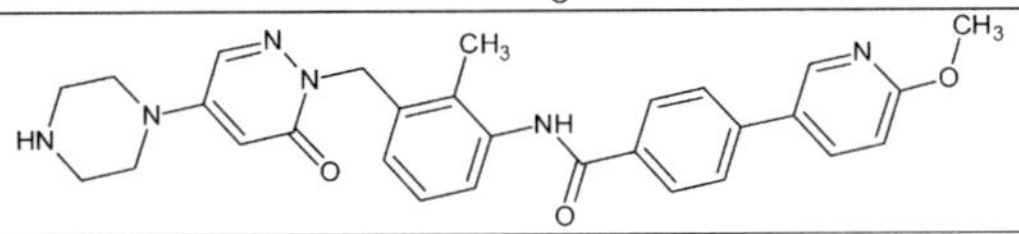	8.2	< 5	6.0	< 5

Table 1 *MrgX1 and aminergic activities for HTS hit **1** and optimised MrgX1 agonist **2***

nerves that transmit the incoming signals through the principal junctions before entering the spinal cord.

An agonist of the MrgX1 receptor **1** was identified by screening of the GlaxoSmithKline compound collection using HEK293 cells stably transfected with MrgX1 and monitoring for the release of calcium by a FLIPR assay.

Compound **1** was found to have equal affinity for aminergic receptors and a medicinal chemistry programme was initiated to eliminate the unwanted binding and improve the activity of the compound at MrgX1.[10] Rational modification of compound **1** gave compound **2** which has more than tenfold greater potency at MrgX1 and negligible binding at the aminergic receptors. Compound **2** was inactive at a further fifty 7-transmembrane receptors and has a drug-like molecular weight (510.58) and a reasonable lipophilicity (cLogP 2.7) for membrane permeability.

The next step for this compound, as for the vast majority of compounds entering a lead optimisation programme, would be testing for its interaction with cytochrome P450 enzymes in *in vitro* assays. Inhibition of P450s reduces metabolism which is unacceptable as it can result in a toxic accumulation of other drugs that a patient may be taking. The other test is *in vitro* clearance which estimates the rate at which the drug will be metabolised by P450 enzymes. The MrgX1 agonist **2** has several points at which metabolism might be expected to occur, with the methoxyl and the unsubstituted piperazine moieties particularly liable to P450 oxidation.

Major pharmaceutical products are dosed orally as this is the method with which patients are most likely to comply and gives a predictable exposure without puncturing the skin - the primary defence against infection. However, oral dosing implies small drug substances that are transported from the gut by the hepatic portal vein to the liver where the more hydrophobic are exposed to metabolism by the hepatic cytochrome P450s and the more hydrophilic are readily excreted. In order to achieve good blood levels, hydrophobic drug substances must resist first pass metabolism by the hepatic cytochrome P450s but, to give a desirable rate of clearance from the blood, they must be metabolised by hepatic and peripheral cytochrome P450 enzymes that aid their excretion. High lipophilicity assists compounds in the transition of membranes either in the gut or the brain, but high lipophilicity also makes them good substrates for cytochrome P450 oxidation.

Several natural product medicines appear to resolve these conflicts and a study of the mechanisms they use may suggest ways in which oral dosing can be achieved without the premature destruction of drug substances by hepatic P450 enzymes. Three of these mechanisms will be presented in the following section and their application to improve the bioavailability of the MrgX1 agonist discussed.

2 MECHANISMS THAT IMPROVE NATURAL PRODUCT BLOOD LEVELS

2.1 Reduced Metabolism of the Drug

FK506 (Figure 2) is known to bind the abundant human enzyme cis-trans prolyl isomerase (FKBP) to form a stable complex and it is this complex which binds the target phosphatase calcineurin and imparts immunosuppressant activity.[11] The complexation of FK506 also reduces its binding to CYP3A4 and this is assumed to protect it from metabolism *in vivo* as unbound FK506 is readily attacked by the enzyme. Both FK506 and rapamycin, which shares the same FKBP binding motif, have long half lives *in vivo* (30 and 60 hours respectively) and the FKBP binding motif or FK506 itself, have been linked to unrelated compounds to form an FKBP complex and protect them from cytochrome P450 oxidation.[12] However, when this technique was applied to amprenavir, HIV protease inhibitory data was not given for the conjugate **3** suggesting limited practical application.

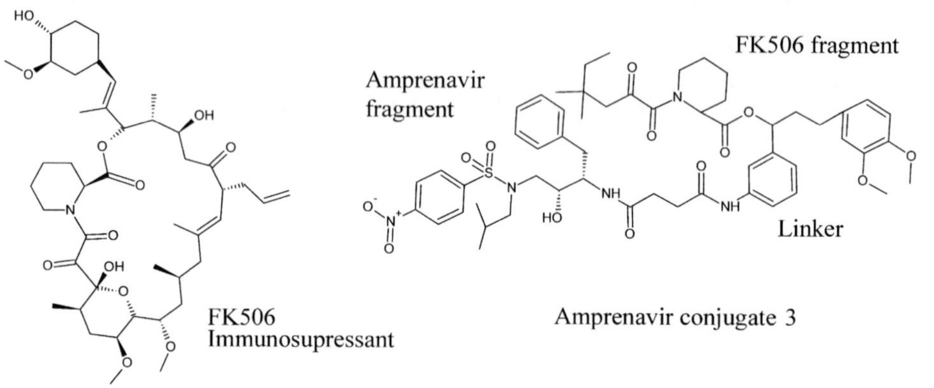

Figure 2 *Molecules containing the FK506 binding motif have some protection against cytochrome P450 oxidation in the presence of FKBP.*

If the MrgX1 agonist **2** were attached to the FKBP binding motif, the resulting conjugate would be expected to bind FKBP and provide steric bulk to protect the agonist from attack by P450 enzymes. However, the complexed agonist might not diffuse freely to the small afferent nerves of the dorsal root ganglion and might not have agonist activity at MrgX1 receptors. The half life of the agonist in the blood could be extended, but four to six hours is usually preferable to the long half lives recorded for FK506 and rapamycin.

2.2 Active Uptake of the Drug from the Gut

The blood levels of macrolides may also be improved by active uptake of the drug from the gut lumen into the epithelial cells forming the gut wall. The antibiotics clarithromycin and azithromycin (Figure 3) exceed the molecular weight and the number of H bond acceptors which Lipinski suggests for good oral absorption but give bioavailabilities of up to 50%. Garver's group at GSK has shown that these compounds are actively taken up into intestinal epithelial cells by Organic Anion Transporter Proteins (OATs).[13] Rifampicin blocks the action of the OATs but does not affect the metabolism or excretion of the macrolides, and in the presence of rifampicin, their blood concentration is halved. The macrolides are also substrates for the pGP transporter which exports the drug from the

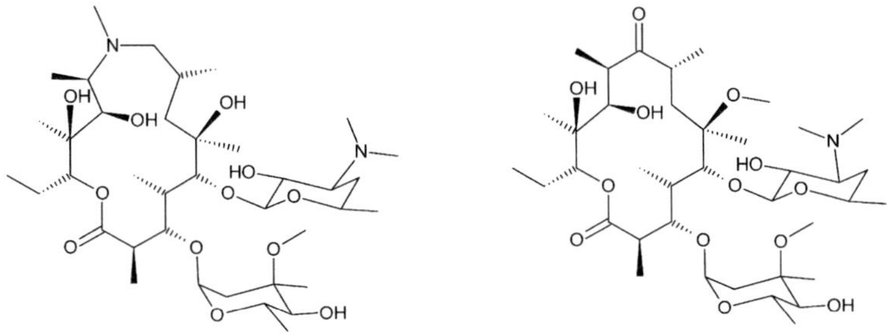

Azithromycin
cLogP 3.3, MW 749
14 H-bond acceptors
Antibacterial

Clarithromycin
cLogP 3.2 , MW 747.95
14 H-bond acceptors
Antibacterial

Figure 3 *Macrolides absorbed from the gut by an active transport mechanism.*

epithelial cells back into the gut lumen and inhibition of this export by rifampicin may contribute to the partiality of the reduction of macrolide blood levels, especially at higher rifampicin concentrations. Other examples, such as the active transport of ketone containing drugs by the intestinal peptide transporter Pep T1 are now being reported.[14]

2.3 Avoidance of First Pass Metabolism by Lymphatic Absorption

Following ingestion, small, hydrophilic molecules pass through the enterocytes forming the gut wall, into the hepatic portal vein and are transported to the liver where they are subjected to first pass metabolism by hepatic enzymes (Figure 4).

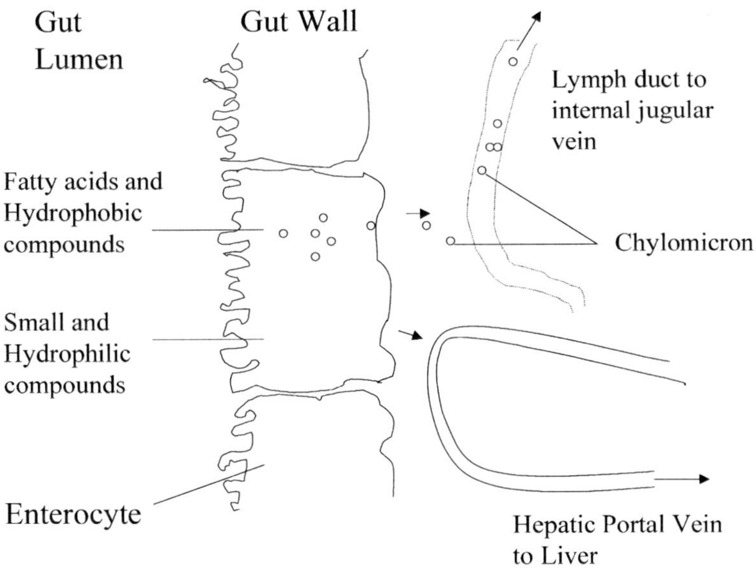

Figure 4 *Diagrammatic representation of modes of absorption from the small intestine*

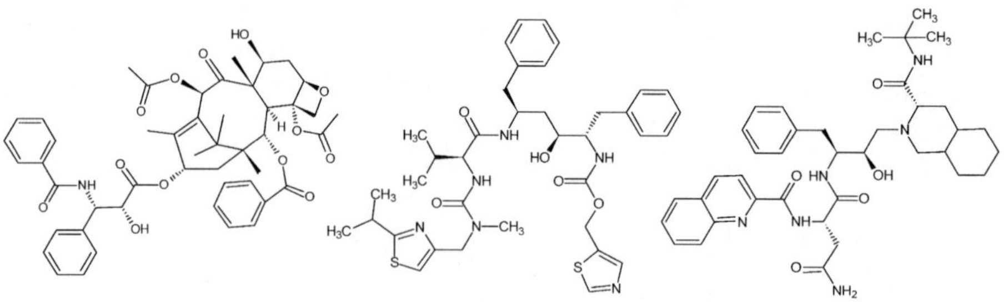

Taxol	Ritonavir	Saquinavir
cLogP 8.6	cLogP 5.3	cLogP 4.4
Microtubule	Protease / Cyp3A4	Protease inhibitor
Stabiliser Cancer	Inhibitor, HIV	HIV

Figure 5 *Hydrophobic drugs absorbed* via *the lymphatic system.*

By contrast, very hydrophobic compounds with a cLogP greater than 5 such as fatty acids, are packaged in the enterocyte into lipoprotein vesicles called chylomicrons. These enter the lymphatic system and are transported in the lymph as a milky emulsion to be discharged into the venous blood flow *via* the subclavian veins. In following this route, fatty acids are distributed throughout the body and do not pass through the liver until they have perfused the periphery. Consequently they avoid first pass metabolism and are not destroyed by hepatic cytochromes[15] before they have an opportunity to exert their physiological action. This mechanism may be used by other hydrophobic substances.

The hydrophobic drug taxol (Figure 5) has a cLogP 8.6 and is an excellent substrate for cytochromes 2C and 3A,[16] but it can be orally absorbed to give therapeutic blood levels when it is formulated with triglycerides which have been etherified with polyethyleneglycol (PEG triglycerides). The lipid-like PEG triglycerides aid solubilisation of the poorly water soluble drug and are believed to enhance its membrane permeability and to direct its absorption *via* the lymphatic system.[17] The same strategy has been used for the hydrophobic HIV protease inhibitors ritonavir and saquinavir (cLogP respectively 5.3 and 4.4) where PEG triglycerides such as Cremophor enhance lymphatic absorption.

In early work, small amounts of polar drugs, such as salicylic acid, isoniazid and caffeine were discovered in the lymph after oral administration to the rat[18] and this has raised the possibility that more polar drugs might be directed towards lymphatic absorption. Amphotericin B (Figure 6), which has a hydrophilic cLogP of 1.65, is an effective antifungal agent that has been given *iv* for 50 years, but recently the use of lipid based formulations to promote lymphatic absorption has enabled therapeutically effective doses to be administered orally.[19,20] The authors of this work note little renal toxicity for amphotericin B in these lipid based formulations and that, in addition to lymphatic absorption, other undiscovered mechanisms may contribute to its improved blood level. Recent unpublished results from our group show that polar molecules with a molecular weight below 500 were able to reach their site of action in the lymph nodes without giving a detectable blood level, suggesting that they were absorbed *via* the lymphatic system.

While lipid formulation is an effective method of getting drugs into the body by the oral route, it has not been perceived by medicinal chemists as a method of avoiding first pass metabolism for drugs which have some susceptibility to cytochrome P450s. Part of the reason for this is the difficulty in sampling lymph when compared to the ease of determining drug concentrations in blood at multiple time points. There is relatively little

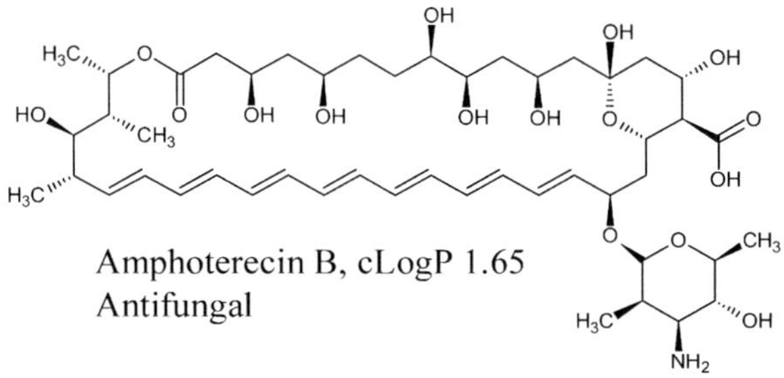

Figure 6 *Amphotericin B can be absorbed from the gut into the lymph*

lymph to sample, it is an inhomogeneous emulsion and its collection requires surgical intervention which can not be achieved in the high throughput manner required in modern medicinal chemistry. However, techniques are now being developed for the repeated sampling of lymph and for indirect methods of assaying drug concentrations in lymph by determining the difference in drug concentration in portal and venous blood.[21]

The MrgX1 agonist **2** has metabolic liabilities and a poor aqueous solubility at pH 7 (1 ug/ml) but these deficiencies can be ameliorated by solubilisation in a lipid medium and by avoiding first pass metabolism. An increase in the lymphatic flow of the MrgX1 agonist **2** should be promoted by an increased flux from the gut into the enterocytes together with an increase in the formation and export of chylomicrons. Progress in both these areas[17] will permit oral dosing of compounds which have more sensitive functionality and a wider range of molecular weight and cLogP; flexibility which will be required as drugs become more complex with demands such as polypharmacology.

3 NATURAL PRODUCT MEDICINES INDICATE DISEASE TARGETS

3.1 Animal Models Fail for MrgX1 and Other Neuroscience Targets

Definition of the physiological role of MrgX1 has been hampered by the lack of a close orthologue in lower mammals, particularly those in which a pain model has been developed. Of the 27 mouse and 9 rat Mrg genes, the rat DRR1 is the closest equivalent to the human, but shares only 60% of its sequence. This is explained by a comparison of the Mrg genes across species, which shows a high ratio of significant changes relative to conservative changes, indicating that the gene is undergoing rapid evolution.[22,23] MrgX1 and rat DRR1 receptors are located in the same relative position and are activated by the same peptide agonist, BAM8-22, but this is only circumstantial evidence for a similarity in function. The lack of a credible equivalent to the human MrgX1 gene in lower mammals shows that higher human nervous functions have diverged from those of lower mammals and this invalidates animal models designed to find a function for the gene.

Similar circumstances recur across the neurosciences; the fate of candidate drugs demonstrating the failure of animal models. Licensed neurological drugs are used to treat illnesses for which they were not developed.[24] In these cases, the determination of efficacy in animal models is bypassed and the drug is effectively tested on humans until a better application than the original is found. Many candidate drugs are abandoned because they are ineffective in man. The drug may be efficacious in animals but does not work in humans because the animal model is not a true reflection of the human disease. Even candidate drugs which reach the licensing authorities for neurological indications require elaborate statistics to show that they have a greater benefit than a placebo. The improvement is frequently marginal and licensing of the drug may only be justified by a reduction in side effects relative to older treatments. The genetic differences between the animals on which drugs are developed and the humans for whom they are designed are far greater than the variation amongst humans that is often blamed for the poor efficacy of drugs across genotypes. Where humans have evolved further than animals,[25] there is little reason to expect animal models to provide a useful guide to disease and the use of such models is driven by the lack of an alternative. Recently evolved areas may be particularly prone to failure because of the untested nature of the new physiology, suggesting that humans have a predilection for neurological ailments and that animals do not possess the machinery which causes some human disease. These considerations apply only where humans have evolved further than lower mammals. Testing in animals for drug safety in shared systems remains an essential step in drug development.

Poor cognition of the targets able to alleviate human disease is a principal impediment in the development of effective drugs, but it is possible that natural product medicines can help in the selection of the best receptors, or the best set of targets, to cure neurological ailments. I would like to spend the rest of this lecture explaining how natural remedies should be used to assist in selecting targets for neurological disorders such as pain, depression or schizophrenia.

3.2 Polypharmacology – A Complication of Pain and Other Neuroscience Diseases

Major pharmaceutical companies are investigating several targets for the relief of neuropathic pain (Figure 7) including ionotropic receptors, voltage gated ion channels and 7-transmembrane receptors. A leading drug, gabapentin and its bioavailable prodrug pregabalin, are believed to act on the calcium channels. Complete analgesia can be achieved by blockage of the sodium channels with the loss of all sensation that is familiar from local anaesthetics. Ion channels are also being studied with compounds that permit normal firing but modulate the activity to prevent rapid firing, which is associated with pain. It is unclear which of these targets or which mode of drug action will provide long lasting and side-effect free pain relief without concomitant addiction, but it is probable that, in common with other neurological diseases, treatments which modulate the action of multiple targets will be pre-eminent.

Polypharmacology is already familiar in neuroscience, for example, the antipsychotic agent quetiapine is an antagonist at dopamine, serotonin and adrenergic receptors. It has been a successful atypical antipsychotic and is used off-label for other conditions such as alcoholism, anxiety and insomnia. Its side effects include sedation, dizziness and nausea, showing that the selection of receptors blocked by quetiapine is not optimal for the disease and it is difficult to see how to improve the selection without using higher mammals as the principal test species. Making a human medicine with the required degree of activity at multiple receptors when the justification for the selection of receptors is theoretical, will be expensive and progress slow. Quetiapine itself was facilitated by having a generally depressive effect on aminergic receptors which, as a group, are readily antagonised by compounds containing a basic nitrogen with a positive charge at physiological pH.

The information necessary to identify the targets that cure human disease already exists in the natural remedies used on humans. Plants, microorganisms, lower animals and humans share many basic pathways and these are controlled by similar signalling molecules, but the variety of the pathways and the range of molecules employed are greater in less specialised organisms. The reduction in complexity may be illustrated by the loss of some amino acid catabolic pathways in higher mammals or the myriad secondary

Pain Targets in Pharmaceutical companies

Ionotropic receptors: *N*-Methyl-D-Aspartate, AMPA, Kainate,
Voltage gated channels: Ca^{2+} (gabapentin), KCNQ K^+, Na^+, Vanilloid (capsaicin)
7-TM receptors: Metabotropic Glutamate 1 to 8, bradykinin, MrgX1

An Example of Polypharmacology

Quetiapine (Seroquel) – Antipsychotic
Antagonist at dopamine (D1, D2), serotonin (5HT-1A, 2) and adrenergic (a1, a2) receptors.

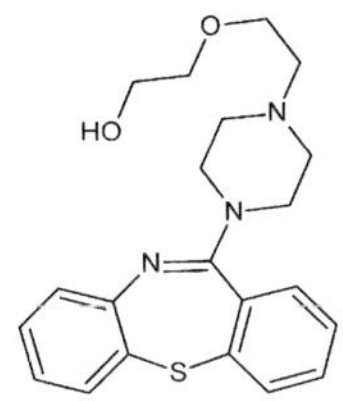

Figure 7 *Multiple targets in neuroscience diseases*

metabolites purified from lower organisms that are absent in mammals. Extracts of lower organisms containing mixtures of these active molecules have been tested on humans in a stochastic manner for generations to determine if they have a beneficial effect on human disease. Where possible, the active component has been isolated and developed for use in western medicine. Instead of isolating the compound responsible for the activity, there may be greater value in discovering upon which human organs and receptors the natural remedy acts, especially if the amelioration of disease is due to several compounds acting on multiple targets.

3.3 Harvesting Information from Natural Remedies

The efficacy of natural remedies can be considered proven when active components such as morphine or artemesinin have been isolated from them but, if no single active species can be discovered, or the disease is difficult to define or is subject to a strong placebo effect, absolute proof of efficacy is elusive.

Recent analyses have highlighted the size of the task. In a WHO / CSIRO study of the value of whole herbal remedies in dementia, May *et al*[26] concluded that their efficacy was equal to that of modern pharmaceuticals and most important, that the herbal remedies had no adverse effects. Strict requirements for inclusion of trials in the study included a high score on the reporting quality scale of Jadad *et al*[27] for randomised controlled studies and clinical outcomes relevant to symptoms of dementia for whole herbal mixtures compared with standard therapy or a placebo.

Of 1668 potentially relevant articles, 14 met the inclusion criteria, representing a large effort both in the trials and in their evaluation. The conclusion that the efficacy of herbal remedies is comparable to that of modern pharmaceuticals exposes the difficulty in drawing definite conclusions from the testing data of neurological remedies from either origin and provides little further proof of efficacy beyond the fact that the herbal medicines continue to be used.

An alternative approach to the collection of natural remedy data might be to analyse the mode of action of efficacious herbal remedies for a particular neurological ailment, such as pain. If several efficacious remedies act through the same mechanisms then those mechanisms merit investigation for curing of the disease.

3.4 Variation in the Composition of Natural Remedies

The rapidity with which natural remedies are being collected and collated from around the world is demonstrated by contemporaneous references gleaned during the preparation of this talk.[28-31] Large pharmacopoeia already exist for the Chinese and Indian herbal traditions[32,33] and the reliability of the preparations is being improved by the development of methods for assaying and standardising the remedies.[34]

Variation in the composition of remedies prepared at different times and in different places is conveniently illustrated by the antibacterial, antiviral, carcinostatic, antifungal and anti-inflammatory preparation called propolis. This is a resin, softened by warmth that is used by bees as a sealant, glue and antibiotic and to embalm dead invaders in the hive. Over 300 components of the mixture have been characterised and include compounds which would be expected to contribute to the various activities such as flavonoids, sesquiterpenes and aromatic acids together with metals, vitamins and other constituents.

The bees make it from collected plant exudates and from their own secreted metabolites and from specially collected ingredients, but these vary with the bees' diet, the variety of the surrounding flora and the season. Individual, purified components have only weak activity and many studies attribute the medicinal properties of propolis to synergy between the components.[35] These observations again suggest that analysis of the actions of the whole propolis on the patient's physiology would be more likely to provide useful information than analysis of the component compounds.

This approach has been used to analyse the antitumour activity of propolis in rats, leading to a better understanding of the suppression of tumour induced angiogenesis by propolis *in vivo* (Figure 8).[36] Whole propolis suppressed extracellular signal-regulated kinase (ERK) activation allowing the caspase cascade to destroy maturing endothelial cells, which would have formed new blood vessels. The authors found that rolling of the endothelial cells to form a tube was also repressed and they are investigating other mechanisms which may contribute to the inhibition of angiogenesis.

The principles of this investigation could be applied to human neurological diseases through the testing of whole herbal remedies on cloned human receptors *in vitro* thereby alleviating the problems in target selection which are inadequately addressed by animal models.

Computation methods of target selection are currently in preparation[37] and have been proven by identifying all the major antibiotic classes from their protein binding data, but these methods require large amounts of detailed data in order to fulfil their potential and this is not yet available.

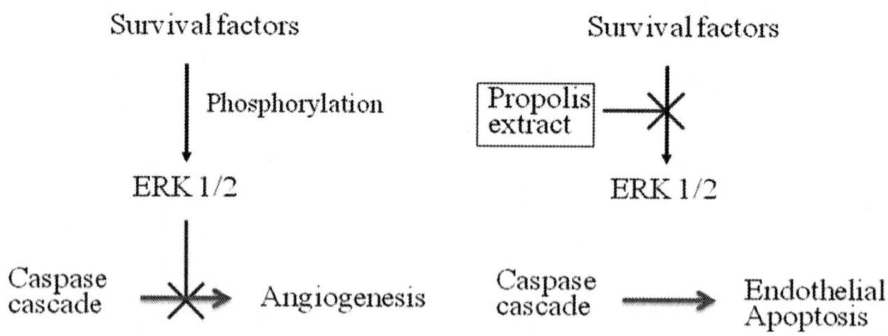

Figure 8 *Suppression of tumour induced angiogenesis by propolis*

3.5 A Method for Identifying Targets in Neurological Diseases

Natural remedies are used to cure humans. Activation of a particular human receptor such as the MrgX1 receptor, by several efficacious natural pain remedies would constitute evidence that an MrgX1 agonist participates in pain relief and that activating the MrgX1 receptor will not cause mechanistic toxicity. This provides a novel pain target and eliminates the unrealistic search for a lower mammal model of the MrgX1 receptor.

If modulation of multiple receptors is required to treat an illness, as adumbrated by the action of quetiapine in psychosis, then the problem of determining which receptors should be activated and which should be repressed becomes challenging by theoretical means. However, whole, natural remedies screened against cloned human receptors *in vitro* will indicate which receptors should be modulated to cure neurological disease. High throughput screens of individual human receptors are relatively cheap and available - for example the French firm CEREP runs a screening service against a continuously expanding range of popular receptors. Whole remedies, prepared in a fashion identical to the herbalist, could be tested in these screens to discover on which receptors the natural remedies act. A record of the variations a herbalist may introduce into his preparations due to the availability of constituents or the physiognomy of his patient would also be desirable.

Naturally derived medicinal compounds such as flavonoids, are metabolized at a relatively high rate and their action in the human body may be due either to the administered compounds or to their metabolites. Consequently, it will be necessary to test the metabolites for activity at human receptors and this can be done by artificial digestion of the natural remedies and testing of the resulting mixtures in the high throughput screens. Isolation of the metabolites is unnecessary as it is the effect of the whole mixture on receptors that is the desired information. This approach will also detect synergistic effects between molecules such as those believed to be responsible for the efficacy of propolis. Once the pattern of receptor modulation required to treat a particular ailment has been determined, a rationale is established for the medicinal chemist to begin the challenging task of designing molecules which have the desired degree of activity at multiple receptors.

Acknowedgements

I am grateful to GlaxoSmithKline for my experience in modern rational design of drugs for neurological disease and for the stimulus to propose an alternative method.

References

1 A. Ganesan, *Curr.Opin.Chem.Biol.*, 2008, **12**, 306.
2 G.M. Cragg, D.G.I. Kingston, D.J. Newman, *Anticancer Agents from Natural Products*, Publisher: Taylor & Francis Group, **2005**, ISBN: 0849318637.
3 A.D. Kinghorn, Y-W. Chin and S.M. Swanson, *Curr.Opin.Drug Discov.Devel.*, 2009, **12**, 189.
4 D.G.I. Kingston, *J.Nat.Prod.*, 2009, **72**, 507.
5 A-T. Hauser, M. Jung, *Planta Med.*, 2008, **74**, 1593.
6 R.R. Hafidh, F. Abas, A.S. Abdulamir, F. Jahanshiri, F.A. Bakar and Z. Sekawi, *International Journal of Cancer Research*, 2009, **5**, 69.
7 Y. Sun, K. Xun, Y. Wang, X. Chen, *Anti-cancer drugs*, 2009, **20**, 757.
8 Ben Gilbert personal communication, Biodiversity: New Leads for the Pharmaceutical and Agrochemical Industries. 1999, St Andrews, Scotland.

9 J.A. Butera, *J.Med.Chem.*, 2007, **50**, 2543.

10 B. Wroblowski, M.J. Wigglesworth, P.G. Szekeres, G.D. Smith, S.S. Rahman, N.H. Nicholson, A.I. Muir, A. Hall, J.P. Heer, S.L. Garland and W.J. Coates, *J.Med.Chem.*, 2009, **52**, 818.

11 K. Stankunas, J.H. Bayle, J.J. Havranek, T.J. Wandless, D. Baker, G.R. Crabtree and J.E. Gestwicki, *ChemBioChem*. 2007, 8, 1162.

12 P.S. Marinec, J.K. Lancia and J.E. Gestwicki, *Mol.Biosyst.*, 2008, **4**, 571.

13 E. Garver, E.D. Hugger, S.P. Shearn, A. Rao, P.A. Dawson, C.B. Davis and C. Han, *Drug Metab.Dispos.*, 2008, **36**, 2492.

14 D. Foley, P. Bailey, M. Pieri and D. Meredith, *Org.Biomol.Chem.*, 2009, **7**, 1064.

15 J.P. Hardwick, *Biochem.Pharmacol.*, 2008, **15**, 2263.

16 K.T. Kivisto, H.K. Kroemer and M.M. Eichelbaum, *Br.J.Clin.Pharmacol.*, 1995, **40**, 523.

17 C.M. O'Driscoll, B.T. Griffin, *Adv.Drug Delivery Rev.*, 2008, **60**, 617.

18 S.M. Sieber, V.H. Cohn, W.T. Wynn, *Xenobiotica*, 1974, **4**, 265.

19 K. Sachs-Barrable, S.D. Lee, E.K. Wasan, S.J. Thornton, K.M. Wasan, *Adv.Drug Delivery Rev.*, 2008, 60, 692.

20 P. Gershkovich, E.K. Wasan, M. Lin, O. Sivak, C.G. Leon, J.G. Clement, K.M. Wasan, *J.Antimicrob.Chemother.*, 2009, **64**, 101.

21 R.J. Jandacek, T. Rider, Q. Yang, L.A. Woollett and P. Tso, *Am. J. Physiol. Gastrointest.Liver Physiol.*, 2009, **296**, G226.

22 S. Yang, Y. Liu, A.A. Lin, L.L. Cavalli-Sforza, Z. Zhao, B. Su, *Gene*, 2005, **352**, 30.

23 M.J. Zylka, X. Dong, A.L. Southwell, D.J. Anderson, *Proc.Natl.Acad.Sci. U.S.A.*, 2003, **100**, 10043.

24 R.S. Stafford, *New England Journal of Medicine*, 2008, **358**, 1427.

25 T.J. Ryan, S.G. Grant, *Nat.Rev.Neurosci.*, 2009, DOI: 10.1038/nrn2717.

26 B.H. May, C.C.L. Xue, A.W.H. Yang, A.L. Zhang, M.D. Owens, R. Head, L. Cobiac, C.G. Li, H. Hugel, and D.F. Story, *Phytother.Res.*, 2009, **23**, 447.

27 A.R. Jadad, R.A. Moore, D. Carroll, C. Jenkinson, D.J.M. Reynolds, D.J. Gavaghan, H.J. McQuay, *Controlled Clinical Trials*, 1996, **17**, 1.

28 R.M. Kunwar, Y. Uprety, C. Burlakoti, C.L. Chowdhary, and R.W. Bussmann, *Ethnobotany Research & Applications*, 2009, **7**, 005.

29 F. Eftekhar, F. Raei, M. Yousefzadi, S.N. Ebrahimi, and J. Hadian, *J. Z.Naturforsch.*, 2009, **64c**, 20.

30 M.A. Signorini, M. Piredda, and P. Bruschi, *Journal of Ethnobiology and Ethnomedicine*, 2009, **5**, 6.

31 R.W. Bussmann and D. Sharon, *Journal of Ethnobiology and Ethnomedicine*, 2009, **5**, 4.

32 *Pharmacopoeia of the People's Republic of China* (2005 English Edition).

33 *Aruyvedic Indian Medicinal Herbs*, Pubdate: 2009/02/10.

34 S. Li, Q. Han, C. Qiao, J. Song, C.L. Cheng, and H. Xu, *Chinese Medicine*, 2008, **3**, 7.

35 M. Marcuccei, *Apidologie*, 1995, **26**, 83. V.S. Bankova, S.L. De Castro, M.C. Marcucci, *Ibid.*, 2000, **31**, 3.

36 K. Kunimasa, M-R. Ahn, T. Kobayashi, R. Eguchi, S. Kumazawa, Y. Fujimori, T. Nakano, T. Nakayama, K. Kaji and T. Ohta, *Evidence-based Complementary and Alternative Medicine*, online, 7 April (eCAM, doi:10.1093/ecam/nep024).

37 E. Bullmore and O. Sporns, *Nature Reviews Neuroscience*, online 4 February 2009; doi:10.1038/nrn2575.

2 Improved Strategies for Natural Product Exploitation

DISCOVERY AND DEVELOPMENT OF ANTIBIOTICS WITH NOVEL MODES OF ACTION

Sheo B.Singh

Merck Research Laboratories, Rahway, New Jersey 07065, USA
E-mail: sheo_singh@merck.com

1 INTRODUCTION

Microbial natural products and screening methods based on inhibiting bacterial growth have played a pivotal role in the discovery and development of antibiotics. This discovery paradigm began with penicillin in 1929 and what followed was nothing less than monumental. Scientific and industrial communities worked hand-in-hand to develop penicillin as a broad-spectrum antibiotic. This is perhaps one of the greatest and most broadly influential discoveries of all time which led to the realization that microbial organisms could be a source not only of antibiotics, but also other therapeutic agents and allowed for the creation of natural products based research operations both in academia and industry. Concerted, world-wide efforts for the next twenty years (1940-1962) culminated in the discovery of essentially all the chemical scaffolds of the antibiotics that are in clinical practice today.[1,2] Unfortunately, higher than expected success led to higher aspirations leading to speculation and the drawing of the naïve conclusion that these antibiotics or their chemical variants would be sufficient to conquer bacterial infections. No sooner had these thoughts gained traction, than fast dividing bacteria were selecting for resistance against all classes of antibiotics including newer variants of penicillin.

Methicillin was one of the more important penicillins, but soon after its introduction to the market, various hospital strains of *Staphylococcus aureus* started to show resistance and now methicillin resistant *Staphylococcus aureus* (MRSA) strains are prevalent, accounting for 60-89% of nosocomial infections.[3-5] MRSA infection leads to a large number of deaths every year both in the United States and other parts of the World.[6] Vancomycin was considered the antibiotic of last resort for the treatment of Gram-positive infections, however 20-30% of *Enterococci* are now found to be resistant to this drug.[3] Linezolid and daptomycin are newer antibiotics that represent options for treatment of Gram-positive infections including MRSA. Like *S. aureus*, other organisms select for resistance against antibiotics both in hospital and community settings and these include Gram-negative bacteria which are occasionally harder to treat. Recent Centers of Disease Control and Prevention (CDC) surveillance data suggests that greater than 20% of *Pseudomonas aeruginosa* are imipenem resistant in the hospital environment and there may be limited or no treatment options available (Figure 1).[3] Treatment of Gram-negative infections is inherently more challenging and the discovery and development of new agents to treat them are equally demanding. Incremental chemical modification of existing

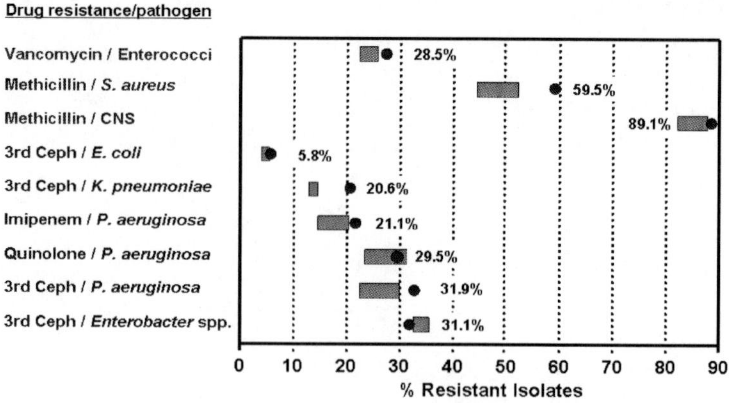

Figure 1 *CDC surveillance of US hospitals for nosocomial rates of resistance. Data from 1998-2002 (grey rectangles). Data from 2003 (black circles). The separation between the end of the grey rectangle and black circle represents the change in one year.*

chemical scaffolds has been the source for a large number of highly potent, broad spectrum antibiotics and continues to deliver new drugs that provide temporary coverage of resistant organisms. However, modification of the existing scaffolds is becoming increasingly difficult and discovery of new antibiotic scaffolds with known modes of action, or preferably, novel scaffolds with completely novel modes of action, is urgently needed.

1.1 Sources of Antibiotics

The majority of antibiotics that are in clinical practice owe their origin to natural products. Exceptions are the oxazolidinones (e.g. linezolid), the quinolones (e.g. ciprofloxacin), and the sulfonamides. Most of the leads for non-antibacterial targets originate from screening of synthetic chemical collections which are the products of medicinal chemistry programs of other disease targets and these often require the removal of antibacterial activity - particularly for an oral drug. Therefore, screening of such collections for antibacterial leads has mostly been a fruitless exercise.[2,7] Unfortunately, recent natural product screens have also failed to yield significant new antibiotic leads. Despite significant advances in analytical methods of detection and characterization of natural products and miniaturization of microbial culturing techniques, rediscovery of known compounds remains the predominant occurrence which leads to significantly increased cost of discovery of each new natural product. This has had a significant negative impact on natural product screening programs and has lead to the abandonment of all but a few in the large pharmaceutical companies. Differentiation of a known from unknown natural product when screening by whole-cell empiric assays remains the biggest challenge, especially when compounds are highly potent and are present in very low concentrations and despite the availability of highly sensitive chemical dereplication methods. The key to overcoming this challenge is to develop innovative, target specific, whole-cell screening technologies that can differentiate the compounds at early screening stages. Antisense technology is one such method.

1.2 Underrepresentation of Targets by Current Antibiotics

Clinically used antibiotics affect about 25 molecular targets, inhibiting DNA, RNA, protein and cell wall synthesis.[2] Genomic analysis and gene knock-out studies of bacteria have indicated that there about 10 fold more (~300) highly conserved essential proteins that could be broad spectrum drug targets.[8-12] Therefore, it is imperative upon us to find innovative methods to design new target-based whole cell assays to discover compounds that act against new targets and allow identification of compounds that would undoubtedly have antibacterial activity *a priori*. It would be ideal if such a technology were general in nature and therefore could be applied to all targets. Antisense based target sensitization is one such approach[12] in which individual targets can be sensitized by expression of antisense against the specific target.

1.3 Approaches for Screening of Antibiotics

Historically, a number of screening approaches have been used for the discovery of new antibiotics and a few key ones are summarized in Table 1. The most productive of these approaches has been empiric screening, using a growth inhibition assay of whole cell bacteria with either wild-type or a resistance phenotype. This approach relies on *a priori* detection of antibacterial activity which is followed by determination of the targets or modes of action. Inherent in this approach is a lack of differentiation of the antibiotics that kill bacteria with a defined mode of action *versus* general cellular poisons. The historical use of this approach for the discovery of most of the existing antibacterial agents, creates a significant challenge in distinguishing novel compounds from already known compounds. In most of the high volume industrial screening operations that use this approach, repeated discovery of the same natural product scaffolds predominates, necessitating implementation of robust dereplication methods. While analytical methods have evolved significantly, the task of dereplication remains a significant challenge primarily due to sensitivity differences between the analytical and biological detection methods - not to mention the sensitivity differences between the compounds and compound classes. This repeated and redundant discovery is quite expensive and is certainly one of the causes of the negative sentiment against natural products discovery.

In the nineteen eighties, following the advent of molecular biology, most organizations either augmented the empiric whole cell screening approaches with cell free enzyme or receptor assays or completely substituted the empiric screening with the cell free assays. Screening with cell free assays was undoubtedly attractive because of the richness of knowledge that came with a defined mode of action.

Approach	Antibacterial	MOA	Benefits and Disadvantages
Whole cell wild type bacteria	√	X	Differentiation of antibiotic from general poison
Cell free enzyme	X	√	Often lack cellular activity
Antisense sensitized strains (target dial down)	√	√	High sensitivity, Universal applicability, targets? Robustness?
Target up-regulation	√	√	Low sensitivity, Universal applicability, targets? Robustness?
Wild-type resistance pair	√	√	Low sensitivity, generally good for discovery known classes?

Table 1 *Benefits, cellular activity and mode of action (MOA) information from modern screening approaches to antibiotic lead discovery.*

The cell free approach discovered a large number of potent enzyme inhibitors but was fraught with difficulty due to the fact that most of the inhibitors lacked whole-cell antibacterial activity because of their inability to reach their intracellular targets due to poor penetration[13] which might be augmented by active efflux.[14,15] On the other hand, cell free assays are highly successful and are commonly used for the discovery of leads for targets other than bacteria and fungi. The pathogenic organisms protect themselves with cell walls which preserve them from many chemicals and allow them to survive in tough environments. No systematic rules exist to supervene the protective mechanisms and impart cellular activity to cell free bacterial enzyme inhibitors. Screening with the cell free assays clearly made sense at the time, but in retrospect, the failure of this approach and the expense associated with it may have contributed to the termination of antibacterial programs in many major pharmaceutical companies. The disappointing result of this approach was nicely summarized by the GlaxoSmithKline group.[7] It is certain that other pharmaceutical companies have had similar disappointing experiences.

2 ANTISENSE SCREENING ASSAYS

2.1 Modes of Target Based Cellular Screening

The obvious advantage of whole cell screening is that it ensures upfront antibacterial activity. Likewise, the obvious advantage of cell free screening is that it ensures knowledge of the mechanism of action. These advantages could be combined if whole cell screens were target based. Over the years, many target-based whole cell screening methods have been used to find target specific inhibitors. Such approaches include comparison of responses of paired strains such as a drug resistant mutant and its parent strain, or an over-expressed target strain and parent strain, or modification of strains by other methods. Whole cell assays have been used that monitor changes in cell morphology or specific phenotypes caused by alteration of a particular pathway or that measured expression of a reporter gene directly or indirectly regulated by essential intracellular proteins. However, most of these approaches lack generality.

Recently, conditional expression of antisense RNA of essential genes in *S. aureus* cells has been found to inhibit cell growth and alter cell sensitivity to selective inhibitors,[10,12] leading to the possibility of developing new hypersensitive target-based whole cell assays. This approach appeared to be general and is applicable for all essential bacterial gene targets. The high sensitivity and universality of the antisense approach made it ideal for screening for antibiotics and it became the main discovery engine at Merck, leading to the discovery of a large number of novel natural products as antibacterial agents (*vide infra*).

As mentioned earlier, target up-regulation or over-expression is another approach that could be applied to screening for new antibiotics of natural or synthetic origin (Table 1) but this approach lacks sensitivity and generality.

Lastly, screening using a drug-sensitive and a drug-resistant pair of organisms is another approach to the discovery of new antibiotics that has been used for a long time. This method also suffers from low sensitivity when compared to the antisense approach, but has potential for the discovery of compounds with novel structural types that bind to the same binding site as the compound that was used for selection of the resistant strain. In practice this has not been the case and it has been a great tool for rapid discovery of compounds belonging to the same structural class and is helpful in the discovery of new congeners or analogues (*vide infra*).

2.2 Antisense Expression Systems and Two-Plate Differential Sensitivity Assays

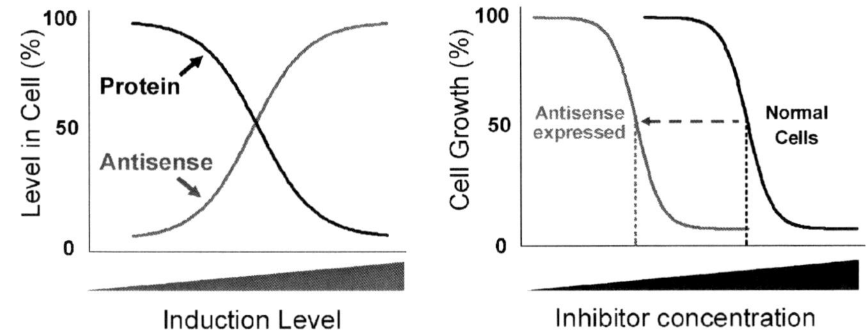

Figure 2 *Schematic representation of how induction of antisense RNA leads to a reduction of the protein levels of the targeted mRNA, leading to differential growth sensitivities between target depleted cells and control cells without antisense induction.*

Several inducible antisense expression systems for *S. aureus* were recently reported.[10,12,16] One of the most important systems, reported by Forsyth et al,[12] involves a xylose inducible promoter and this became very useful for screening purposes since xylose does not have inherent antibiotic properties. Its application to drug discovery was recently reviewed.[17] Briefly, xylose induction of target specific antisense leads to a proportional decrease of the target mRNA and hence the target protein, leading to increased sensitivity to the target specific inhibitors. In order to determine target specificity, the activity of a given compound or extract is tested for its ability to inhibit growth of the antisense sensitized *S. aureus* strain and then compared with its ability to inhibit growth of the control *S. aureus* strain under identical conditions. If the compound or extract has target specific activity the antisense strain will show hypersensitivity relative to the control strain and there will be a left shift of the growth inhibition curve (Figure 2). The hypersensitivity to target based inhibitors allows for the discovery of compounds even when they are present in relatively lower concentrations or low titres.

While this liquid assay format was great for precise testing of compounds, it requires titration of each compound and extract which is not conducive to high throughput primary screening on an industrial scale. It was reasoned that a similar outcome could be obtained by changing the liquid assay to an agar diffusion assay format which would be akin to a standard primary screening assay model in which the compound could be tested at a single concentration.[18] The concept was based on the hypothesis that, when an antibiotic is applied to a well of an agar plate seeded with growing bacterial cells, it diffuses into the agar forming a concentration gradient (natural titration). Close to the well the concentration of an antibiotic is higher resulting in the inhibition of cell growth, whereas further from the well the concentration is lower leading to no inhibition and continued cell growth. Hence, a clear circular zone is formed around the well due to the inhibition of cell growth. The zone size depends on the dose, potency, diffusion rate of inhibitors and the growth rate of the bacteria.

With these principles in mind, the Merck team reported the development of the first agar-diffusion two-plate differential sensitivity assay (two-plate assay)[18] in which one plate was seeded with *S. aureus* cells expressing *fabF* AS-RNA (AS plate) and the other with cells lacking AS-RNA expression (control plate). When a *bona fide* FabH and/or FabF inhibitor is applied to both plates, it should selectively inhibit the growth of the *fabF* AS

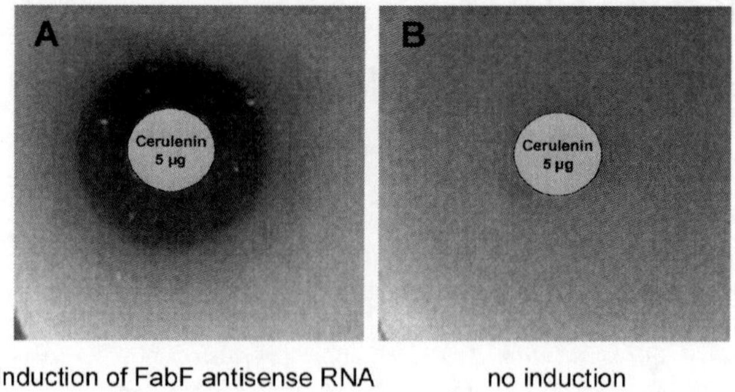

Figure 3 *Two-plate differential sensitivity assay. A: a plate seeded with S. aureus fabF AS-RNA stain where antisense expression was induced with 50 mM xylose. B: a control plate seeded with S. aureus with no expression of fabF antisense RNA. A filter disc containing 5 µg cerulenin was placed on the top of agar of each plate.*

strain and form a larger inhibition zone on the AS plate relative to the control plate. To validate the two-plate assay 5 µg of cerulenin, a known FabF selective inhibitor, was spotted on each plate. As expected, cerulenin produced a large zone on the *fabF* AS plate (18 mm) and no zone on the control plate (Figure 3). As expected the compounds that do not work by inhibition of FabF/H enzymes and work by other modes of action, show no zone differentials between the plates. This was demonstrated using fourteen known antibiotics and the results of screening of natural products extracts, by Singh et al.[17]

A similar differential sensitivity assay was applied to the screening for inhibitors of, or interactions with ribosomal protein S4 (rpsD) leading to the discovery of a number of antibiotics. This screening method provides a higher level of sensitivity which allows for the detection of antibiotics that are present in lower concentrations and would otherwise have been missed.

2.3 Discovery of Inhibitors of Condensing Enzyme (FabH/F) of Fatty Acid Synthase

Fatty acids are an integral part of phospholipids and are essential for bacterial survival. They are synthesized by fatty acid synthases which comprise a large number of enzymes representing a series of molecular targets (Figure 4). Of these, elongation condensing enzyme (FabF) is the most attractive due its higher conservation across the bacterial species. Fatty acid synthesis has been validated as a clinical target by triclosan and isoniazid, which are both inhibitors of FabI. Despite these clinical validations, essentiality of fatty acid synthesis as a target in certain bacterial species has recently been challenged.[19]

Screening of over 250,000 microbial extracts using the FabH/F two-plate assays at a single concentration produced a 0.3% hit rate which was reduced to 0.1% by triaging through a FASII biochemical assay.[18,20] The bioassay guided fractionation allowed rapid identification of all known inhibitors of condensing enzymes such as cerulenin, thiolactomycin, and thiotetromycin. These compounds are rather poor antibiotics (MIC > 64 µg/mL) but were easily detected by the assay, even when their titres in fermentation

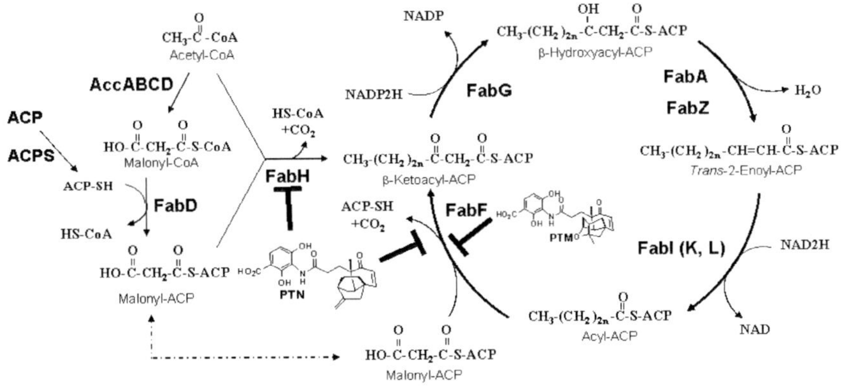

Figure 4 *Biosynthetic pathway of bacterial fatty acid synthesis*

broths were very low and did not show any zone of inhibition on the control plate. They would, therefore, not have been detected by wild-type strain screening alone. These findings validated the screening process and continued screening allowed for the identification of Tü3010,[21] an amide analogue of thiotetromycin. While Tü3010 showed no difference in activity in the cell free FASII assay[22] compared to thiolactomycin and thiotetromycin, it showed more than 15 fold greater antibacterial activity.[18] Similarly the screening efforts led to the isolation of a series of novel acetyllenic allenes and phomallenic acids A-C (Figure 5).[18,23]

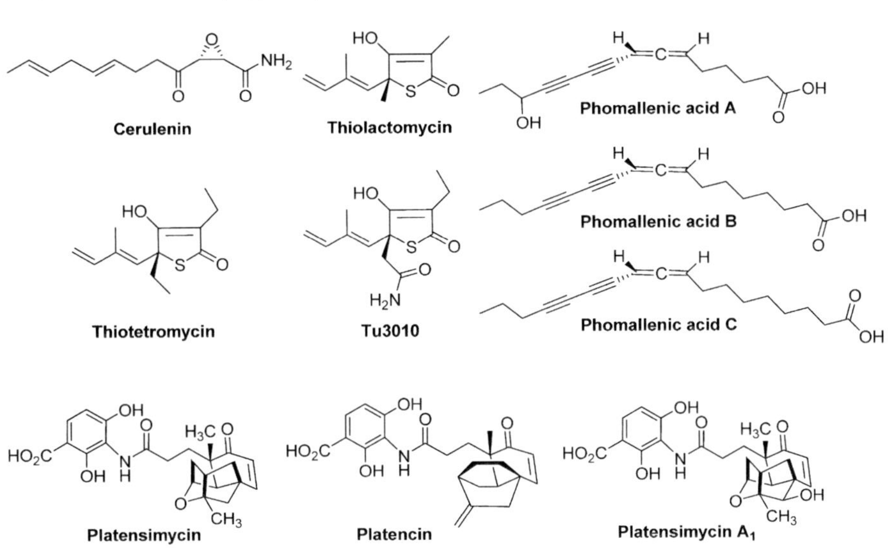

Figure 5 *Chemical structures of natural product inhibitors of bacterial condensing enzymes (FabH/F)*

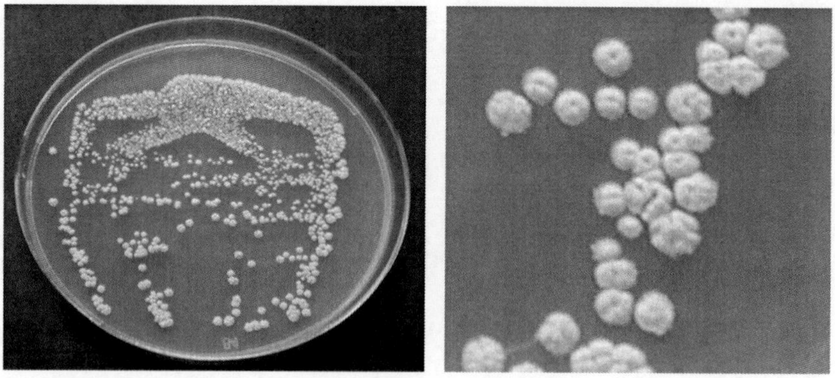

Figure 6 *Photograph of platensimycin producing culture of Streptomyces platensis MA7327*

These discoveries helped in the validation of the antisense strategy, provided a significant confidence boost and allowed us to expand our screening efforts. Screening of new sources from substrates collected from diverse geographical locations led us to the discovery of platensimycin (PTM) and platencin (PTN) as selective inhibitors of fatty acid biosynthesis or, more specifically, inhibitors of condensing enzymes.[20,24-26] Platensimycin was originally isolated from a strain of *Streptomyces platensis* (Figure 6) recovered from soil samples collected in South Africa and subsequently from another strain of *S. platensis* recovered from a soil sample collected in Spain.[20,24] Platencin was isolated from another strain of *S. platensis* recovered from a soil sample also collected in Spain.[25,26] Structurally both platensimycin and platencin comprise a common 3-amino-2,4-dihydroxybenzoic acid unit and different C-17 acids. The C-17 acid in platensimycin is a tetracyclic enone and in platencin it is a tricyclic enone. Biosynthetic studies of platensimycin and platencin by precursor feeding experiments suggested that the aminobenzoic acid unit is derived from a C7N unit involving the TCA cycle and the C-17 acid portion is derived from a non-mevalonate terpenoid pathway.[27,28]

2.4 Platencin And Platensimycin

2.4.1 Biological Activities. These novel and unprecedented natural products show potent broad spectrum Gram-positive antibiotic activities with low (0.06-2 μg/mL) MIC against clinically relevant bacteria including major drug resistance organisms such as MRSA, VRE, and VISA.[20,25] The compounds derive their antibacterial activity from selective inhibition of fatty acid biosynthesis in bacteria. However, they differ in their precise modes of action. Platensimycin selectively inhibited the *S. aureus* elongation condensing enzyme FabF (IC_{50} 48 nM) and was essentially inactive against initiation condensing enzyme (FabH) (IC_{50} 67 μM) thus showing over 800 fold specificity for FabF. In a FabF direct binding assay platensimycin gave a binding affinity of 19 nM[20] and inhibited cellular fatty acid biosynthesis (inhibition of synthesis of phospholipids) with an IC_{50} 0.1 μg/mL as measured by inhibition of macromolecular synthesis showing over 10,000 fold selectivity over other macromolecules (Figure 7). Platencin showed equivalent potency to platensimycin in all cellular assays including inhibition of *S. aureus* fatty acid

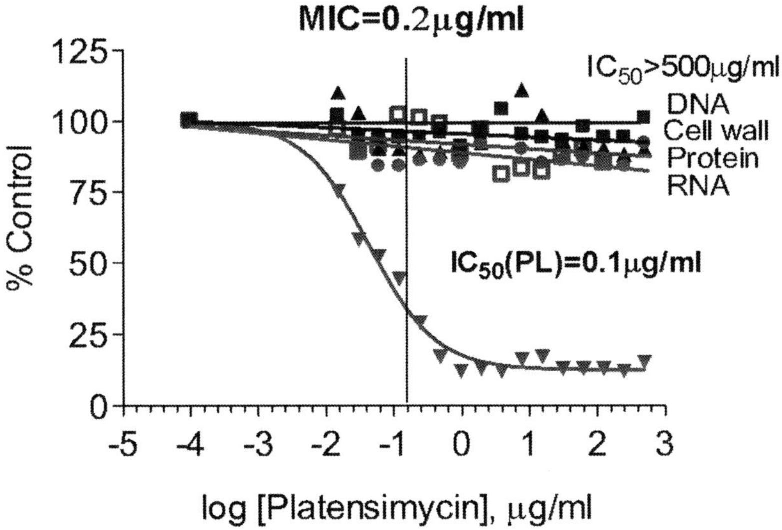

Figure 7 *Inhibition of macromolecular synthesis by platensimycin (PL)*

synthesis (IC_{50} 0.23 μg/mL), but was six fold less active in the FabF direct binding assay (IC_{50} 0.11 μM) and four fold more active against *S. aureus* FabH (IC_{50} 16.2 μM). This apparently paradoxical difference in the cellular and cell free activities could be better explained by comparing the activities of the two compounds in a gel elongation assay using cell lysate in which FabH and FabF inhibitory activities were measured simultaneously in the same assay.[25] In this assay, platensimycin exhibited IC_{50} values against FabF (0.29 μM) and FabH (247 μM) and platencin showed IC_{50} values of FabF (4.58 μM) and FabH (9.17 μM). Comparison of FabF and FabH activities of platensimycin indicated 850 fold selectivity for FabF whereas platencin showed only two fold selectivity for FabF (Table 2). Cross comparison of the activity of platensimycin and platencin showed a 15 fold better *S. aureus* FabF activity in favour of platensimycin and 30 fold better *S. aureus* FabH activity in favour of platencin. The balanced, albeit lower, FabF/H potencies of platencin appear to be responsible and to account for, the potent fatty acid synthesis inhibition which is essentially equal to the fatty acid synthesis inhibition of platensimycin.

Enzyme	Platensimycin (IC_{50}, μM)	Platencin (IC_{50}, μM)	Fold selectivity
FabF (*S. aureus*)	0.048	ND	-
FabF binding (*E. coli*)	0.019	0.113	6x for PTM
FabH (*S. aureus*)	67	16.2	4x for PTN
Gel elongation cell lysate			
FabF (*S. aureus*)	0.29	4.58	15x for PTM
FabH (*S. aureus*)	247	9.17	30x for PTN

Table 2 *Summary of key biochemical data of platensimycin and platencin*

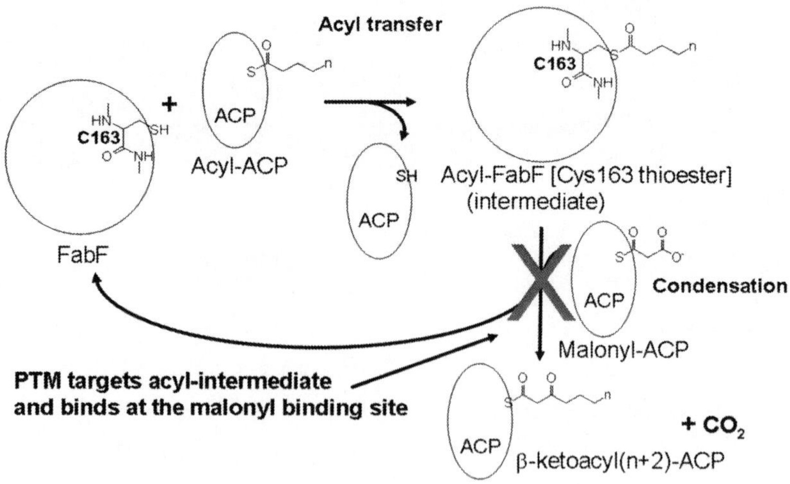

Figure 8 *Enzymatic reactions catalyzed by FabF and target of platensimycin*

2.4.2 Molecular Modes of Action. Condensing enzymes FabH and FabF are bi-functional enzymes and catalyze two discrete steps. The first reaction catalyzed by both of these enzymes is acylation of the active site cysteine followed by binding of the second substrate malonyl CoA. Claisen condensation of these two substrates is then followed by decarboxylation (Figure 8). It has been shown by binding experiments that platensimycin will not bind to the enzyme in the absence of the fatty acid substrate, but does bind in its presence, confirming that platensimycin binds and inhibits only the acylated enzyme.[20] This phenomenon was confirmed by X-ray co-crystallographic studies. All attempts to co-crystallize platensimycin with native FabF enzyme were unsuccessful and this was attributed to the closed nature of the active site of the unacylated enzyme. The acylated enzyme was not amenable to crystallization, however, it has been postulated that mutation of the active site cysteine to a glutamine leads to a protein conformation that mimics the acyl enzyme intermediate, though it is catalytically incompetent. Thus, we mutated Cys163 to Gln163 in *E. coli* FabF, co-crystallized it with platensimycin and determined the crystal structure at 2.5Å resolution.[20] In this structure Phe400, the gate keeper residue, is in an open conformation at an angle of 45° to the catalytic pocket. The movement of this residue caused by acylation or substitution with Gln163 is critical for the opening of the malonyl CoA binding pocket which permits the entry and binding of platensimycin.[20]

The co-crystal structure showed that all but one hetero atom of platensimycin are hydrogen bonded. A significant portion of the binding energy is generated by a salt bridge interaction of the acid group of the aminobenzoic acid with two active site histidines (His340 and His303) (Figure 9). The C-4 hydroxy group of the aminobenzoic acid forms a water mediated H-bond with Glu265. The C-17 tetracyclic enone acid portion of platensimycin showed a number of H-bond all with backbone residues which are summarized in Figure 9. The aminobenzoic acid unit of platencin showed a similar

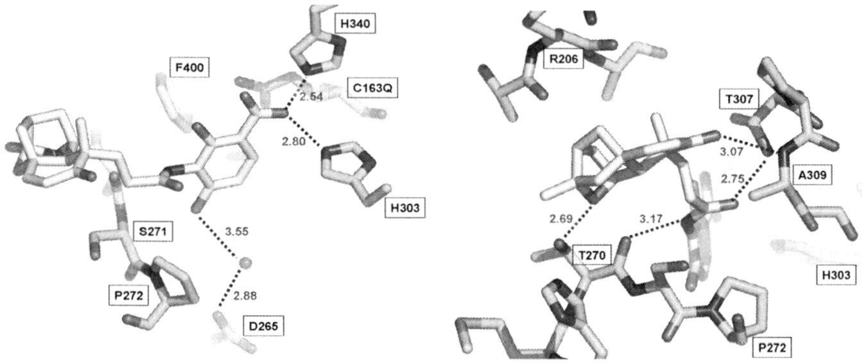

Figure 9 *Crystal structures of ecFabF (C163Q) bound to platensimycin showing hydrogen bond interactions*

network of salt bridges and H-bond interactions, but the C-17 tricyclic enone acid portion showed slightly different interactions (Figure 10).[29] The key difference was a lack of H-bond interaction with Thr270 due to absence of the ether bridge in the C-17 acid of platencin. The other H-bonds were identical except that the distances were longer compared to platensimycin. The lack of an ether H-bond and increased H-bond distances could account for the relatively lower FabF inhibitory and binding activity of platencin. No FabH crystallographic data is available to rationalize the difference in the FabH activity of the two compounds. However, examination of the amino acid residues of the two enzymes suggest that the binding sites of the terpenoid units contain larger numbers of hydrophilic amino acid residues in the FabF enzyme and more hydrophobic residues in FabH enzyme.

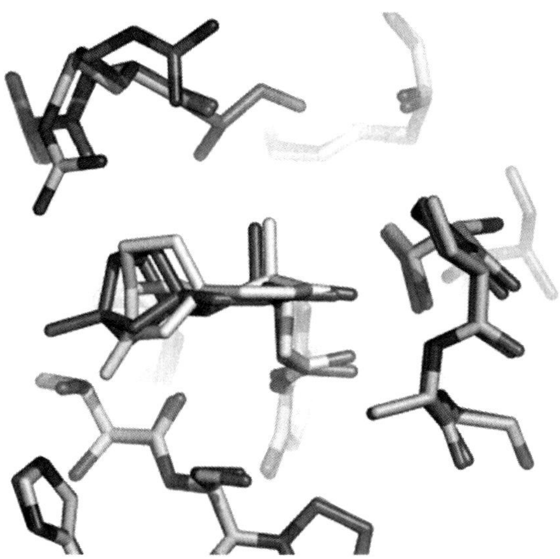

Figure 10 *Superposition of crystal structures of ecFabF (C163A) bound to platencin (light) and platensimycin (dark)-only C-17 terpenoid potion is shown.*

Clearly the terpenoid portion of platencin is much more hydrophobic than the corresponding terpenoid portion of platensimycin suggesting a positive hydrophobic-hydrophobic interaction for platencin but not for platensimycin.[26] These data provide persuasive evidence that platensimycin and platencin inhibit condensing enzymes by binding to the malonyl CoA binding site of the acyl-enzyme intermediates, inhibiting the binding of malonyl CoA, stopping decarboxylation and consequently blocking chain elongations. The binding of platensimycin and platencin to acyl-enzyme intermediates to give this inhibition mechanism is unprecedented.

Platensimycin and platencin exhibited excellent *in vivo* efficacy in an *S. aureus* mouse model of infection when infused intravenously, without causing any overt toxicity. They showed 3-5 $\log_{10}$ reduction of *S. aureus* colony forming units (CFU's) at 100-150 µg/h infusion.[20,25] *In vivo* efficacy of platensimycin was affected by its high clearance (~100 mL/min/kg), but it did not show any toxicity when infused at 50µg/h for 9 days followed by 100 µg/h for 8 days. In another experiment, platensimycin was continuously infused 600µg/h for 3 days and no overt toxicity was observed.

The antisense based screening strategy enabled the discovery of the novel antibiotics, platensimycin and platencin, which exert their activity by inhibiting the acyl enzyme intermediate of the fatty acid condensing enzyme, a novel mode of action. These discoveries clearly demonstrated the value of this technology and screening strategy in the discovery of novel antibiotics. Since platensimycin and platencin are potent antibiotics, they could be easily detected in crude extracts by screening with a wild type bacterial strain. Indeed, the extract showed strong activity in the control plate (equivalent to wild type). However, had it not been for antisense target based whole cell screening, the extracts producing platensimycin and platencin could not have been differentiated from thousands of other active extracts observed during standard whole cell screening. In other words, these extracts were no different from many other antibiotic producing extracts when tested in bacterial growth assay using wild type strains. Therefore, what this assay provided was the ability to differentiate the extracts that contained the potent antibiotics, e.g., platensimycin and platencin, which exerted their activity by inhibiting condensing enzymes, from other extracts that contained antibiotics that did not target these enzymes. Chemical screening of the extracts of *S. platensis* MA7327 led to isolation of a series of congeners of platensimycin and platencin.[29-33]

2.5 Inhibition of Ribosomal Protein S4 (RPSD)

Inhibition of protein synthesis has been established as one of the highly effective and clinically validated approaches to control bacterial growth and eliminate bacterial infections.[2,34] Protein synthesis is catalyzed by the ribosome which comprises two asymmetric macromolecular units, the large (50S) and small (30S) subunits. The small subunit is composed of 16S rRNA and 21 r-proteins, S1-S21.[35-37] Ribosomal proteins are known to help the ribosome maintain its quaternary structure. Most of the drugs that work by inhibiting protein synthesis bind not only to rRNA but also to one or more ribosomal proteins and the inhibition of protein synthesis is the composite result of both interactions. Alteration of rRNA binding to ribosomal proteins inactivates protein synthesis. Therefore, selective alteration of the conformation of a particular ribosomal protein, or inhibiting the synthesis of a ribosomal protein, would result in a loss of function, leading to the inhibition of bacterial protein synthesis. Small ribosomal protein S4 (RPSD) is one of these proteins which appeared to be a suitable target for antibacterial discovery. It is encoded by the *rpsD*

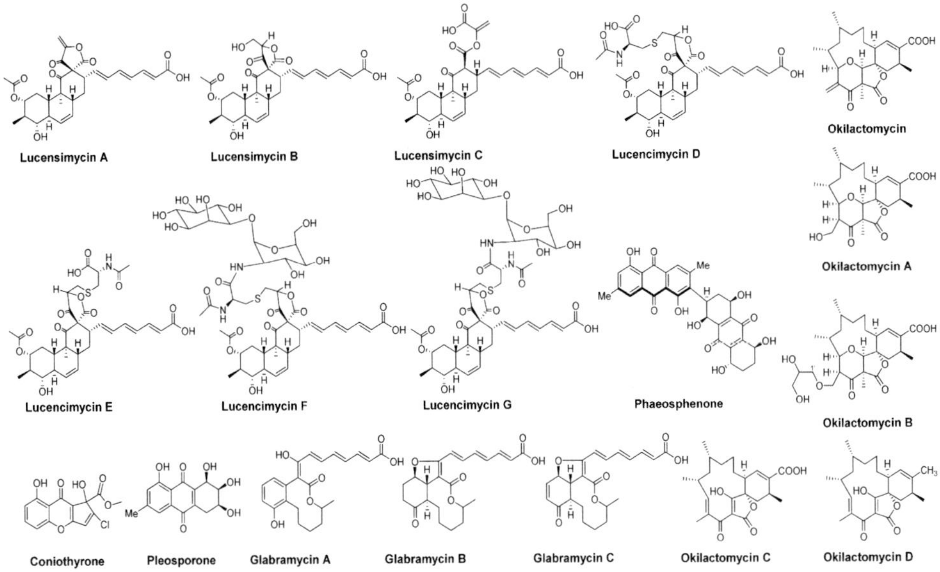

Figure 11 *Natural products discovered by rpsD antisense assay*

gene in both Gram-positive and Gram-negative bacteria, is conserved across bacterial species and is essential for bacterial growth.[12,38]

The same extracts that were used for screening against the FabF two plate assays described earlier, were screened using antisense sensitized *rpsD* two plate whole cell assays and gave a < 0.2% hit rate. Bioassay guided fractionation of a number of these extracts led to purification of six classes of novel antibiotics shown in Figure 11. These are seven members of lucensimycins A-G,[39-41] phaeosphenone,[42] coniothyrione,[43] pleosporone,[44] glabramycins A-C,[45] and okilactomycins.[46] Many of these compounds are not very potent antibiotics and did not show significant growth inhibition of the wild type strains of *S. aureus* so it is unlikely that they would have been discovered if this screening technology had not been used and this demonstrates the power of hypersensitivity for screening and the discovery of novel natural products.

2.6 Bacterial Target Array - A Simultaneous Screen for All Essential Bacterial Gene Targets using Antisense

Based on the success of individual target antisense screening results for the discovery of antibiotics, Merck scientists recently extended this approach to a chemical genomics strategy based on 245 *S. aureus* antisense RNA strains, each engineered for reduced expression of target genes essential for *S. aureus* growth.[47] As has been demonstrated earlier, attenuation of gene expression sensitizes cells to compounds that inhibit the activity of a gene product or associated process. Pools of strains grown competitively in the presence of bioactive compounds generate characteristic profiles of strain sensitivities reflecting the compound's mechanism of action. This tool was validated with a structurally and mechanistically diverse set of reference antibiotics.[47] The method described by Merck represents the first use of an integrated array of hypersensitized strains to systematically

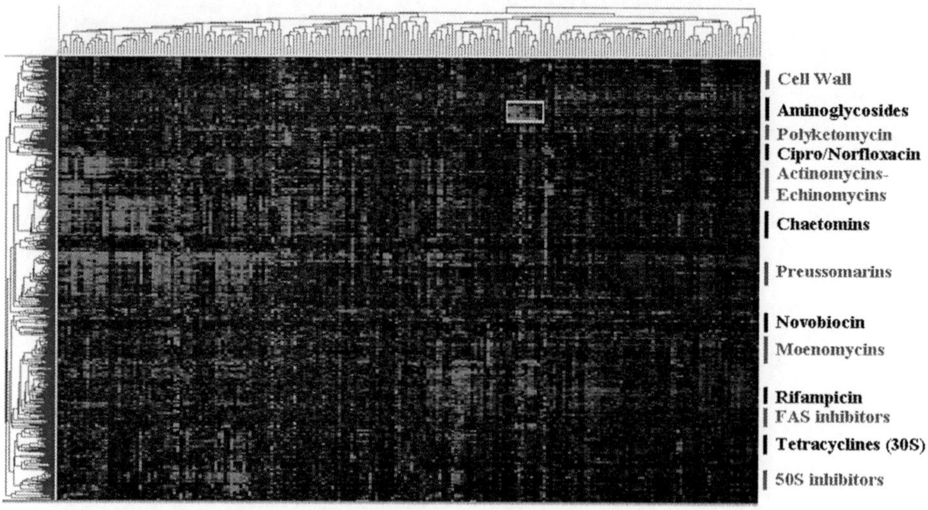

Figure 12 *Two-dimensional cluster analysis of AISS profiles of various compound classes belonging to defined mechanisms and 245 strains showing depletion (lighter areas) and resistance (darker areas). The compounds representing a specific mechanism cluster together. For detailed explanation and more precise profiling data, see Donald et al.*[47]

profile the modes of action of small molecule antibiotics in a clinically significant bacterium, *S. aureus*. The array consists of 245 strains that correspond to highly conserved essential genes which show growth phenotypes after induction by xylose-inducible antisense RNA expression. The system is semi-automated and is capable of generating antisense induced strain sensitivity (AISS) profiles indicative of the mechanistic imprint of a structurally and mechanistically diversified set of antibiotics. The process and the data from a reference set of 59 antibiotics was described by Donald et al.[47] This method easily differentiates compounds which inhibit bacterial growth by different mechanisms and provides visual differentiation when visualized by two-dimensional hierarchical clustering of AISS profiles and modes of action (Figure 12). This method allows for assessment of modes of action. For example, analysis of the AISS profile produced by novobiocin, a DNA synthesis inhibitor, showed significant depletions of the two DNA gyrase strains, *gyrA*-AS and *gyrB*-AS and both topoisomerase IV strains *parE*-AS and *parC*-AS, faithfully identifying the known multimeric enzyme complexes that are inhibited by novobiocin.[47] While profiling of known compounds provides validation of the methodology and mechanistic information on the action of a pure compound whether synthetic or a natural product, the key question is whether this technology can be useful for the discovery of novel natural products by profiling crude extracts. This was easily demonstrated by screening of the crude extract of *Streptomyces platensis*, the producer of platensimycin. Two dimensional cluster analyses of AISS profiles of the extracts of *Streptomyces platensis* MA7327 indicated significant depletion of *fabF*-AS along with *accA*-AS and *accC*-AS (Figure 13). Purified platensimycin showed an identical profile to that observed with the extract, confirming that the method is easily applicable to the profiling of crude extract and the discovery of new natural products.

Similar results were obtained with various other antibiotic producing strains such as novobiocin, moenomycin and many others that showed depletion of strains relevant to the

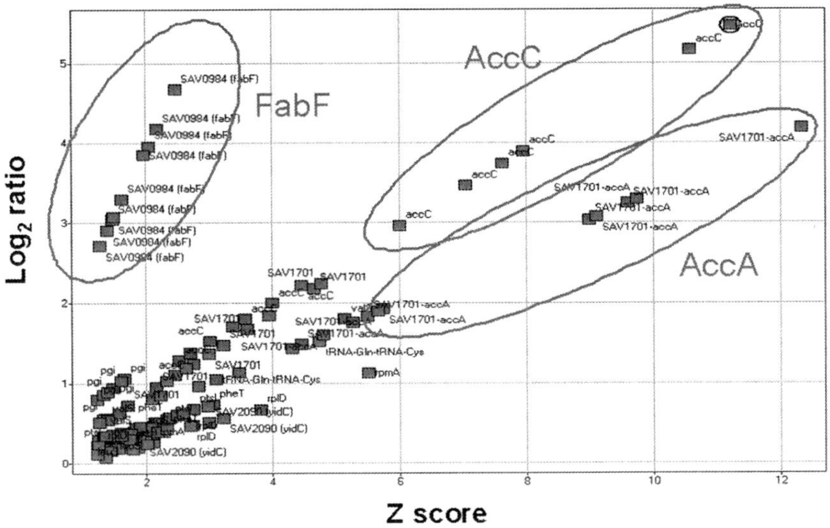

Figure 13 *Two-dimensional AISS profiles of Streptomyces platensis extracts containing platensimycin (3 mg/L), the natural titre. Data points represent independent doses across a minimum of two profiling experiments. All the data is represented. FabF, the enzyme target for platensimycin is significantly depleted and is clustered away from the baseline noise. The Acetyl CoA carboxylases (ACC) A and C, the enzymes early in the pathway are also depleted perhaps due to pathway effect. Platensimycin does not inhibit purified ACC enzymes. Purified platensimycin produced similar profiles suggesting the utility of the profiles in the screening.*

mode of action of the profiled compounds and corresponding extracts. In fact, the increasing number of compound profiles and capture of the data in a database, made the bacterial array an extremely powerful tool for comparing the activity of an unknown with a known compound allowing dereplication of natural products relatively easily; particularly when combined with highly sensitive high-resolution liquid chromatographic mass spectral (HRLCMS) analysis. This approach allowed for dereplication of natural products from first small scale (2-10 mL) fermentations without any preparative chromatographic fractionations and lead to the exclusive discovery of novel natural products. This is a great tool for the discovery of novel antibacterial agents.

3 SENSITIVE-RESISTANT PAIR SCREENING

Sensitive-resistant pair assays compare the activity of an antibiotic or active extract against a sensitive strain of a bacterium with the activity against a strain of the same bacterium that harbours resistance to the antibiotic. The design of the bioassay consists of seeding a control, antibiotic sensitive bacterial strain (e.g., *S. aureus* MB2865) on one agar plate and an antibiotic resistant strain (e.g., thiazomycin resistance *S. aureus* MB5832) on another. The compounds that exert their activity by binding to the site at which the resistance is mapped will be more active against the control antibiotic sensitive strain and form a larger zone of clearance compared to the plate containing the antibiotic resistant strain. The

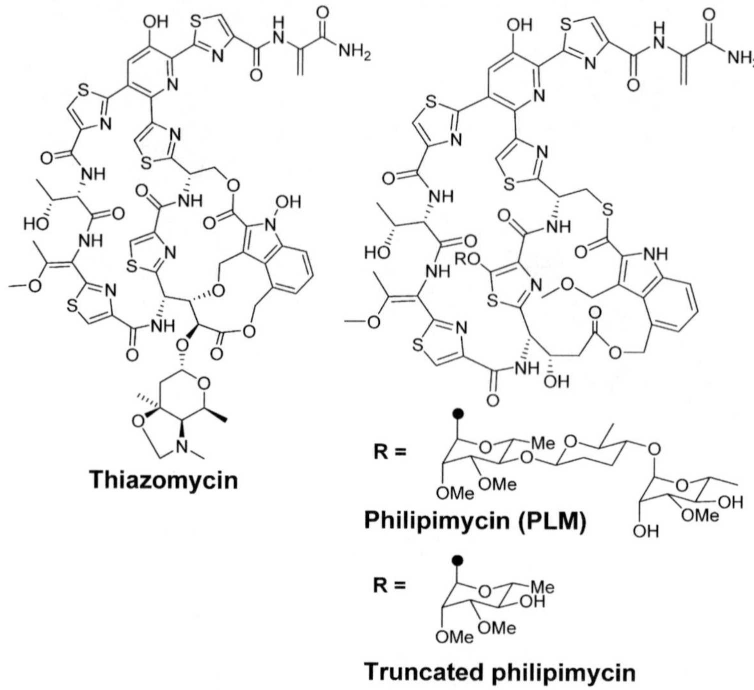

Figure 14 *Chemical structures of thiazomycin, philipimycin and truncated philipimycin*

compounds that inhibit bacterial growth by different mechanisms will show no differential in their zone of clearance between the two plates. This method was elegantly applied to the discovery of a novel thiazolyl peptide, philipimycin (PLM) and an acid catalysed hydrolytic product, truncated philipimycin (Figure 14).[48] Philipimycin showed a potent Gram-positive antibacterial activity like many thiazolyl peptides and displayed *in vivo* activity in a mouse infection model.[48]

Philipimycin was discovered by screening a large collection of natural product extracts against sensitive and thiazomycin resistant pairs of *S. aureus* strains (e.g., see Figure 15).[48] Thiazomycin is a protein synthesis inhibitor and binds to the ribosome and L11 protein.[49] The thiazomycin resistant strain was generated by growing *S. aureus* under a selective pressure of thiazomycin leading to a 12 base pair deletion in the L11 protein which was mapped to four amino acids (Pro_{23}-Val_{24}-Gly_{25}-Pro_{26}).[49,50] This screening method turned out to be extremely powerful for detection and targeted dereplication (by HRFTMS) of thiazolyl peptides. By random screening, we were able to detect 65 of 89 known thiazolyl peptides in a very short time, including several new compounds such as philipimycin. This method can be used for the discovery of novel members of a known class of natural products if so desired. While discovery of novel structural classes of compounds that bind to the same L11 protein binding site as thiazolyl peptides was the real goal, we did not discover compounds other than thiazolyl peptides in this assay. However, this method remains a viable option for such discoveries and for the discovery of compounds that work by other modes of action using other drug resistant phenotypes, such as macrolides, tetracyclines or coumarins.

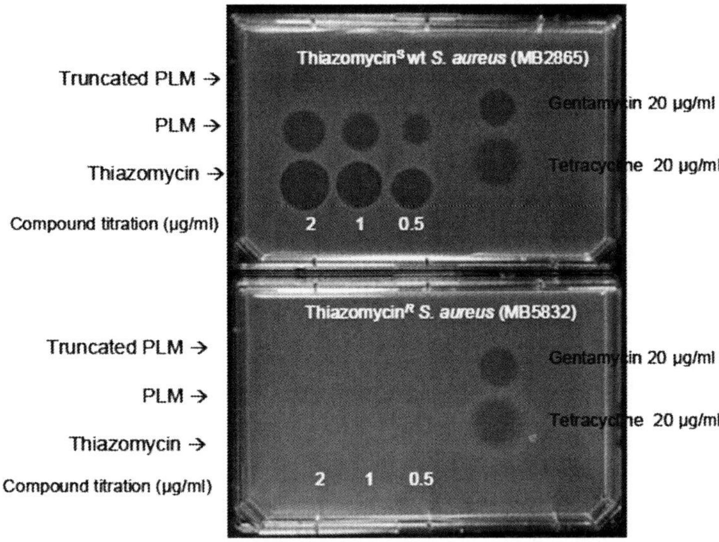

Figure 15 *Two plate sensitive-thiazomycin resistant assay. Top plate is seeded with thiazomycin sensitive strain of S. aureus (MB2865) and the bottom plate is seeded with thiazomycin resistant strain of S. aureus (MB5832). The resistant strain harbours resistance due to deletion of (Pro_{23}-Val_{24}-Gly_{25}-Pro_{26}) in L11 protein. Thiazomycin, philipimycin and truncated philipimycin show larger zones of inhibition in the top plate in a dose dependent manner compared to the bottom plate. The corresponding zone sizes in the two plates are the same for compounds that work by different modes of action (e.g, gentamycin and tetracycline)*

4 SUMMARY AND CONCLUSIONS

Discovery of new antibiotics with novel mechanisms of action is critical in the fight against drug resistant bacteria. While classical and contemporary discovery technologies have a role to play, they have resulted in less than desirable results. Discovery, development and aggressive implementation of new technologies for uncovering new antibacterial chemotypes that inhibit bacterial growth with new modes of actions hold the key for the future success to fight bacterial resistance.

Some of the approaches summarized in this article, particularly antisense based selective target depletion and differential sensitivity screening technology are very powerful whole cell-target based approaches to the discovery of novel natural product antibiotics. This approach is general in nature and applicable for all essential gene products and could be applied to screening for all targets simultaneously in a single test when used in an array format. This technology provides the opportunity for easy targeted dereplication of known natural products and fast discovery of new natural products. The technology is poised for pairing with novel natural products sources for discovery of novel antibiotics with novel modes of action. Drug discovery is a very tedious and unpredictable process with low success rates. Therefore, it is imperative to discover a large numbers of lead-target pairs for further preclinical and clinical studies to develop clinically useful antibiotics.

Acknowledgements

The author is indebted to his Merck colleagues who contributed so immensely for this program. Most notables are: Drs. Jun Wang, Katherine Young, Dennis Schmatz, Stephen Soisson, Sookhe Ha, John Phillips, Robert Donald, Prakash Masurekar, Francisca Vicente, Olga Genilloud, Fernando Pelaez, Gerald Bills, Mary Motyl, Hiranthi Jayasuriya, Kithsiri Herath, John Ondeyka and Chaowei Zhang without whom this work would have been not possible.

References

1 C.T. Walsh, *Antibiotics: actions, origin, resistance*; ASM Press: Washington, DC, 2003.
2 S.B. Singh and J.F. Barrett, *Biochem. Pharmacol.* 2006, **71**, 1006-1015.
3 A report from the NNIS System *Amer. J. Infection control* 2004, **32**, 470-485.
4 G.H. Talbot, J. Bradley, J.E. Edwards, Jr., D. Gilbert, M. Scheld, and J.G. Bartlett, *Clin. Infect. Dis.* 2006, **42**, 657-68.
5 B. Spellberg, R. Guidos, D. Gilbert, J. Bradley, H.W. Boucher, W.M. Scheld, J.G. Bartlett and J. Edwards, Jr., *Clin. Infect. Dis.* 2008, **46**, 155-64.
6 R.M. Klevens, M.A. Morrison, J. Nadle, S. Petit, K. Gershman, S. Ray, L.H. Harrison, R. Lynfield, G. Dumyati, J.M. Townes, A.S. Craig, E.R. Zell, G.E. Fosheim, L.K. McDougal, R.B. Carey and S.K.Fridkin, *J. Am. Med. Assoc.* 2007, **298**, 1763-71.
7 D.J. Payne, M.N. Gwynn, D.J. Holmes and D.L. Pompliano, *Nat. Rev. Drug Discov.* 2007, **6**, 29-40.
8 B.J. Akerley, E.J. Rubin, V.L. Novick, K. Amaya, N. Judson and J.J. Mekalanos, *Proc. Natl. Acad. Sci. (USA)* 2002, **99**, 966-971.
9 C.A. Hutchinson, III, S.N. Peterson, S.R. Gill, R.T. Cline, O. White, C.M. Fraser, H.O. Smith, J.C. Venter, *Science* 1999, **286**, 2165-2169.
10 Y. Ji, B. Zhang, S.F. Van, W.P. Horn; G. Woodnutt, M.K. Burnham, M. Rosenberg, *Science* 2001, **293**, 2266-9.
11 K. Kobayashi, S.D. Ehrlich, A. Albertini, G. Amati, K.K. Andersen, M. Arnaud, K. Asai, S. Ashikaga, S. Aymerich, P. Bessieres, F. Boland, S.C. Brignell, S. Bron, K. Bunai, J. Chapuis, L.C. Christiansen, A. Danchin, M. Debarbouille, E. Dervyn, E. Deuerling, K. Devine, S.K. Devine, O. Dreesen, J. Errington, S. Fillinger, S.J. Foster, Y. Fujita, A. Galizzi, R. Gardan, C. Eschevins, T. Fukushima, K. Haga, C.R. Harwood, M. Hecker, D. Hosoya, M.F. Hullo, H. Kakeshita, D. Karamata, Y. Kasahara, F. Kawamura, K. Koga, P. Koski, R. Kuwana, D. Imamura, M. Ishimaru, S. Ishikawa, I. Ishio, D. Le Coq, A. Masson, C. Mauel, R. Meima, R.P. Mellado, A. Moir, S. Moriya, E. Nagakawa, H. Nanamiya, S. Nakai, P. Nygaard, M. Ogura, T. Ohanan, M. O'Reilly, M. O'Rourke, Z. Pragai, H.M. Pooley, G. Rapoport, J.P. Rawlins, L.A. Rivas, C. Rivolta, A. Sadaie, Y. Sadaie, M. Sarvas, T. Sato, H.H. Saxild, E. Scanlan, W. Schumann, J.F.M.L. Seegers, J. Sekiguchi, A. Sekowska, S.J.

Seror, M. Simon, P. Stragier, R. Studer, H. Takamatsu, T. Tanaka, M. Takeuchi, H.B. Thomaides, V. Vagner, J.M. van Dijl, K. Watabe, A. Wipat, H. Yamamoto, M. Yamamoto, Y. Yamamoto, K. Yamane, K. Yata, K. Yoshida, H. Yoshikawa, U. Zuber and N.Ogasawara, *Proc. Natl. Acad. Sci. (USA)* 2003, **100**, 4678-4683.

12 R.A. Forsyth, R.J. Haselbeck, K.L. Ohlsen, R.T. Yamamoto, H. Xu, J. D. Trawick, D. Wall, L. Wang, V. Brown-Driver, J.M. Froelich, G.C. Kedar, P. King, M. McCarthy, C. Malone, B. Misiner, D. Robbins, Z. Tan, Z.-Y.; Zhu, G. Carr, D.A. Mosca, C. Zamudio, J.G. Foulkes and J.W. Zyskind, *Mol. Microbiol.* 2002, **43**, 1387–1400.

13 K. Young and L.L Silver,. *J. Bacteriol.* 1991, **173**, 3609-14.

14 K. Poole, *Current Opinion in Microbiology* 2001, **4**, 500-508.

15 P.N. Markham and A.A. Neyfakh, *Curr. Opin. Microbiol.* 2001, **4**, 509-14.

16 Y. Ji, A. Marra, M. Rosenberg and G. Woodnutt, *J. Bacteriol.* 1999, **181**, 6585-90.

17 S.B. Singh, J.W. Phillips and J. Wang, *Curr. Opin. Drug Disc. Dev.* 2007, **10**, 160-166.

18 K. Young, H. Jayasuriya, J.G. Ondeyka, K. Herath, C. Zhang, S. Kodali, A. Galgoci, R. Painter, V. Brown-Driver, R. Yamamoto, L.L. Silver, Y. Zheng, J.I. Ventura, J. Sigmund, S. Ha, A. Basilio, F. Vicente, J.R. Tormo, F. Pelaez, P. Youngman, D. Cully, J.F. Barrett, D. Schmatz, S.B. Singh and J.Wang, *Antimicrob. Agents Chemother.* 2006, **50**, 519-526.

19 S. Brinster, G. Lamberet, B. Staels, P. Trieu-Cuot, A. Gruss and C. Poyart, *Nature* 2009, **458**, 83-6.

20 J. Wang, S.M. Soisson, K. Young, W. Shoop, S. Kodali, A. Galgoci, R. Painter, G. Parthasarathy, Y. Tang, R. Cummings, S. Ha, K. Dorso, M. Motyl, H. Jayasuriya, J. Ondeyka, K. Herath, C. Zhang, L. Hernandez, J. Alloco, Á. Basilio, J.R. Tormo, O. Genilloud, F. Vicente, F. Pelaez, L. Colwell, S.H. Lee, B. Michael, T. Felcetto, C. Gill, L.L. Silver, J. Hermes, K. Bartizal, J. Barrett, D. Schmatz, J.W. Becker, D. Cully and S.B. Singh, *Nature* 2006, **441**, 358-361.

21 C. Rapp, G. Jung, C. Isselhorst-Schart and H. Zahner, *Liebigs Ann. Chem.* 1988, 1043-1047.

22 S. Kodali, A. Galgoci, K. Young, R. Painter, L.L. Silver, K.B. Herath, S.B. Singh, D. Cully, J.F. Barrett, D. Schmatz and J. Wang, *J. Biol. Chem.* 2005, **280**, 1669-1677.

23 J.G. Ondeyka, D.L. Zink, K. Young, R. Painter, S. Kodali, A. Galgoci, J. Collado, J.R. Tormo, A. Basilio, F. Vicente, J. Wang and S.B. Singh, *J. Nat. Prod.* 2006, **69**, 377-380.

24 S.B. Singh, H. Jayasuriya, J.G. Ondeyka, K.B. Herath, C. Zhang, D.L. Zink, N.N. Tsou, R.G. Ball, A. Basilio, O. Genilloud, M.T. Diez, F. Vicente, F. Pelaez, K. Young and J. Wang, *J. Am. Chem. Soc.* 2006, **128**, 11916-20 and 15547.

25 J. Wang, S. Kodali, S.H. Lee, A. Galgoci, R. Painter, K. Dorso, F. Racine, M. Motyl, L. Hernandez, E. Tinney, S. Colletti, K. Herath, R. Cummings, O. Salazar, I. Gonzalez, A. Basilio, F. Vicente, O. Genilloud, F. Pelaez, H. Jayasuriya, K. Young, D. Cully and S.B. Singh, *Proc. Natl. Acad. Sci. (USA)* 2007, **104**, 7612-7616.

26 H. Jayasuriya, K.B. Herath, C. Zhang, D.L. Zink, A. Basilio, O. Genilloud, M.T. Diez, F. Vicente, I. Gonzalez, O. Salazar, F. Pelaez, R. Cummings, S. Ha, J. Wang and S.B. Singh, *Angew. Chem. Int. Ed.* 2007, **46**, 4684-4688.

27 K.B. Herath, A.B. Attygalle and S.B. Singh, *J. Am. Chem. Soc.* 2007, **129**, 15422-3.

28 K. Herath, A.B. Attygalle and S.B. Singh, *Tetrahedron Lett.* 2008, **49**, 5755-5758.

29 S.B. Singh, J.G. Ondeyka, K.B. Herath, C. Zhang, H. Jayasuriya, D.L. Zink, G. Parthasarathy, J.W. Becker, J. Wang and S.M. Soisson, *Bioorg. Med. Chem. Lett.* 2009, **19**, 4756-4759.

30 K.B. Herath, C. Zhang, H. Jayasuriya, J.G. Ondeyka, D.L. Zink, B. Burgess, J. Wang, and S.B. Singh, *Org. Lett.* 2008, **10**, 1699-702.

31 H. Jayasuriya, K.B. Herath, J.G. Ondeyka, D.L. Zink, B. Burgess, J. Wang and S.B. Singh, *Tetrahedron Lett.* 2008, **49**, 3648-3651.

32 C. Zhang, J. Ondeyka, D.L. Zink, B. Burgess, J. Wang and S.B.Singh, *Chem. Commun. (Camb)* 2008, 5034-6.

33 S.B. Singh, H. Jayasuriya, K.B. Herath, C. Zhang, J. Ondeyka, D.L. Zink, S. Ha, G. Parthasarathy, J.W. Becker, J. Wang and S.M.Soisson, *Tetrahedron Lett.* 2009, **50**, 5182-5185.

34 J. Poehlsgaard and S. Douthwaite, *Nature Rev. Microbiol.* 2005, **3**, 870-881.

35 V. Ramakrishnan, *Cell* 2002, **108**, 557-72.

36 G.M. Culver, *Biopolymers* 2003, **68**, 234-249.

37 J.M. Ogle, A.P. Carter and V. Ramakrishnan, *Trends Biochem. Sci.* 2003, **28**, 259-66.

38 F.J. Grundy and T.M.Henkin, *J. Bacteriol.* 1991, **173**, 4595-602.

39 S.B. Singh, D.L. Zink, J. Huber, O. Genilloud, O. Salazar, M.T. Diez, A. Basilio, F. Vicente and K.M. Byrne, *Org. Lett.* 2006, **8**, 5449-52.

40 S.B. Singh, D.L. Zink, K.B. Herath, O. Salazar and O. Genilloud, *Tetrahedron Lett.* 2008, **49**, 2616-2619.

41 S.B. Singh, D.L. Zink, K. Dorso, M. Motyl, O. Salazar, A. Basilio, F. Vicente, K.M. Byrne, S. Ha and O. Genilloud, *J. Nat. Prod.* 2008, **72**, 345-352.

42 C. Zhang, J.G. Ondeyka, D.L. Zink, A. Basilio, F. Vicente, J. Collado, G. Platas, G. Bills, J. Huber, K. Dorso, M. Motyl, K. Byrne and S.B. Singh, *J. Nat. Prod.* 2008, **71**, 1304-1307.

43 J.G. Ondeyka, D. Zink, A. Basilio, F. Vicente, G. Bills, M.T. Diez, M. Motyl, G. Dezeny, K. Byrne and S.B. Singh, *J. Nat. Prod.* 2007, **70**, 668-70.

44 C. Zhang, J.G. Ondeyka, D.L. Zink, A. Basilio, F. Vicente, J. Collado, G. Platas, J. Huber, K. Dorso, M. Motyl, K. Byrne and S.B. Singh, *Bioorg. Med. Chem.* 2009, **17**, 2162-6.

45 H. Jayasuriya, D. Zink, A. Basilio, F. Vicente, J. Collado, G. Bills, M.L. Goldman, M. Motyl, J. Huber, G. Dezeny, K. Byrne and S.B. Singh, *J. Antibiot. (Tokyo)* 2009, **62**, 265-9.

46 C. Zhang, J.G. Ondeyka, D.L. Zink, A. Basilio, F. Vicente, O. Salazar, O. Genilloud, K. Dorso, M. Motyl, K. Byrne and S.B. Singh, *J. Antibiot. (Tokyo)* **2009**, *62*, 55-61.

47 R.G. Donald, S. Skwish, R.A. Forsyth, J.W. Anderson, T. Zhong, C. Burns, S. Lee, X. Meng, L. LoCastro, L.W. Jarantow, J. Martin, S.H. Lee, I. Taylor, D. Robbins, C. Malone, L. Wang, C.S. Zamudio, P.J. Youngman and J.W. Phillips, *Chem. Biol.* 2009, **16**, 826-36.

48 C. Zhang, J. Occi, P. Masurekar, J.F. Barrett, D.L. Zink, S. Smith, R. Onishi, S. Ha, O. Salazar, O. Genilloud, A. Basilio, F. Vicente, C. Gill, E.J. Hickey, K. Dorso, M. Motyl and S.B. Singh, *J. Am. Chem. Soc.* 2008, **130**, 12102-10.

49 S.B. Singh, J. Occi, H. Jayasuriya, K. Herath, M. Motyl, K. Dorso, C. Gill, E. Hickey, K.M. Overbye, J.F. Barrett and P. Masurekar, *J. Antibiot. (Tokyo)* 2007, **60**, 565-71.

50 H. Jayasuriya, K. Herath, J.G. Ondeyka, C. Zhang, D.L. Zink, M. Brower, F.P. Gailliot, J. Greene, G. Birdsall, J. Venugopal, M. Ushio, B. Burgess, G. Russotti, A. Walker, M. Hesse, A. Seeley, B. Junker, N. Connors, O. Salazar, O. Genilloud, K. Liu, P. Masurekar, J.F.;Barrett and S.B. Singh, *J. Antibiot. (Tokyo)* 2007, **60**, 554-64.

MARINE MICRO-ORGANISMS: A SOURCE OF NOVEL BIO-ACTIVES

Andrew Mearns Spragg

Aquapharm Biodiscovery Ltd, European Centre for Marine Biotechnology, Dunstaffnage, Oban, PA37 1QA, Scotland, UK. E-mail: enquiries@aquapharm.co.uk

1 INTRODUCTION

Aquapharm is a leading marine biotechnology company pioneering the discovery, isolation and subsequent development of novel bio-chemicals isolated from the under-explored biological diversity of marine microbes.

The company was founded in 2000 financed by a number of grants, a Royal Society of Edinburgh (RSE) Enterprise fellowship and awards. In 2004, the company received £1.25 million financing from a London based angel network, NESTA, Scottish Enterprise and Highlands and Islands Enterprise Argyll and the Islands allowing the business to re-locate to facilities within the European Centre for Marine Biotechnology near Oban, Scotland. In 2007 the company closed a further financing of £4.2 million lead by the Dutch venture capitalist group Aescap Venture and UK based Tate & Lyle Ventures with contributions from existing investors. This funding allowed the company to increase its resources so that it now employs 20 people at its research headquarters in Oban. In March 2010, the company announced a further £4.2 million investment from existing investors Aescap Venture, Tate & Lyle Ventures, Highlands and Islands Enterprise (HIE) and NESTA.

The cornerstone of the company's discovery platform is a collection of >7,500 marine micro-organisms, which it continues to expand. Through the application of its know-how and proprietary screening technologies to this collection, Aquapharm can boost the production of secondary metabolites, many of which have new scaffold diversity that is not present in synthetic or terrestrial natural product (NP) compounds. Natural products from this unique source of new biological and chemical diversity are already being exploited commercially with major partners such as Croda International Plc and Dr Reddy's Laboratories in the personal care and pharmaceutical fields respectively

Aquapharm will use its proprietary microbial platform to build a large marine derived compound library containing small molecule and peptide compounds with the size and structural characteristics of potential new drugs. Upon its completion, the compound library will accelerate Aquapharm's drug discovery efforts, principally in the anti-microbial field targeting Gram negative and Gram positive infections.

In addition, Aquapharm's development pipeline is also fed by an active discovery portfolio of anti-oxidants, anti-inflammatories for applications in personal care and functional food ingredients.

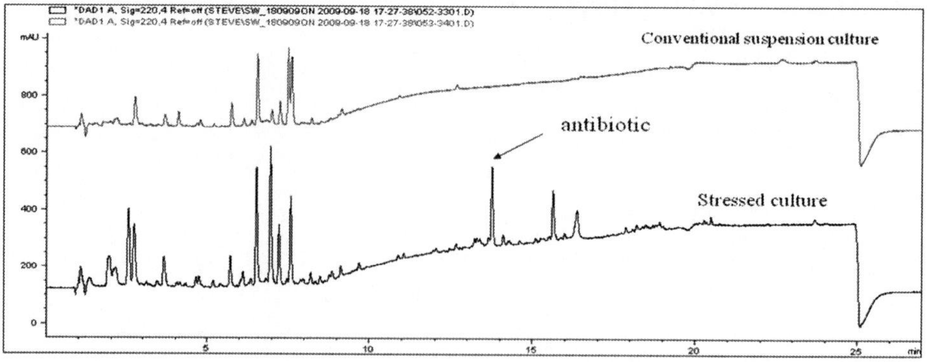

Figure 1 *Aquapharm SeaRch™ technology demonstrating differences in chemical expression profiles produced in bacteria grown under conventional suspension (upper trace) vs a stressed culture system (lower trace).*

To fully exploit the natural product potential of its microbial resource, the company employs a unique fermentation technology to this culture collection; SeaRch™ which allows for the production of important secondary metabolites by switching on sleeping biosynthetic pathways present in micro-organisms. Utilising this technology it is possible to induce certain stress mechanisms within the organism causing the production of secondary metabolites not previously seen from a traditional fermentation. Induction of key stress mechanisms with microbes can lead to changes in DNA methylation status causing changes to the chemical expression profile (as illustrated in Figure 1) which can lead to the recovery of compounds that are not normally expressed under standard culture conditions.

Secondary metabolites such as these are often induced in response to predators, disease etc. and are thus more likely to contain biologically active compounds e.g. antibacterials than primary metabolites.

Using these extracts from its collection of marine micro-organisms, Aquapharm is currently developing its compound library for high throughput screening (HTS) and subsequent lead generation using a structured program of growth, extraction, fractionation and distribution into a multi-well format.

When completed, the semi-purified compound library will complement Aquapharm's in house structural elucidation and product development capabilities, in order to help identify leads for specific commercial applications.

2 MARINE MICROBIOLOGY AS A SOURCE OF NOVEL CHEMISTRY

Nature's pharmacy has long been utilised to treat disease and improve the quality of life. Well-known examples include the analgesic, aspirin, derived from the bark of the willow tree, the heart medicine, digoxin, (digitalis glycosides) derived from foxgloves and most famously the antibiotic penicillin produced from the *Penicillium* fungus. Indeed, modern pharmaceutical drug discovery owes much to natural product research as more than 61% of small molecule chemical entity (NCE) drugs marketed between 1981 and 2002 originated from natural products.[1] The majority of these products have been derived from so called 'secondary metabolites' of terrestrial microorganisms, however the rate of new chemistry discovered from land based organisms has slowed and there is growing recognition of the importance of the marine environment for the provision of new biological and chemical diversity.

Interest in marine natural products as a source of novel chemistry really took off in the 1950s once the technology for humans to access marine habitats improved with advanced scuba technologies to facilitate sample collection.[2]

It is not unsurprising that marine natural products have provided much interest as a source of new drug compounds, as this is where the earth first gave birth to life some 3.5 billion years ago. It was only about 500 million years ago that a small fraction of this biodiversity managed to adapt to live on land, giving rise to the familiar terrestrial life forms we see around us today. However, a much greater pool of biodiversity evolved in the oceans creating millions more species compared to those on land. Marine micro-organisms have developed unique physiological and chemical abilities to survive and adapt to the diverse and sometimes extreme habitats found in the marine environment. Adaptations to extremes in pH, salt, temperature and pressure have provided marine microbes with the ability to produce rare and unusual secondary metabolites. All this translates into the oceans containing the largest abundance of species found anywhere on the planet. "Members of this "rare biosphere" are highly divergent from each other and may represent a nearly inexhaustible source of genomic innovation."[3]

Marine natural product research over the years has built up a remarkable legacy since the first symposium on the subject was organised on Rhode Island in 1967 entitled "Drugs from the Sea". The higher than expected hit rates of marine natural compounds in screening for drug leads was often a justification to search this environment and many notable discoveries have been made in areas of infectious disease, cell cycle modulation and cellular metabolism. Werner Bergmann published three reports in the early 1950's reporting the isolation of novel arabino- and ribo-pentosyl nucleosides obtained from marine sponges off the Florida coast which lead to the development of the anti-cancer derivative drugs vidarabine and cytarabine.[2,4] Other examples include macrolactin A,[5] a novel macrolide discovered in the early 1980s from a deep sea bacterium by William Fenical's group at the Scripps Institute of Oceanography. This molecule had potent antiviral activity against HIV and Herpes Simplex Virus (HSV). The first marine-derived compound to be approved in the United States was Ziconotide (Prialt; Elan Pharmaceuticals), a peptide isolated from a tropical cone snail for use in pain relief. It was approved for sale under the name Prialt by the U.S. Food and Drug Administration on December 28, 2004 and by the European Commission on February 22, 2005. Since then, Trabectedine (Yondelis®; PharmaMar) was approved in 2007 for the treatment of patients with advanced soft tissue sarcoma, after failure of anthracyclines and ifosfamide, or who are unsuited to receive these agents. Yondelis is credited as being the first marine anti-cancer drug to be approved by the EU.

Marine microorganisms are expected to serve as new sources of chemically interesting and biologically significant secondary metabolites, serving as lead compounds for potential drug development or pharmacological tools for basic research in life sciences.[6,7] Most marine microorganisms are of the Gram negative and positive eubacteria, cyanobacteria, and the myxobacteria groups, which are generally thought to produce many medically useful substances.

The ocean environment is massively complex, consisting of extreme variations in pressure, salinity, temperature, and biological habitats. The biota of marine microorganisms has developed unique metabolic and physiological functions that not only ensure survival in extreme habitats but also offer a potential for the production of novel enzymes and bioactive metabolites for potential exploitation. It is estimated that less than

5% of the bacteria observed by microscopy are culturable under standard laboratory conditions. Out of the large number of species examined, only a fraction of marine bacteria have been isolated and cultured. This is exemplified in the following Figure 2 that used principal component analysis (PCA) as a mathematical procedure to illustrate the global patterns of bacterial diversity isolated from saline and non saline environments.[8]

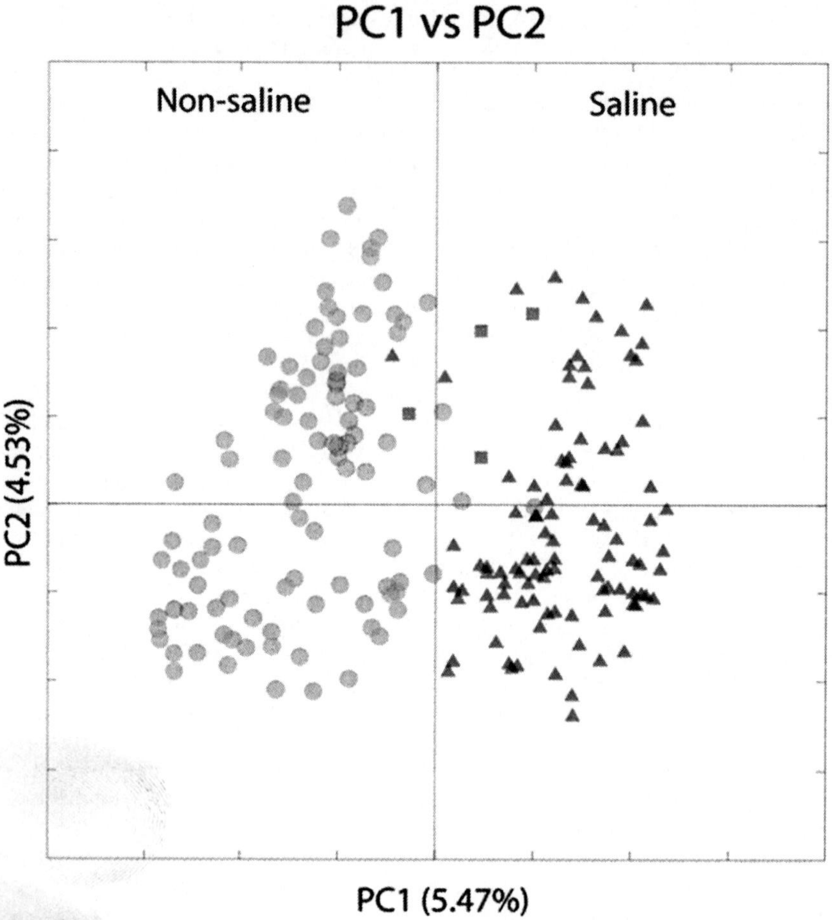

Figure 2 *Graph showing principle component analysis illustrating the divergence of microbial biodiversity isolated from saline vs. non saline environments with a clear taxonomic divide (circles indicate non-saline environments; triangles indicate saline environments; squares indicate mixed environments).*

These data were generated by Catherine A. Lozupone and Rob Knight based their analyses of 21,752 16S rRNA sequences compiled from 111 studies of diverse physical environments.[8] Surprisingly, they found that the major environmental determinant of microbial community composition and diversity was based on salinity rather than extremes of temperature, pH, or other physical and chemical factors. Marine sediments in particular were shown to be more phylogenetically diverse than any other environment type. By comparison, terrestrial soils were shown to contain below-average phylogenetic diversity.

Many types of marine habitats have created niches for the evolution of diverse life forms. The surface and internal spaces of the sessile marine organisms provide a unique micro-habitat and several symbiotic microorganisms are reported to occupy up to 40% of their tissue volume. These microbes are proven to be highly specific in their relationship with filter-feeding organisms like sponges, alcoynarians, ascidians, and marine plants and are often responsible for the production of interesting secondary metabolites.

For example the dolastatins, a series of cytotoxic peptides originally isolated from the Indian Ocean sea hare *Dolabella auricularia* by Arizona State University in 1972 have now recently been isolated from marine cyanobacteria living symbiotically with the host organism.[10] In fact cyanobacteria may be the true biosynthetic origin of many bioactive molecules isolated from marine invertebrates that sequester cyanobacteria through diet or by symbiosis. Further study in this area may solve important supply issues to developing many bioactive natural products as clinical drug candidates through fermentation.

The chemistry of Dolastatin is striking: it is an unusual depsipeptide containing 4 unusual amino acids *N,N*-dimethylvaline, dolaisoleucine, dolaproline and dolaphenine. Dolastatin 10 demonstrated *in vitro* inhibitory activity against a number of human cell lines including melanoma, sarcoma and ovarian cancer cell lines and had progressed to Phase II clinical studies.[9,11,12] Although the toxicity profile of Dolastatin 10 was acceptable, no clinically significant activity could be detected in any of the conducted trials and did not justify continued study of this molecule as a drug candidate. However, a new synthetic derivative TZT-1027 is being evaluated in Phase I trials in Japan, the USA and Europe.[13,14] The synthetic derivative was altered by the replacement of the terminal dolaphenine amino acid residue with phenethylamine, but the two compounds share the same mechanism of action; they both bind near the vinca peptide site on tubulin and inhibit tubulin polymerization with equal potency.

Marine bacteria offer a prolific source of highly novel chemical structures with the potential as drug leads.[15] Recent examples include, salinosporamide A (NPI-0052) the p26 proteasome inhibitor. This compound was discovered by Fenical, Jensen and co-workers at the Scripps Institution of Oceanography, California, USA, in a bacterium recovered from deep-sea sediment in 2003. The mechanism of action of salinosporamide A (Figure 4, left) is similar to that of the proteasome inhibitor bortezomib (Velcade; Millennium/Janssen-Cilag), which is approved for the treatment of multiple myeloma, but it overcomes resistance to bortezomib in patients with relapsed or refractory multiple myeloma. Salinosporamide A has completed Phase I clinical trials for multiple myeloma under the sponsorship of Nereus Pharmaceuticals in La Jolla, California.[16,17]

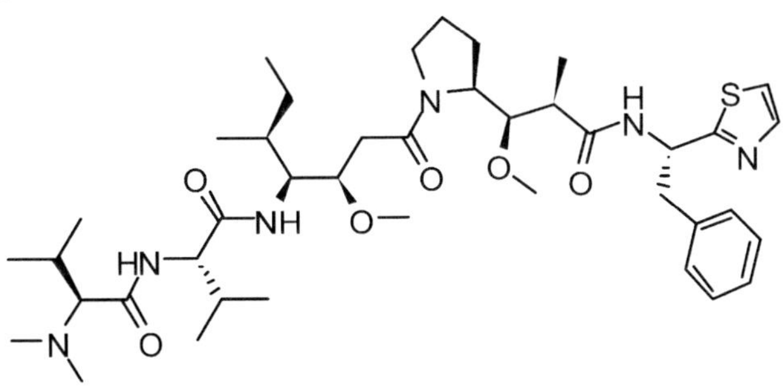

Figure 3 *Dolastatin-10.*

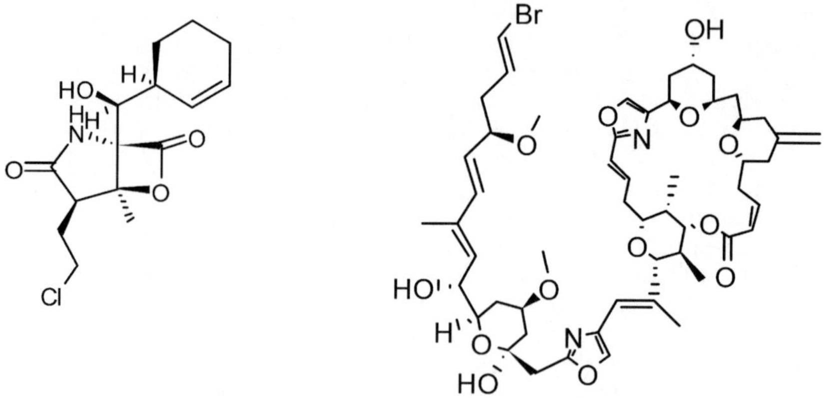

Figure 4 *Structures of salinosporamide A (left) and phorboxazole A (right).*

The potent cytostatic agent phorboxazole A was isolated in 1995 by Searle and Molinski from the marine sponge *Phorbas sp.*[18] Phorboxazole was found to inhibit cell cycle in S-phase at subnanomolar concentrations. This molecule may be produced by cyanobacteria that either live within the host, or expressed from microbial genes that have been integrated into the host sponge (Figure 4, right).

3 NOVEL ANTIBACTERIALS FROM MARINE MICRO-ORGANISMS

It is widely known that many types of microorganisms are capable of producing novel antibiotics and the subject has been widely reviewed.[19] Antimicrobial activity is thought to be commonplace within the marine environment and may be due to excessive competition dynamics between different micro-organisms for "living" space, assimilation of nutrients and control of population dynamics within micro-environments (sediments, surface structures, biofilms etc.).[20] A growing number of marine fungi are also providing new sources of novel and potentially life-saving bioactive secondary metabolites e.g. novel antibacterial, antiviral and antiprotozoal compounds.[21]

3.1 Zopfiellamide

The facultative marine ascomycete *Zopfiella latipes* was grown in submerged cultures, and two novel antimicrobial compounds, zopfiellamide A and zopfiellamide B (Figure 5) were isolated from both the culture fluids and from the mycelia.[21] These two compounds display moderate antibacterial effects towards the Gram positive *Arthrobacter citreus, Bacillus brevis, B. subtilis, B. licheniformis, Corynebacterium insidiosum, Micrococcus luteus, Mycobacterium phlei,* and *Streptomyces sp.* and towards the Gram negative *Acinetobacter calcoaceticus*, with minimal inhibitory concentrations between 2 and 10 μg/ml for zopfiellamide A.

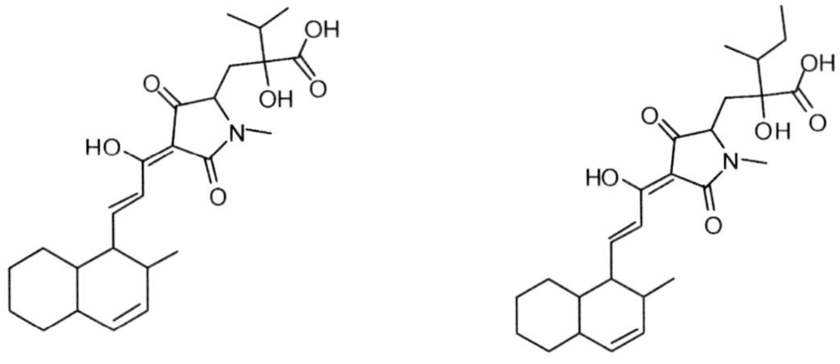

Figure 5 *Zopfiellamides A and B.*

3.2 Abyssomicin

Collaborative efforts by Riedlinger's group at the Mikrobiologisches Institut, Universität Tübingen in Germany and the Universities of Newcastle and Kent, in the UK resulted in the discovery of three novel polycyclic polyketides named as abyssomicins. These molecules were isolated from a deep sea marine actinomycete strain, *Verrucosispora* AB-18-032.[22] The compounds were isolated by employing a novel screening method to detect inhibitors of the biosynthetic pathways of aromatic amino acids and para-aminobenzoic acid, the precursor of folic acid, using an agarplate diffusion assay modified as an antagonism test. The main component abyssomicin C was found to inhibit the pathway between chorismate and para-aminobenzoic acid and was observed to be strongly active against Gram positive bacteria, including multi-resistant clinical isolates of *Staphylococcus aureus* with MICs in the range of 4μg/ml and 13μg/ml.

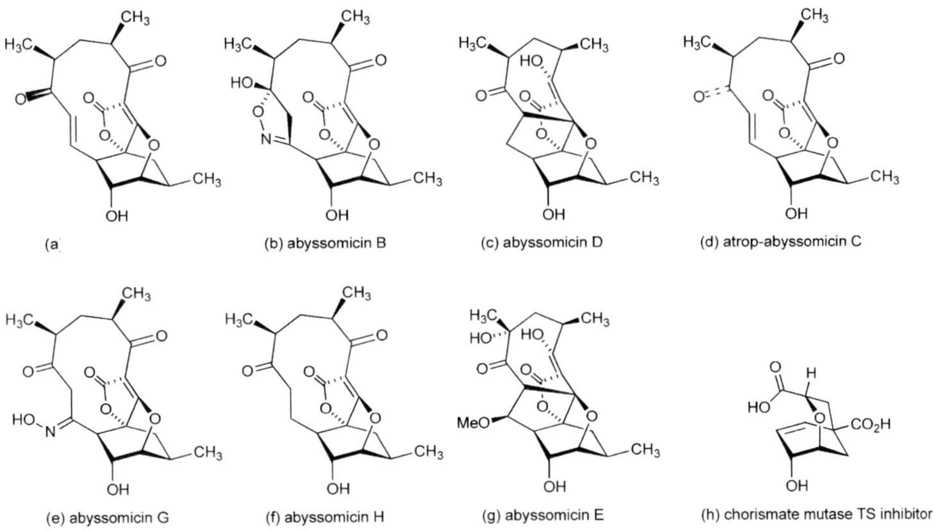

Figure 6 *The abyssomicins (structure (h) represents putative abyssomicin pharmacophore).*

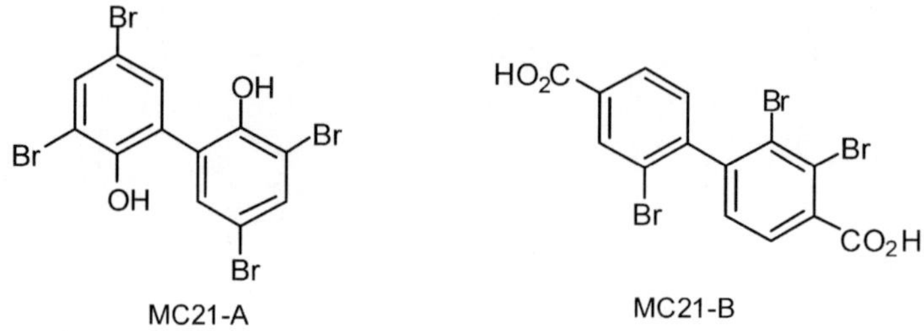

Figure 7: *MC21-A & MC21-B.*

3.3 MC21-A

Researchers at the Marine and Highland Bioscience Center, Saga University, Japan recently isolated a new marine bacterium, *Pseudoalteromonas phenolica* sp. nov. O-BC30T that was shown to produce a bactericidal antibiotic MC21-A active against methicillin-resistant *Staphylococcus aureus* (MRSA).[23]

The purified active ingredient (MC21-A) from the methanol extract of the cells of *P. phenolica* O-BC30T was determined to be 3,3′,5,5′-tetrabromo-2,2′-biphenyldiol by spectrometric analyses. Its anti-MRSA activity against 10 clinical isolates of MRSA was comparable to that of vancomycin (MC21-A MICs, 1 to 2 µg/ml; vancomycin MICs, <0.25 to 2 µg/ml). This compound was also high active against *Enterococcus serolicida, Enterococcus faecium*, and *Enterococcus faecalis* but was less active against *Streptococcus* spp. Time-kill studies also demonstrated that MC21-A was bactericidal and that its killing rate was much higher than that of vancomycin. Furthermore, MC21-A was not cytotoxic to human normal fibroblast, rat pheochromocytoma, and Vero cells at concentrations up to 50 µg/ml. These results suggest that MC21-A might be useful as a lead compound in the development of new types of anti-MRSA substances with modes of action different to those of vancomycin and teicoplanin.

4 NOVEL ANTIVIRAL COMPOUNDS PRODUCED BY MARINE MICRO-ORGANISMS

The morphological diversity of marine viruses is high, varying in capsid size as well as tail length and structure. In the deep ocean around hydrothermal vents spindle forms are found, which are typical for some groups of viruses infecting Archaea. In surface waters, very large viruses can be found, often with capsid diameters of 200 - 400 nm (0.0002 - 0.0004 mm). These viruses are larger than some prokaryotic cells and are able to infect single-celled photosynthetic organisms (phytoplankton). Even larger virus-like particles (700 nm) have been found in the food vacuoles of radiolarians (skeleton-forming protists).[24]

It is widely accepted that viruses are inactivated in the marine environment[33,34] and it is reasonable to argue that marine micro-organisms living in the oceans will have evolved key mechanisms to protect themselves against viral infection / competition.

Early studies demonstrated that compounds produced by the marine bacterium *Vibrio marinus*, were able to inactivate enteroviruses and that extracts of a *Flavobacterium* sp. *Pseudomonas* and *Vibrio* all displayed antiviral activities against poliovirus.[25] The viral assays used in these early studies used the Vero cell line propagated in Eagle's minimal essential medium with Earle's salts, 10% (v/v) newborn calf serum, 50 pg/ml gentamicin,

100 IU/ml penicillin and 100 pg/ml streptomycin. It was deduced that *Pseudomonas* and *Vibrio* displayed the prominent antiviral (antipolio virus) activity. The source of this antiviral activity was found to be secreted in culture supernatants, but no work was carried out to determine the exact chemical composition of such inhibitory agent(s).

Antiviral activity produced by marine micro-organisms has now been shown to be widespread.[26] Studies have demonstrated that marine and estuarine bacteria are able to produce antiviral activity against infectious hematopoietic necrosis virus (INHV).[26] Among isolates taken from marine environments, 32% of strains showed a 50% plaque reduction against this specific viral assay, including 6% showing a 90% plaque reduction. Most of the bacteria which exhibited plaque reduction of more than 90% belonged to the genera *Pseudomonas, Achromobacter*, and *Vibrio*.[26,27]

Currently, one of the most promising approaches to antiviral therapy is the development of inhibitors of the multifunctional non-structural (NS)3/NS4A heterocomplex serine protease and offers hope as an effective target for anti-HCV therapy. Studies conducted at the University of British Columbia reported on the application of high-throughput/high-content biochemical and cell-based fluorescence screening to identify and characterize *in vitro* and *in cellulo* nano-molar noncompetitive allosteric inhibitors of the HCV NS3/4A heterocomplex serine protease from marine-based compounds.[28] Although the source of this molecule was not disclosed, this underpins the importance of marine natural compounds as potent, naturally occurring antiviral agents able to targeted viral protease-associated replication complexes.

Other reports describe the isolation of a novel extracellular polysaccharide (EPS-2), with immunomodulatory and antiviral properties produced by a strain of *Geobacillus thermodenitrificans* isolated from a shallow marine vent of Vulcano Island (Italy).[29] The study reported on the ability of EPS-2 to hinder HSV-2 replication in human peripheral blood mononuclear cells (PBMC) but not in WISH cells in a dose-dependent manner. Of specific interest was the author's concluding remarks that treatment with EPS-2 had the potential to partially restore immunological disorders determined by HSV-2.

4.1 Macrolactin

Macrolactin A (Figure 8) is a good example of a novel antiviral compound isolated from an unusual Gram positive deep sea bacterium. The bacterium was shown to produce a series of new macrolide natural products, macrolactins A-F.[5] The major metabolite, macrolactin A was shown to inhibit mammalian herpes simplex virus (type I and II) and to protect T lymphocytes against human immunodeficiency virus.

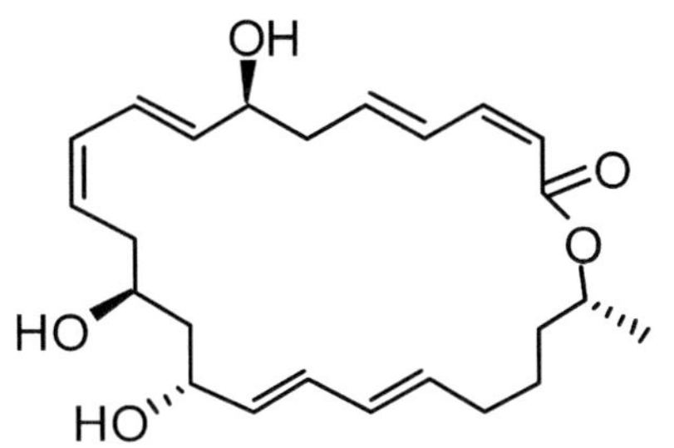

Figure 8 *Macrolactin A.*

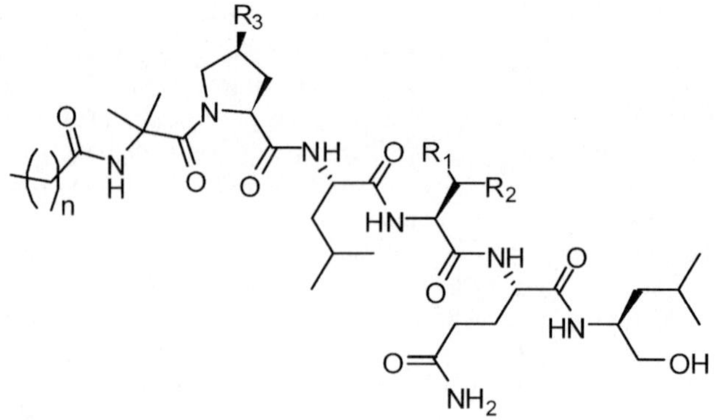

Figure 9: *Halovirs*

4.2 Halovir

A series of peptides designated halovirs A–E were isolated during the saline fermentation of a marine-derived fungus of the genus *Scytalidium*. These lipophilic, linear peptides were shown to be potent *in vitro* inhibitors of herpes simplex viruses I and II. It was found that the halovirs directly inactivated herpes viruses, a mechanism of action that could be applicable in the prevention of HSV transmission. Salient structural features of the halovir hexapeptides include a nitrogen terminus acylated by myristic (C14) or lauric (C12) acids, an unusual aminoisobutyrate-hydroxyproline dipeptide segment, and a carboxyl terminus reduced to a primary alcohol.[30]

5 FUTURE PROSPECTS

With increasing government restrictions on the registration of new drugs coupled with the current framework of HTS in pharmaceutical companies, it is possible that the rate of new drug discovery from natural products could reach zero. However, this is likely to be short lived as the potential that natural products offer for new drug discovery in the future is enormous. With the estimated total number of living species ranging from 2 million to 100 million, the number of biosynthetic products and enzymes remaining to be examined is enormous.[7]

To fully mine this resource, advances in scientific techniques such as genome sequencing (454 sequencing or single-molecule real-time (SMRT) methods will fully enable metagenomics for unculturable organisms.[32] This coupled with techniques to uncover "silent pathways" in plants and micro-organisms will open up science to a range of new chemical scaffolds and biocatalysts and importantly will allow preservation of any threatened species, through cataloguing of its genetic blueprint.

Systems biology could eventually lead to the metabolome mapping of most species. Such a library of biochemical information could act as a powerful tool for the design and generation of new chemistry and drug leads. Just as synthetic chemists today plan total syntheses of a target compound using established reagents and well characterised transformations, synthetic biologists will one day be able to call on vast arrays of enzymes to rationally make complex molecules.

Although the current industry model for drug discovery does not favour natural products, the biochemical resource of nature is vast. Emerging tools in biochemistry will enable humanity to fully capture this opportunity of exciting discoveries that will in time lead to new medicines.

References

1 W. Sneader, in *Drug discovery; A History* ed John Wiley & Sons Ltd., West Sussex, 2005 Ch 1 p. 1; Section 3 Ch 21 p.287.

2 T.F. Molinski, D.S. Dalisay, S.L. Lievens and J.P. Saludes, *Nature Reviews Drug Discovery,* 2009, **8**, 69.

3 M. L. Sogin, H.G. Morrison, J.A. Huber, D.M. Welch, S.M. Huse, P.R. Neal, J. Arrieta and G.J. Hernd, *Proc. Natl. Acad. Sci. USA*, 2006, **103**, 12115.

4 H. J. Field and E. De Clercq, *Microbiology Today,* 2004, **31**, 58.

5 K. Gustafson, M.Roman and W.Fenical, *J. Am. Chem. Soc.*, 1989, **111**, 7519

6 J.W. Blunt, B.R. Copp, W.P. Hu, M.H. Munro, P.T. Northcote and M.R. Prinsep, *Nat. Prod. Rep.*, 2009, **26**,170.

7 G. Blunden, *Phytotherapy Research,* 2001, **15**, 89.

8 C.A. Lozupone and R. Knight., *Proc. Natl. Acad. Sci. U.S.A.*, 2007, **104**, 11436.

9 T. Madden, H. T. Tran, D.Beck, R. Huie, R.A. Newman, L. Pusztai, J.J. Wright and J. L. Abbruzzese, *Clinical Cancer Research.*, 2000, 6, 1293.

10 G.G. Harrigan and G. Goetz, *Journal of Applied Phycology.,* 2002, 14, 103.

11 H. Zhen-Bo, S. M. Gignaca, H.Quentmeiera, G. R. Pettita and H. G. Drexler, *Leukemia Research,* 1993, **17**, 333.

12 H.Kindler, P. Tothy, R. Wolff, R. McCormack, J. Abbruzzese, S. Mani, K. Wade-Oliver and E. Vokes, *Investigational New Drugs.*, 2005, **23**, 489.

13 P. Schöffski, B. Thate, G. Beutel, O. Bolte, D. Otto, M. Hofmann, A. Ganser, A. Jenner, P. Cheverton, J. Wanders, T. Oguma, R. Atsumi and M. Satomi, *Annals of Oncology.,* 2004, **15**, 671.

14 D.S. Everardo, E.H. Kraut, P.M. Hoff, D.F. Moore, D. Jones, R. Pazdur and J.L. Abbruzzese, *American Journal of Clinical Oncology.*, 2002, **25**, 451.

15 W. Fenical, P.R. Jensen, M.A. Palladino, K. S. Lamb, G. K. Lloyd and B.C. Potts, *Bioorganic & Medicinal Chemistry.*, 2009, **17**, 2175.

16 R.H. Feling, G.O. Buchanan, T.J. Mincer, C.A. Kauffman, P.R. Jensen and W. Fenical, *R.H., Angew. Chem. Int.,* 2003, **42**, 355.

17 D. Chauhan, L. Catley, G. Li, K. Podar, T. Hideshima, M. Velankar, C. Mitsiades, N. Mitsiades, H. Yasui, A. Letai, H. Ovaa, C. Berkers, B. Nicholson, T.H. Chao, S.T. Neuteboom, P. Richardson, M.A. Palladino and K.C. Anderson, *Cancer Cell.,* 2005, **8**, 407.

18 P.A. Searle and T.F. Molinski, *J. Am. Chem. Soc.,* 1995, **117**, 8126.

19 H. Rahman, B. Austin, W. J. Mitchell, P. C. Morris, D. J. Jamieson, D. R. Adams, A. Mearns Spragg and M. Schweizer, *Mar. Drugs,* 2010, **8**, 498.

20 E. Armstrong, L. Yan, K.G. Boyd, P. Wright and J. G. Burgess, *Hydrobiologia.*, 2001, **461**, 37.

21 P. Bhadury, B.T. Mohammad and P. C. Wright, *Journal of Industrial Microbiology & Biotechnology.,* 2006, **33**, 325

22 J. Riedlinger, A. Reicke, H. Zahner, B. Krismer, A.T. Bull., L.A. Maldonado, A.C. Ward, M. Goodfellow and B. Bister, *J. Antibiotics.,* 2004, **57**, 271.

23 A. Isnansetyo and Y. Kamei, *Int. J. Systematic & Evolutionary Microbiol.*, 2003, **53**, 583.

24 M.G. Weinbauer, *Encyclopaedia of Earth*, 2007, http://www.eoearth.org/article/Marine_viruses

25 Chun-Tao Che, *Drug Development Research,* 2004, **23**, 201.
26 Y. Kamei, M. Yoshimizu, Y. Ezura, and T. Kimura, *Bulletin Japanese Soc. Scientific Fisheries*, 1987, **53**, 2179.
27 S. S. Kumar, P. Rosamma, and C. T. Achuthankutty, *Current Science.*, 2006, **91**, 807.
28 F. Jean, *CHI Conference.,* 2008, April 28th.
29 A. Arena, C. Gugliandolob, G. Stassi, B. Pavonea, D. Iannelloa, G. Bisignanoc and T. Luciana Maugeri, *Immunology Letters.*, 2009, **123**, 132.
30 D.C. Rowley, S. Kelly, C.A. Kauffman, P. R. Jensen and W. Fenical, *Bioorganic and Med. Chem.,* 2003, 4263.
32 M.J. Levene, J. Korlach, S.W. Turner, M. Foquet, H.G. Craighead and W.W. Webb, *Science,* 2003, **299**, 682.
33 E.W. Akin, W.F. Hill, G.B. Cline and W.H. Benton, *Water Res,* 1976, **10,** 59.
34 R.L. LaBelle and C.P. Gerba *Appl Environ Microbiol.* 1980 **39**, 749.

A UNIQUE NATURAL PRODUCT PLATFORM APPLIED TO THE DISCOVERY OF NOVEL COMPOUNDS WITH ACTIVITY AGAINST THE MALARIA PARASITE, *PLASMODIUM FALCIPARUM*

Michael Kemmler, Sandra Duffy, Vicky M. Avery, Robert Witzig, Kim Beyer, Ian Bathurst and Peter Eckard.
BioFocus DPI AG, Gewerbestrasse 16, CH-4123 Allschwil, Switzerland. Eskitis Institute for Cell and Molecular Therapies, Eskitis Builiding 75, Griffith University, Brisbane Innovation Park, Don Young Road, Nathan QLD 4111, Australia. Medicines for Malaria Venture, ICC, CP 1826, Route de Pre-Bois 20, CH-1215 Geneva, Switzerland.
E-mail: michael.kemmler@glpg.com, www.biofocusdpi.com

1 INTRODUCTION

1.1 Natural products

Natural products have been a successful source for drugs in the past, accounting for approximately 50% of the drugs introduced to the market in the last 20 years. Nevertheless, their use for hit discovery dramatically decreased in the 1990's due to the advent of high throughput screening (HTS) and combinatorial chemistry combined with disadvantages in natural product work such as long timelines, a lack of reproducibility, the loss of activity upon fractionation and false positives due to the use of complex samples.

The need for novel compounds combined with technological advances in natural product chemistry has nourished the recent renaissance of natural products in drug discovery. Here we will present BioFocus DPI's approach.

1.2 Malaria

The four malaria species that produce human disease are *P. vivax*, *P. falciparum, P. malariae* and *P. ovale. Plasmodium falciparum*, the most lethal strain, is the most prevalent species throughout the tropics and subtropics. Malaria is usually transmitted when a person is bitten by an infected female *Anopheles* mosquito. Clinical symptoms of the disease include fever and flu-like illness, such as chills, headache, muscle aches, and tiredness. These symptoms may be accompanied by nausea, vomiting, and diarrhoea. Malaria can also cause anaemia and jaundice (yellow colouring of the skin and eyes) due to the loss of red blood cells.

Infection with one type of malaria, *Plasmodium falciparum*, if not promptly treated, may cause kidney failure, seizures, mental confusion, coma, and death. Malaria is a major public health problem in more than 90 countries, inhabited by more than 2.4 billion people - 40% of the world's population.

1.3 Aim

The Medicines for Malaria Venture (MMV), in collaboration with BioFocus DPI and the Eskitis Institute for Cell and Molecular Therapies (Eskitis), has established a drug discovery program that will enable the identification of small molecules derived from natural sources to become new therapies for fighting malaria. This collaboration is based on key strengths of BioFocus DPI and Eskitis, expertise in natural products drug discovery and anti-malarial HTS.

2 HIGH THROUGHPUT SCREENING OF NATURAL PRODUCT LIBRARIES

2.1 The Microbial Strain Collection

The sources of microorganisms for the collection are shown in Figure 1 together with an indication of the stress, separation and cultivation techniques that the collected organisms are subjected to so as to obtain a single, pure culture. The collection of microbial species is made across as wide a spectrum of environments as is reasonably practical and the resulting homogenous cultures are classified and stored. A summary of some of the phyla available in the collection is given in Figure 2.

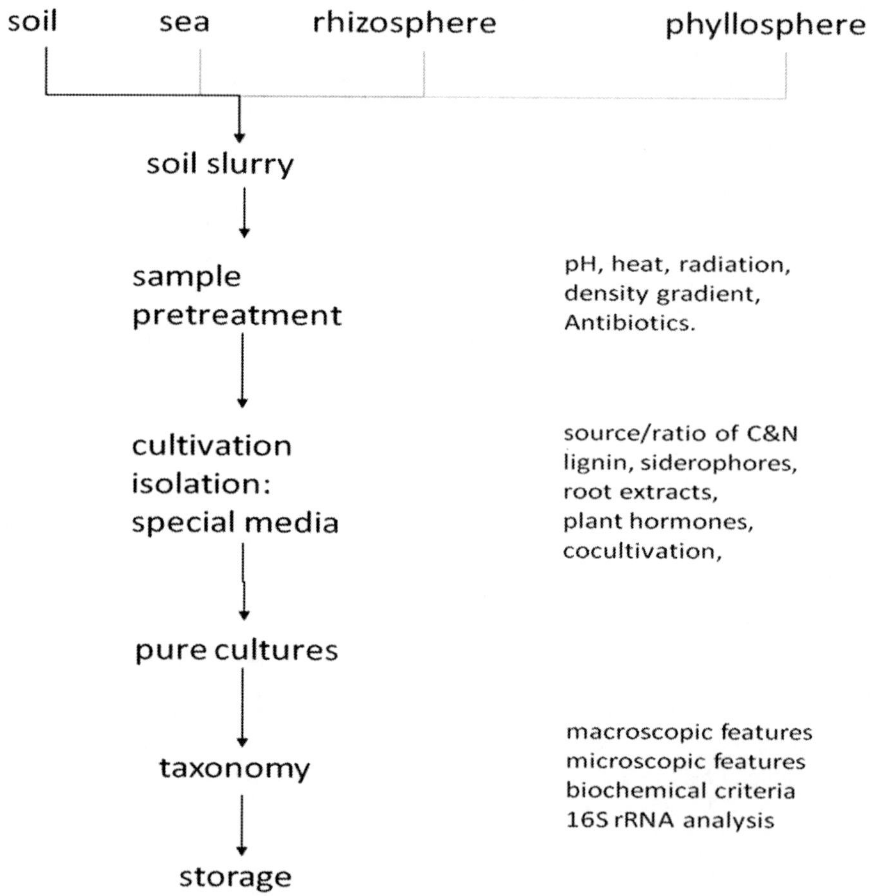

Figure 1 *The sources and treatment techniques employed to build the microbial strain collection.*

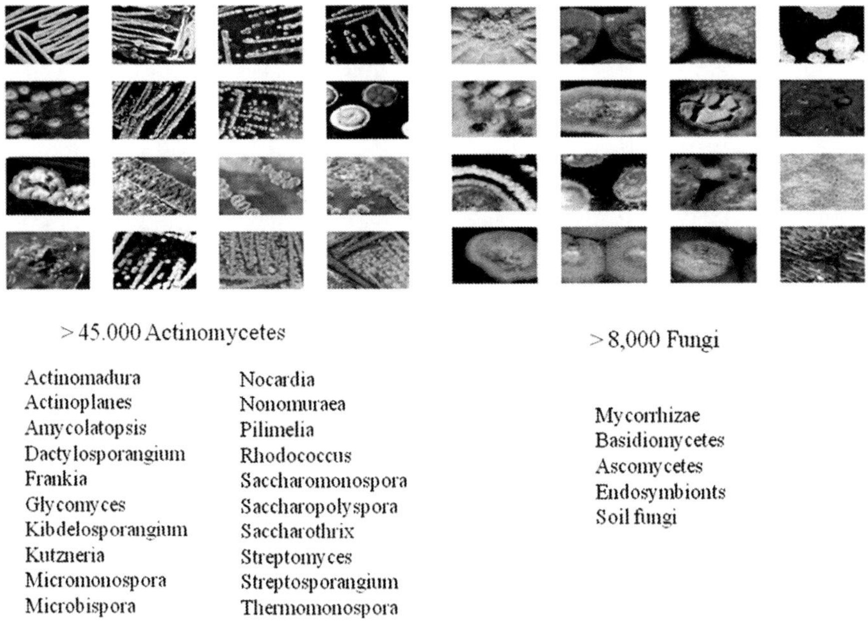

Figure 2 *Examples of the organisms comprising the strain collection.*

2.2 Generation of Metabolite and Mycelial Subfractions

Microbial strains from the collection are fermented to produce a spectrum of secondary metabolites. The mycelia are then separated by filtration and the filtrate fractionated to give nine subfractions per strain. These are dissolved in DMSO to give 2mg/ml solutions that are stored with the dry film of mycelium extract under argon at -20°C.

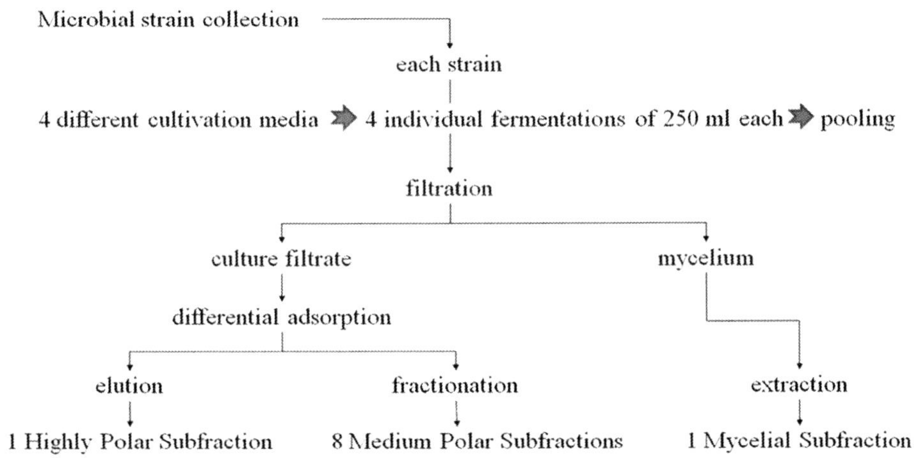

Figure 3 *Flow chart depicting the production of fractionated microbial metabolites.*

2.3 Screening the Culture Subfractions.

The High Throughput Screening and biological testing of these samples were performed at the Eskitis Institute and natural product studies, including the fractionation of the culture media, were carried out by BioFocus DPI.

The natural product library employed in this project consisted of 140,172 pre-purified subfractions. Each subfraction contained on average 10 to 20 different compounds, yielding an HTS compatible library with huge chemical diversity.

The test organisms used were the chloroquine sensitive *Plasmodium falciparum* strain 3D7 and the chloroquine resistant *Plasmodium falciparum* strain Dd2. The cytotoxicity of the subfractions was determined using the non-cancerous HEK293 human cell line.

The assay utilises the DNA-intercalating dye 4',6-diamidino-2-phenylindole (DAPI) to monitor changes in parasite number in infected red blood corpuscles. Images of the stained parasites within the test well were analysed using the Accapella spot detection programme associated with Opera™. The HTS imaging assay was developed in 384-well format for use on the Opera™ confocal imaging system with Twister arm and is capable of screening over 70,000 assay wells (200 plates) per day. The basic assay can also be used with Total Intensity data output from a standard microplate reader such as a VICTOR II. The assay typically yields Z'-factor values of 0.5-0.6 with signal-to noise ratios of 10:1.

Pictures showing the effect of artemisinin on *P. falciparum* at varying concentrations are shown in Figure 4. This data is converted into a numerical format in the table below based on the number of spots which have a defined fluorescence and shows the decline in numbers of *P. falciparum* parasites with increasing concentrations of artemisinin.

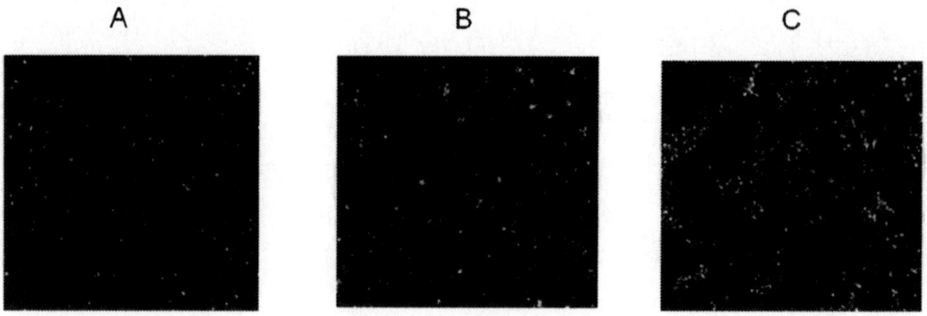

Image	Condition	Spots	% Inhibition
A	500nM Artemisinin	159	100
B	20nM Artemisinin	332	78
C	0.5nM Artemisinin	954	0

Figure 4 *Pictures of* Plasmodium falciparum *treated with A 500 nM, B 20 nM and C 0.5 nM artemisinin. The table below contains the same data converted into a numerical format.*

3 HIGH THROUGHPUT SCREENING FOR *P.FALCIPARUM* INHIBIORS

3.1 Selection of Active Subfractions

In the primary HTS, 140,172 natural product subfractions were screened against *P. falciparum* 3D7 at a concentration of 0.7µg/ml. The HTS assay consistently gave Z' values of 0.5-0.8 during screening, demonstrating the high quality of the assay. Of these samples, 4,050 gave greater than 50% inhibition of *P. falciparum* and these were further subdivided into 1774 subfractions which gave 50 – 79% inhibition and 2276 subfractions which gave greater than 80% inhibition.

The less active fractions were retested against the sensitive 3D7 strain at concentrations of 0.7 and 0.35 ug/ml and against HEK293 cells at 0.7 ug/ml. The more active fractions were tested against *P. falciparum* 3D7 and Dd2 at a range of five concentrations below 0.7 ug/ml and also against HEK293 cells.

A final selection of 1075 active subfractions was made comprising subfractions which gave better than 50% inhibition at concentrations of 0.35 ug/ml against both strains or against 3D7 if only one strain was tested. Active subfractions were more than 10 fold more active against *P. falciparum* than HEK293 cells and comparable activity against both strains of parasite was favoured. This workflow is portrayed in graphic form in Figure 5.

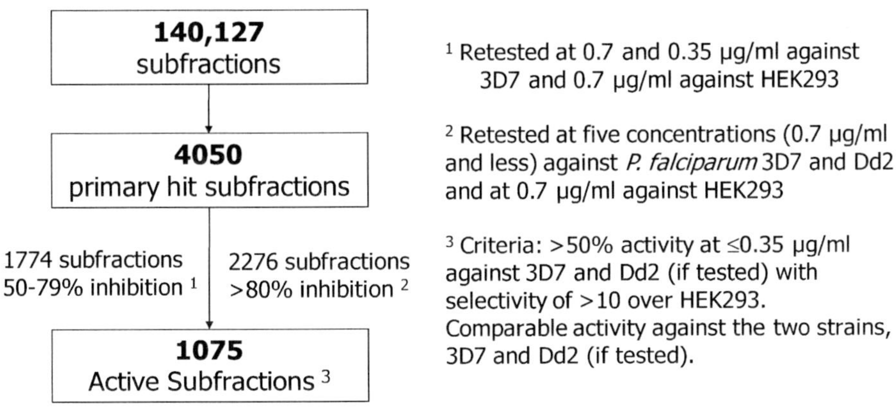

Figure 5 *HTS and secondary screening workflow.*

3.2 Distribution of *P. falciparum* Activity

Analysis of the origin of the active subfractions (Figure 6) showed that activity was

	In Collection	In Active Subfractions	%
Number of strains:	14312	878	6.1
Number of fungal strains:	1997	24	1.2
Number of bacterial strains:	12315	854	6.9
Number of different collection sites:	2448	655	26.8

Figure 6 *Analysis of the origin of the active subfractions.*

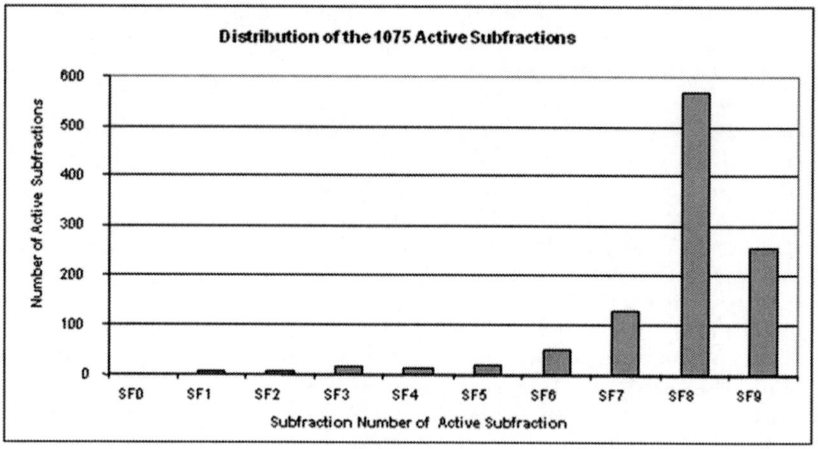

Figure 7 *Analysis of the subfractions from which the active subfractions were derived.*

restricted to a relaitively small proportion of the bacterial strains and an even smaller set of fungal strains. A similar analysis of the subfractions from which the active subfractions were derived (Figure 7), showed that the active compounds were present predominantly in the less polar fractions 7, 8 and 9.

3.3 Fine Fractionation of the Active Subfractions

Analysis of the microbial source organism, the fraction order and the observed activity and selectivity of the subfractions allowed the number of fractions for further investigation to be reduced to 149 without significant sacrifice of diversity. In order to identify the active components, 10-20 mg of each subfraction were further fractionated using preparative reverse phase HPLC with a water/acetonitrile gradient and this resulted in each subfraction being divided into 88 Fine Fractions.

An example of this fractionation is shown in Figure 8. Peaks showing significant

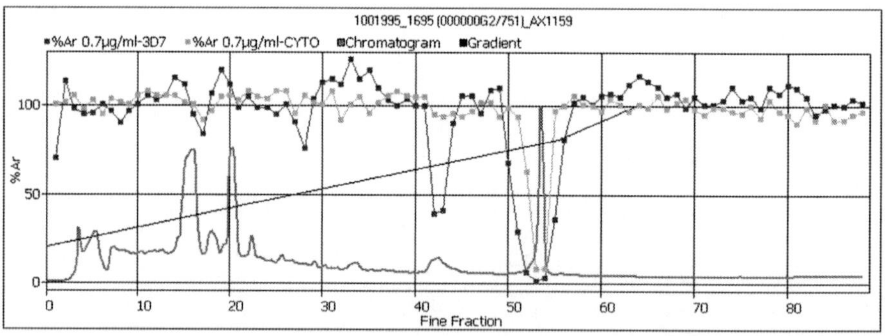

Figure 8 *An example of fine fraction (at 0.7µg/m)l testing results [Residual activity (Ar),* P. falciparum *3D7 (top grey), HEK293 (top black)]. The UV-chromatogram of the semipreparative fractionation is shown below with the water/acetonitrile gradient.*

activity against *P. falciparum* 3D7 and selectivity for the parasite relative to HEK293 cells, such as that in fractions 42-43, are subjected to dereplication whereas those showing little activity or poor selectivity were treated with a lower priority.

3.4 Dereplication of the Fine Fractionation

Active fine-fractions and neighbouring inactive fine-fractions are analyzed by LC-MS and ^{1}H-NMR. Correlation of specific properties such as the molecular mass, UV-absorption, ^{1}H-NMR and biological activity are used to define an active compound and to search in-house and public databases (DNP, Chapman&Hall/CRC and Antibase) to determine whether the compound is new. Dereplication is currently ongoing.

Known compounds of interest can be produced in larger quantities by fermentation and isolation. The producer strain of novel compounds is fermented in a larger scale, usually 10 - 100 litre and the active compound isolated using the properties identified during dereplication. New compounds are usually obtained in milligramme quantities which allows for structure elucidation and further biological testing.

3 CONCLUSION

MMV, in collaboration with Eskitis and BioFocus DPI, has performed a high-throughput screen against the malaria parasite *Plasmodium falciparum*. At this point in the joint project BioFocus DPI's prorietary subfraction collection has been successfully screened against the chloroquine sensitive 3D7 and chloroquine resistant Dd2 strains of *P. falciparum*.

The screen resulted in the identification of 1,075 active subfractions, corresponding to 0.8% of the library. 149 Subfractions with greater than 10 fold selectivity for *P. falciparum* over HEK 293 have been identified for further investigation, which is continuing.

Acknowedgements

Financial support: This work is supported by MMV.

HIGH CAPACITY COUNTERCURRENT CHROMATOGRAPHY FOR FAST ISOLATION OF NATURAL PRODUCTS

Ian J.Garrard and Derek Fisher

Advanced Bioprocessing Centre, Brunel University, Uxbridge, West London, UK
E-mail: ian.garrard@brunel.ac.uk

1 INTRODUCTION

Countercurrent chromatography (CCC) is a technique that was first invented in the mid 1960's.[1] After a brief period of interest by many scientists, the technique declined in popularity, being effectively out-competed by rival techniques that were faster and more reliable. To many, CCC was known as a slow technique with separations measured in hours or days, with poor capacity allowing injections of only tens of milligrams, poor reliability and difficult scale-up because of a poor understanding of the factors required to scale up or instruments on which to do so. However, recent engineering improvements and advances in the understanding of the process have transformed the technique to one which provides separations in minutes, injection loadings in the scale of grams to hundreds-of-grams, equipment that is considerably more robust and a scale up from analytical to pilot level that is quick and easy.[2]

These developments have led to a resurgence of interest in CCC, especially for natural products, with the technique offering numerous advantages over both solid-phase chromatography and traditional liquid-liquid extraction. The current range of CCC instruments is a worthy inclusion in the arsenal of techniques necessary for the purification of challenging natural product molecules. This article describes the technique and its advantages over other purification methods and demonstrates the purification potential with a number of natural product examples.

2 COUNTERCURRENT CHROMATOGRAPHY

2.1 Mechanism of Action

CCC is a liquid-liquid purification technique comprising a liquid stationary phase and a liquid mobile phase. Separation of the components of a mixture is based on differential distribution between the two liquid phases. The immobilisation of the liquid stationary phase and the multiple mixing and separation of phases are achieved by centrifugal forces. The instruments are consequently termed CCC-centrifuges.

A CCC centrifuge consists of a drum on which tubing is wound. When any coil is filled with liquid and rotated, every object within, whether it is lighter (e.g. an air bubble) or heavier (e.g. a glass bead), tends to move towards one end of the coil as a consequence of trying to maintain its original position. The end towards which the bubble and bead move is called the head of the coil, and the other end is called the tail.

If the air bubble represents the upper solvent phase and the glass bead the lower solvent phase, it can be seen that both phases "screw" themselves towards the head of the coil. Since they cannot both occupy the same space, one phase dominates and thus displaces the other towards the tail. So if the dominant phase is selected as the stationary phase of the system, then it will be retained in the coil by this screwing action, making the displaced phase the mobile one being pushed by a pump through the coil in the direction it is tending to go. In countercurrent chromatography terms, the mobile phase is being pumped from head to tail.

It can be seen that, if the coil were rotated in the opposite direction, the other phase is selected as the stationary one which is retained in the coil. In countercurrent chromatography terms, the mobile phase is being pumped from tail to head. In practical terms, the coil is always rotated in the same direction but the pump is switched to the other lead from the coil. The coil thus contains a liquid stationary phase, held in place purely by the "screwing" motion of the coil and a liquid mobile phase being pumped through the coil in the direction in which it is naturally being displaced. This gives a hydrodynamic countercurrent flow.

However, if simply spun on its own axis, the two phases inside the coil would not mix due to the high $r\omega^2$ rotational force (the "centrifugal" force). There would be laminar flow of the mobile phase over the stationary one and extremely poor partitioning of solutes. Therefore, the coil is spun in planetary motion, revolving around the central axis of the sun gear while simultaneously rotating about its own axis at the same angular velocity (Figure 1).

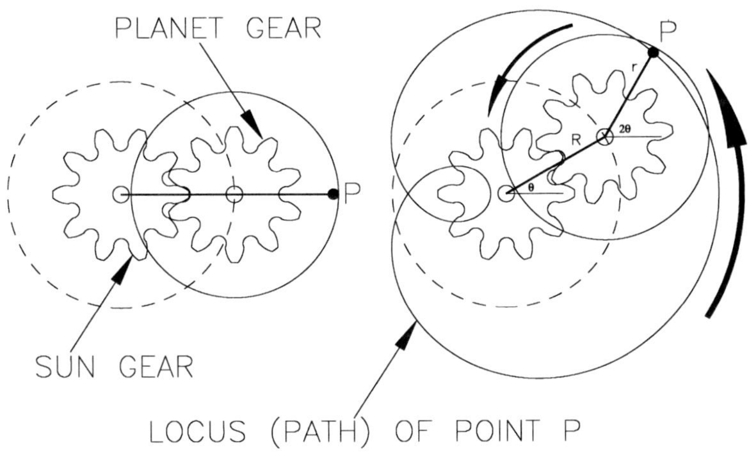

Figure 1 *Motion of the bobbin in the CCC centrifuge.*

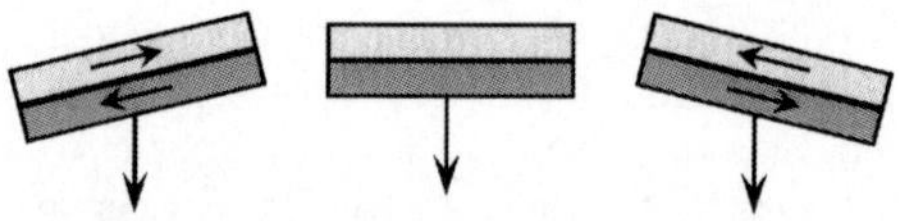

Figure 2 *Tilting tube effect that produces the wave mixing in the planetary motion centrifuge.*

This creates a mixing effect akin to the "swish-swosh" motion that occurs when a tube of two liquids is tilted from side to side (Figure 2). The net result of this motion is the formation of waves and a wave mixing effect (Figure 3).

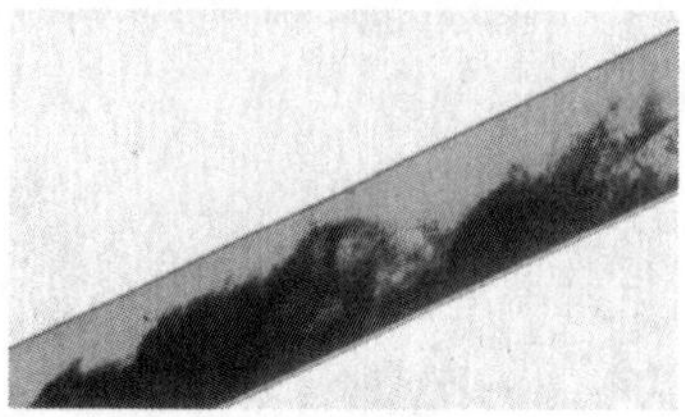

Figure 3 *A model of the wave mixing that occurs in the CCC centrifuge.*

When a sample is introduced into the centrifuge, it will experience a series of mixing and settling steps according to the low or high acceleration vectors (Figure 4). The components of the mixture introduced into the coil will separate according to their distribution ratio between the two solvent phases (KD = concentration of solute in stationary phase / concentration of solute in mobile phase).

It will be noticed in the diagram that there is one mixing and one settling zone per coil loop. Furthermore, the coil in Figure 4 is simply wound in an axial direction, producing a helical coil. In practice, a typical CCC coil will be multilayer wound and may have 30 loops spinning at 1200 rpm. It will therefore undergo 2.16 million partitioning steps per hour (30 x 1200 x 60). This provides for high resolution separations within the instrument.

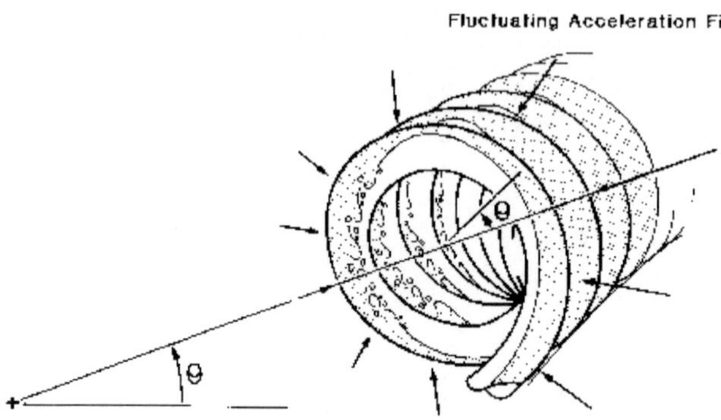

Figure 4 *The mixing and settling that takes place in a J-type centrifuge.*

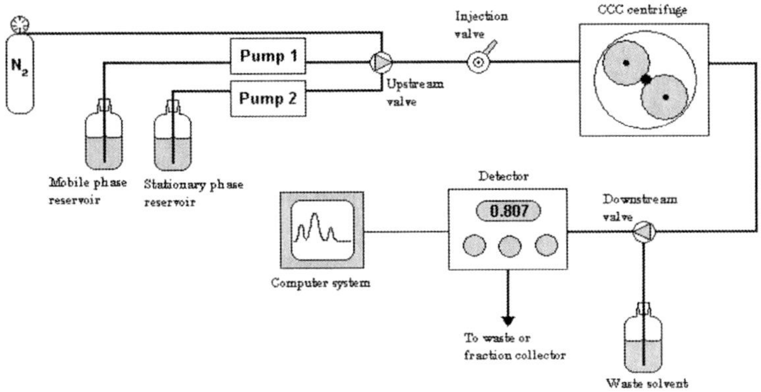

Figure 5 *Schematic layout of a typical preparative CCC setup.*

The synchronous planetary motion of the CCC coil has a special feature significant in the design of instruments: there is no twisting of the flow tubes linking the coil to the pump and detector. This is because the rotation about its own axis unwinds the twist produced by its motion around the sun gear. Thus, there are no rotating seals to wear out, leak or provide the risk of cross-contamination.

A schematic layout for a typical CCC setup is shown in Figure 5. In effect, the CCC coil is equivalent to a preparative HPLC column that does not wear out or need replacing. All the ancillary equipment is the same - pumps, injection valves, detector, fraction collector and chromatography software. Two pumps allow the easy switching of the liquid stationary to the liquid mobile phase, while a nitrogen cylinder allows the final extraction of the stationary phase from the coil if that is desired. The injection valve can be a standard Rheodyne type, but when the sample is particularly crude, viscous or may precipitate, direct injection from a syringe may be necessary. In the diagram a UV detector has been shown since this form of detection is non-destructive. However, in natural product purifications, CCC is commonly used with a bleed to a light scattering detector (ELSD) and has even been used with mass spectrometry detection.[3]

Since both phases are liquid, CCC can be operated in either normal or reverse phase mode, depending on which solvent phase is selected to be stationary. It is even possible to switch over mid-run to elute slow moving compounds. Alternatively, components that elute very slowly can be recovered by pushing out the stationary phase without any compound losses whilst maintaining resolution.[4]

At the Brunel Advanced Bioprocessing Centre, we have a range of CCC centrifuges running from analytical to pilot scale, the latter capable of processing several kilograms of crude material per day (Table 1). Known as Mini, Midi, Maxi and Super-Maxi, the Mini centrifuge is the size of a microwave oven, the Midi fits into a normal chemistry fume cupboard, and the Maxi instruments are pilot machines about 2 metres cubed (Figure 6). Commercially available machines are now produced by Dynamic Extractions Ltd (www.dynamicextractions.com). Importantly they have now developed a composite machine, the Spectrum, which includes both analytical and preparative coils in one machine, providing a versatile instrument.

Instrument	MINI	MIDI	MAXI	"SUPER" MAXI
Scale of application	Analytical	Preparative	Pilot	Pilot
Processing rate	mg/hr	g/hr	kg/day	kg/hr
Rotor radius (mm)	50	110	300	300
Volume of coils (ml)	17	930	4600	18000
Coil bore (mm)	0.8	3.6	10.0	10.0
Max rotor speed (rpm)	2100	1400	850	850
Typical flow rate (ml/min)	0.25-2	5-80	100-1000	400-3000
Typical elution time (min) for K_D=1 peak	20	20	20	20
Typical loading (g)	0.005-0.050	1-30	50-300	250-1000

Table 1 *The range of CCC centrifuges at the Brunel Advanced Bioprocessing Centre with their specifications.*

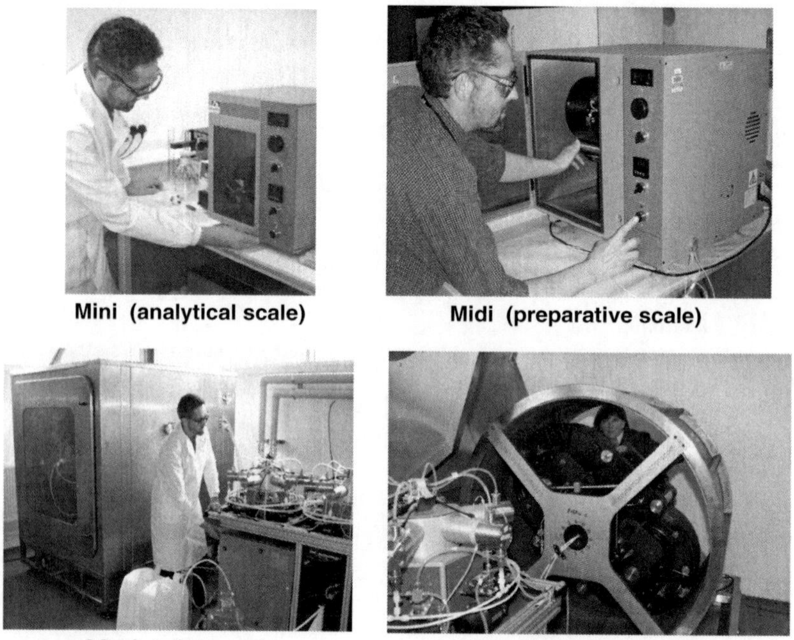

Mini (analytical scale) **Midi (preparative scale)**

Maxi (pilot scale) **"Super" Maxi (pilot scale)**

Figure 6 *The Mini, Midi, Maxi and "Super" Maxi range of CCC centrifuges.*

2.2 Operational Advantages of CCC

The key advantages of countercurrent chromatography in the purification of natural products:

1. Far lower solvent usage compared with solid phase chromatography systems operating at the same scale. Furthermore, a simple analysis of solvent composition allows the recycling of the solvents, reducing the usage still further.[5]
2. 100% recovery of the sample components. Since there is no solid phase, there is no possibility of losses arising from irreversible adsorption onto the solid matrix. The target compound will be in one liquid phase or the other so it can always be recovered.
3. Particulates, such as cell debris are tolerated, so filtering a crude sample is not necessary and precipitation of the sample on meeting the solvent system does not adversely affect the purification process. A direct extraction of compounds from a crude natural mixture is possible.
4. High loading capacity with short processing times. The smallest Mini-CCC centrifuge, with a 17 ml coil and running at 1 ml/min flow rate, can routinely accept injections of 5-20 mg and is capable of loadings of up to 50 mg crude material.
5. The scale up from milligram to kilogram level is quick and predictable. Optimisation runs performed on the Mini CCC are immediately transferred to the Maxi machines (1000 times scale up) and show almost identical chromatograms.
6. The technique can be operated in normal batch injection, chromatography mode or as a continuous extraction process for better throughput.[6]
7. A wide range of polarities can be processed due to the range of solvents that may be used, from highly polar molecules such as peptide antibiotics[7] to highly non-polar molecules such as lycopene.[8] A recent review of over 200 papers on the purification of Chinese herbal medicines by CCC showed an ACD logP polarity range from -4 to +12 with the majority falling in the -2 to +8 range (Figure 7).[9] In this review, 354 natural product compounds were isolated in total from 108 different plant species and 56 plant families.

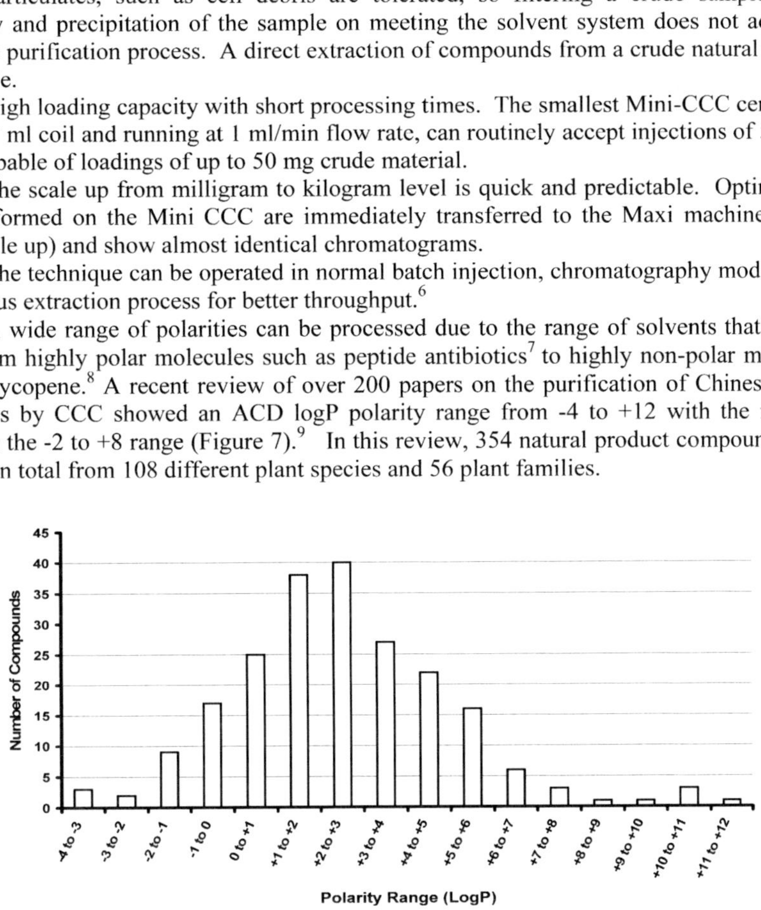

Figure 7 *The natural product compounds purified by CCC sorted by LogP polarity (taken from a review of 200 papers).*[9]

8. The coils are tough and the machinery robust. A set of coils would be expected to last the lifetime of the centrifuge and so maintenance and running costs are low.
9. Unlike solid phase chromatography, there is no change to component retention over time (no column aging effects) as a freshly-filled coil of solvents is used for each run. This makes it easier consistently to satisfy current regulatory requirements when performing purifications under a GLP or GMP environment.

2.3 Natural Product Application Examples

Approximately 30 natural product compounds have been purified using the Brunel CCC-centrifuges in a number of collaborations with academic groups active in natural products. A major focus has been on developing scaled up separations, currently up to kilogramme scale.

A typical CCC natural product purification performed at Brunel Advanced Bioprocessing Centre involved the separation of the isomers, honokiol (ACD logP 4.199) and magnolol (ACD logP 3.938) from the traditional Chinese medicine, Houpu, an extract of *Magnoliae Officinalis*.[10] This was done in collaboration with the State Key Lab of Biotherapy in Chengdu, China. Honokiol has potent anti-angiogenesis activity and material was required for pre-clinical trials and further chemical modification. A suitable two phase solvent system (hexane/ethyl acetate/methanol/water 1/0.4/1/0.4 v/v) was identified using the selection table described in reference[11] operated on a liquid handling robot. The upper, organic phase was chosen as the mobile phase (normal phase mode) in order to make the subsequent drying down of fractions more rapid. Loading studies were then performed on the small scale Mini CCC centrifuge (17.2 ml coil, 0.8 mm bore, 1800 rpm spin speed) to determine both the optimum injection concentration and volume. The chosen conditions were then directly scaled-up to the 4.6 litre Maxi centrifuge and the same run performed on a preparative scale (75 g injection loading in 230 ml phase i.e. 5% of column volume, spin speed 600 rpm, 600 ml/min flow rate, 30 minute run time) to yield 30 g of honokiol per injection with a purity greater than 99%.

This material was then used by the Chengdu group in a series of *in vivo* studies to improve its therapeutic effectiveness by formulation in liposomes[12-14] or in hydrogels.[15,16] In a programme to synthesise a series of novel derivatives of honokiol as potential analogues for anti-proliferative and anti-tumour treatment, they have produced three C-formyl derivatives of honokiol (3'-formyl, 5-formyl and 3',5-diformyl).[17] Not only was CCC used to provide the honokiol starting material, but it was also used to isolate these derivatives, including the 3'- and 5'-isomers.

In another example, the highly polar glucosinolate, glucoraphanin (ACD logP 0.53) has been isolated from extracts of broccoli seeds, using a phase system of 1-propanol/acetonitrile/saturated ammonium sulphate/water (1/0.5/1.2/1 v/v) to separate it from glucoiberin (ACD logP -0.81). Using a Midi centrifuge, 50 g was obtained in three days with *ca.* 30 runs of 30 min.[18] In partnership with Dynamic Extractions, this was scaled to 1 kg using a Maxi CCC centrifuge, again with a series of 30 min. runs over three days.

The flavonoids baicalein (ACD log P 0.314) and its 7-O glucoside, diglucoside and glucuronide, chrysin (ACD logP 2.879) and its 7-O-glucuronide and diglucoside, biochanin A (ACD logP 3.139), have all been separated by CCC from extracts of the seeds of *Oroxylum indicum*; analytically with a Mini CCC centrifuge with inline MS monitoring,[19] preparatively with a Midi CCC centrifuge[20] and also from the leaves.[21] Chloroform/methanol/water (9.5/10/5 v/v) was the solvent system used.

The terpenoid triptolidelode (ACD logP 1.268) and the alkaloids peritassine A, wilforgine and wilforine (ACD logP 6.244) have been isolated from extracts of *Tripterygium wilfordii* Hook f using pre-fractionation by medium pressure liquid chromatography followed by preparative CCC on a Midi-CCC centrifuge.[22] In a novel way of using a conventional twin column CCC centrifuge, intermittent CCC extraction has been used to enrich triptolide from 2% in a crude extract to 98% by retaining the material in the column while washing away all other components, either as lower or higher partitioning material.[23]

Another novel approach developed by the Brunel team, has been to optimise the resolution of an extract from *Milletia pachycarpa*[24] using model compounds, thereby preserving valuable starting material. In this way, tephrosin, pyranoisoflavone, dehydrodeguelin and deguelin have been purified using a hexane/ethyl acetate/methanol/ water (1/0.8/1/0.8v/v) system.[25]

Two recent examples of scaling up from the analytical scale Mini CCC centrifuge to the preparative scale Midi centrifuge are provided by the isolation of the biflavonoids amentoflavone, robustoflavone, bilobetin, hinkoflavone, isocryptomerin and an apigen-diglucide from *Selanginella tamarisca*[26] using heptane/ethyl acetate/methanol/water (2/3/2/3/v/v) and salvianolic acid B (ACD log P 2.14) from *Salvia mitiorrhiza* Bunge[27] using hexane/ethyl acetate/methanol/acetic acid/water (1/5/1.5/0.006/5 v/v).

Finally, in a very recent purification study, two new minor coumarin compounds, alsaticol and alsaticocoumarin A have been isolated from extracts of *Peucedanum alsaticum* L fruits using a heptane/ethyl acetate/methanol/water (1/1/1/1/v/v) phase system.[28]

3 CONCLUSION

On its invention back in the mid 1960's, CCC was a technique that showed great promise. Although it has been slow to deliver that promise, both engineering improvements and developments in the understanding of the process over the last ten years have developed the technique into a powerful purification application. The complexity of natural product purifications is such that a range of suitable purification techniques is required and CCC is now a technique that is wonderfully suited to large-scale preparative separations, with advantages over solid-phase techniques such as the ability to accept particulates, no columns to replace, and high speed and high resolution relative to current liquid-liquid techniques.

References

1 Y. Ito, M. Weinstein, I. Aoki, R. Harada, E. Kimure and K. Nunogaki, *Nature*, 1966, **212**, 985.

2 I.A. Sutherland and D. Fisher, *Innovations in Pharmaceutical Technology*, October 2004, 68-71.

3 J. Jones, H. Kidwell and D.E. Games, Rapid *Communications in Mass Spectrometry*, 2003, **17**, 1565.

4 A. Berthod, M.J. Ruiz-Angel and S. Carda-Broch, *Anal. Chem.*, 2003, **75**, 5508.

5 I.J. Garrard, L. Janaway and D. Fisher, *J. Liq. Chromatogr. Relat. Technol.*, 2007, **30**, 151.

6 H. Ye, S. Ignatova, H. Luo, Y. Li, A. Peng, L. Chen and I.A. Sutherland, *Journal of Chromatography A*, 2008, **1213**, 145.

7 Y. Ikai, H. Oka, J. Hayakawa, N. Kawamura, K.I. Harada, M. Suzuki, H. Nakazawa and Y. Ito, *J. Liq. Chromatogr. Relat. Technol.*, 1998, **21**, 143.

8 Y. Wei, T. Zhang, G. Xu and Y. Ito, *J. Chromatogr., A,* 2001, **929**, 169.

9 I.A. Sutherland and D. Fisher, *J. Chromatogr., A,* 2009, **1216**, 740.

10 L. Chen, Q. Zhang, G. Yang, L. Fan, J. Tang, I. Garrard, S. Ignatova, D. Fisher and I. Sutherland, *J. Chromatogr., A,* 2007, **1142**, 115.

11 I.J. Garrard. *J. Liq. Chromatogr. Relat. Technol.*, 2005, **28**, 1923.

12 Q. Jiang, L. Fan, G. Yang, W. Guo, W. Hou, L. Chen and Y. Wei, *BMC Cancer*, 2008, **8**, 242.

13 Y. liu, L. Chen, X. He, L. Fan, G. Yang, X. Chen, X. Lin, L. Du, Z. Li, H. Ye, Y. Mao, X. Zhao and Y. Wei, *International J. Gynecological Cancer*, 20007, **18**, 652.

14 H. Luo, Q. Zhong, L. Chen, X. Qi, A. Fu, H. Yang, F. Yang, H. Lin, Y. Wei and X. Zhao, *J. Cancer Res. Clin. Oncol.*, 2008, **134**, 937.

15 C. Gong, S. Shi, P. Dong, B. Kan, M. Gou, X. Wang, X. Li, F. Luo, X. Zhao, Y. Wei and Z. Qian, *Internat. J. Pharmaceutics*, 2009, **365**, 89.

16 M. Gou, C. Gong, J. Zhang, X.Wang, X. Wang, Y. Gu, G. Guo, L. Chen, F. Luo, X. Zhao, Y. Wei and Z. Qian, *J. Biomedical Material Research Part A*, 2009 doi 10.1002/jbm.a.32546.

17 Y. Luo, Y. Xu, L. Chen, H. Luo, C. Peng, J. Fia, H. Cheng, A. Peng, H. Ye, D. Xie, A Fu, J. Shia, S. Yang and Y. Wei, *J. Chromatogr., A,* 2008, **1178**, 160.

18 D. Fisher, I.J. Garrard, R. van den Heuvel, I.A. Sutherland, F.E. Chou and J.W. Fahey, *J. Liq. Chromatogr. Relat. Technol.*, 2005, **28**, 1913.

19 L.J. Chen, H. Song, D.E. Games and I.A. Sutherland, *J. Liq. Chromatogr. Relat. Technol.*, 2005, **28**, 1993.

20 L.J. Chen, H. Song, X.Q. Lan, D.E. Games and I.A. Sutherland, *J. Chromatogr., A,* 2005, **1063**, 241.

21 Y. Yuan, W. Hou, M. Tang, H. Luo, L. Chen, Y.H. Guan and I.A. Sutherland, *Chromatographia*, 2008, **68**, 885.

22 H. Ye, S. Ignatova, H. Luo, Y. Li, A. Peng, L. Chen and I. Sutherland, *J. Chromatogr., A*, 2008, **1213**, 145.

23 P. Hewitson, S. Ignatova, H. Ye, L. Chen and I. Sutherland, *J. Chromatogr., A,* 2009, **1216**, 4187.

24 H. Ye, L. Chen, Y. Li, A. Peng, A. Fu, H. Song, M. Tang, H. Luo, Y. Luo, Y. Xu, J. Shi and Y. Wei, *J. Chromatogr., A,* 2008, **1178**, 101.

25 H. Ye, S. Ignatova, A. Peng, L. Chen and I. Sutherland, *J. Chromatogr., A,* 2009, **1216**, 5101.

26 Y. Yuan, B. Wang, L. Chen, H. Luo, D. Fisgher and I.A. Sutherland, *J. Chromatogr., A,* 2008, **1194** 192.

27 M.,Zhang, S. Ignatova, Q. Long, F.W. Jun, I. Sutherland, Y. Wang and G. Luo, *J. Chromatogr., A,* 2009, **1216**, 3869.

28 K. Skalicka-Wozniak, T. Mroczeck, I. Garrard and K. Glowniak, *J. Chromatogr., A,* 2009, **1216**, 5669.

3 Functional Molecules and Their Enhancement

STRATEGIES TO ENHANCE THE NEUROPROTECTIVE PROPERTIES OF RAPAMYCIN ANALOGS

Guy T. Carter

A Wyeth Research, Pearl River, NY 10965 USA

1 INTRODUCTION

Rapamycin **1** has been shown to exert its inhibitory effect on cell proliferation through a two-stage process in which it first binds with FKBP-12, and this binary complex then blocks growth factor-dependent signalling by binding to mTOR.[1] The two distinct regions of the rapamycin macrocycle that are involved in these binding events are highlighted in Figure 1. The FKBP-binding region, containing the pipecolyl residue and "tricarbonyl" moiety, is a structural feature conserved in a few related natural products also illustrated in Figure 1. Meridamycins **2**,[2] FK-506 **3**,[3] and FK-520 **4**[3] have all shown neuroprotective properties in vitro, while FK-506 was also found to have beneficial effects in a rodent model of stroke.[4] In the rapamycin series, compound **5**, in which the triene system has been disrupted through a Diels-Alder addition reaction, showed neurotrophic activity without immunosuppression.[5] These studies appeared to indicate that FKBP binding was necessary for neuroprotective activity. There have been a number of different approaches taken to exploit the FKBP binding motif found in these compounds. Synthesis of simplified analogs that mimic this region has yielded compounds such as **6** that have shown some neuroprotective activity.[6] In preliminary work focused on the non-immunosuppressive meridamycin **2**, we have shown that the naturally produced prolyl derivative **7** retains neuroprotective activity.[2] While in the case of FK-520 **4**, biosynthetic engineering yielded a product **8** that maintains potent FKBP binding but with greatly reduced immunosuppressive activity.[7] In the case of rapamycin we have further explored the potential of Diels-Alder chemistry to provide non-immunosuppressive analogues with potent neuroprotective properties.

2 METHODS AND RESULTS

2.1 Diels-Alder adducts of rapamycin and nitrosobenzenes

A set of Diels-Alder adducts between rapamycin and nitrosobenzene derivatives was prepared as shown in Scheme 1. Neuroprotective activity (CTX) was gauged by a survival assay using primary cortical and spinal cord neuronal cultures,[8] these results are shown as

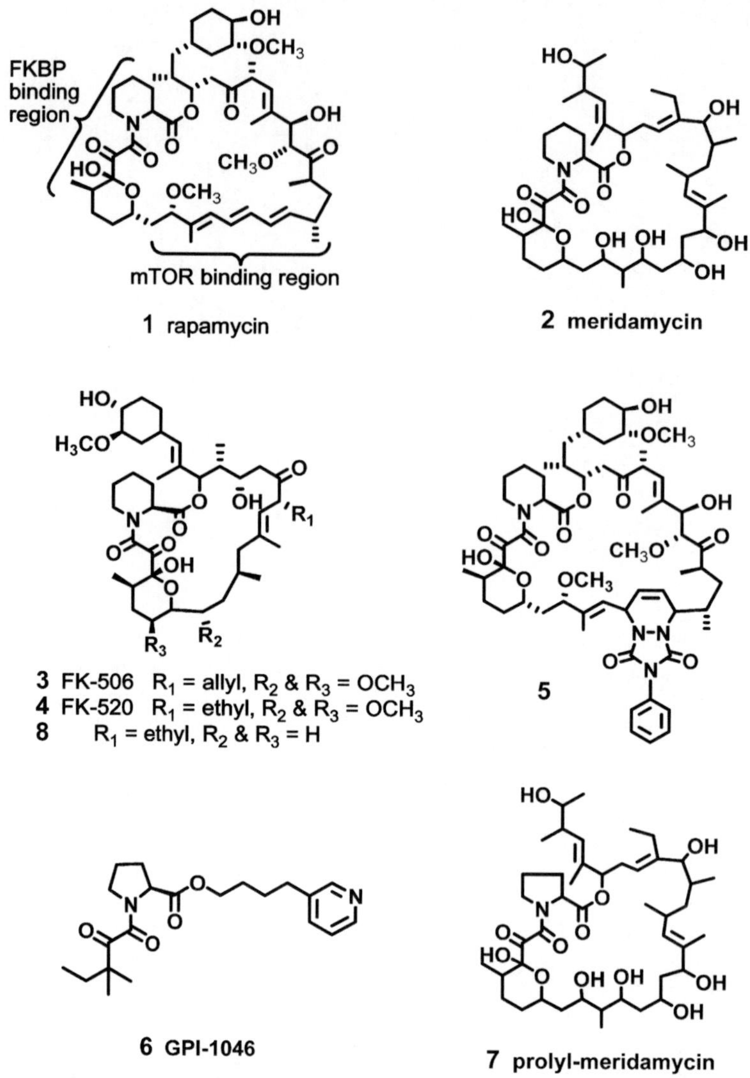

Figure 1 *Structures of representative immunophilin ligands*

EC_{50} values in the scheme. Owing to the small number of analogues prepared only very limited structure-activity correlations were possible. One critical aspect of the chemistry was the observed reversibility of the cycloaddition.

Thus in the case of WYE-592, rapamycin was regenerated in the bioassay medium employed in the T-cell proliferation assay to the extent of about 3%. Even this small amount of free rapamycin was sufficient to inhibit proliferation of the cell line. In order to avoid the confounding effects of intact rapamycin on the subsequent biological assays, the olefinic bond formed in the Diels-Alder process was selectively reduced yielding ILS-920.

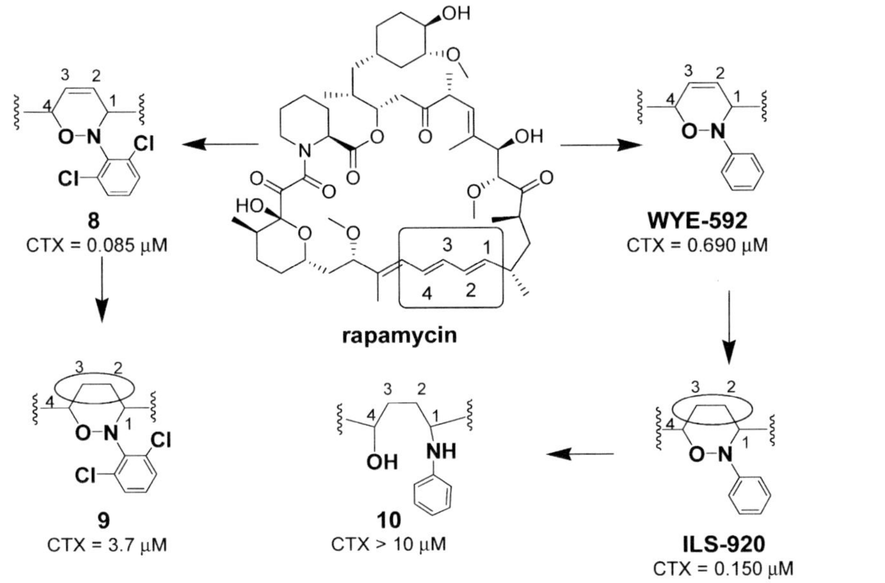

Scheme 1 *Diels-Alder products of rapamycin with nitrosobenzene dienophiles; CTX data (EC_{50}) indicate neuronal protection in cell culture*

In this instance the reduced product was not only chemically stable in solution, but also showed enhanced potency in the neuroprotection assay. Such improvement in activity through reduction of the olefin was not entirely general however, since in the analogous selective reduction of **8** to **9** a substantial loss of potency was observed in the CTX assay. Over reduction resulting in the scission of the nitrogen-oxygen bond and opening of the heterocyclic ring yielding **10**, resulted in a diminution of biological potency. ILS-920 was selected as a compound for further development as it met necessary criteria for advancement.

2.2 Biological properties of ILS-920

Stroke remains one of the major causes of death and long-term disability worldwide, and effective treatments for the debilitating effects of stroke are a major unmet medical need.[9] Thrombolysis with rt-PA (recombinant tissue plasminogen activator) is effective in reducing damage and improving recovery in a small percentage of stroke patients if given within the narrow 3-hour time window following the ischemic event. Often the delay in recognition that a stroke has occurred obviates rt-PA treatment and therefore one of the goals of this work was to discover an agent that was effective a minimum of 6 hours post-stroke. ILS-920 has been evaluated in a number of rodent models of ischemic stroke. In our first report[8] ILS-920 demonstrated efficacy in a transient model in which the middle cerebral artery is occluded for 90 minutes and then reperfused. When administered 4 hours after the insult, both 10 mg/kg and 30 mg/kg doses significantly reduced the ischemia-derived infarct volume while enhancing functional recovery. Subsequent studies in the more rigorous permanent occlusion model, in which there is no reperfusion, saw striking enhancements in recovery after 6-hour intervals (10 and 30 mg/kg) and up to 24-hours post occlusion (30 mg/kg). These beneficial effects were stable and long lasting as revealed through monitoring the neurological deficit scores of treated animals over 90 days. These results provided the foundation for the further development of ILS-920 which has now transitioned to clinical evaluation.

Initial studies aimed at determining the mechanism of action of ILS-920 employed affinity matrices to selectively precipitate proteins from F-11 cell (mouse neuroblastoma/rat dorsal root ganglia hybrid) lysates.[8] Two proteins showed the strongest avidity in these experiments. FKBP-52 which had been recognized previously[10] as a factor in neuroprotective activity of FK-506 was the predominant FKBP identified. The other main target protein was the β-subunit of the L-type voltage-gated calcium channel.[11] It is notable that FK506 also showed strong affinity for FKBP-52 but not for the calcium channel subunit. Rapamycin itself has lower affinity for FKBP-52 than for FKBP-12 and does not appear to bind the β-subunit of the L-type voltage gated calcium channel. How the effects of FKBP-52 inhibition and modulation of calcium channel activity relate to the neurotrophic activities of ILS-920 remain significant research objectives.

3. CONCLUSION

Repurposing the dual binding modes of rapamycin has been demonstrated by disrupting the mTOR binding region of the macrocycle through formation and reduction of Diels-Alder adducts. ILS-920 shows 200-fold preferential binding to FKBP-52 over FKBP-12, and FKBP-52 inhibition enhances neurite outgrowth in cortical neurons. In one of the most challenging models of stroke in rodents, ILS-920 has stimulated sustained enhancement in recovery following treatment 6 to 24 hours post occlusion of the middle cerebral artery.

References

1. J. Choi, J. Chen, S.L. Schreiber, and J. Clardy, *Science*, 1996, **273**, 239.
2. M.Y. Summers, M. Leighton, D. Liu, K. Pong, and E.I. Graziani, *J. Antibiot.*, 2006, **59**, 184-189.
3. W.E. Lyons, E.B. George, T.M. Dawson, J.P. Steiner, and S.H. Snyder, *Proc. Natl. Acad. Sci. U.S.A.*, 1994, **91**, 3191-3195.
4. J. Sharkey and S.P. Butcher, *Nature*, 1994, **371**, 336-339.
5. J.P. Steiner, M.A. Connolly, H.L. Valentine, G.S. Hamilton, T.M. Dawson, L. Hester, and S.H. Snyder, *Nature Medicine,* 1997, **3**, 421428.
6. V.L. Marshal and D.G. Grosset, *Curr. Opin. Invest. Drug,* 2004, **5**, 107-112.
7. W.P. Revill, J. Voda, C.R. Reeves, L. Chung, A. Schirmer, G. Ashley, J.R. Carney, M. Farndis, C.W. Carreras, Y. Zhou, L. Feng, E. Tucker, D. Robinson, and B.G. Gold, *J. Pharmacol. Exp. Ther.*, 2002, **302**, 1278-1285.
8. B. Ruan, K. Pong, F. Jow, M. Bowlby, R.A. Crozier, D. Liu, S. Liang, Y. Chen, M.L. Mercado, X. Feng, F. Bennett, D. von Schack, L. McDonald, M.M. Zaleska, A. Wood, P.H. Reinhart, R.L. Magolda, J. Skotnicki, M.N. Pangalos, F.E. Koehn, G.T. Carter, M. Abou-Gharbia, and E.I. Graziani, *Proc. Natl. Acad. Sci. U.S.A.,* 2008, **105**, 33-38.
9. M.M. Zaleska, M.L.T. Mercado, J. Chavez, G.Z. Feurstein, M.N. Pangalos, and A. Wood, *Neuropharmacology*, 2009, **56**, 329-341.

10. B.G. Gold, V. Densmore, W. Shou, M.M. Matzuk, and H.S. Gordon, *J. Pharmacol. Exp. Ther.*, **289**, 1202-1210.
11. Y. Opatowsky, C.C. Chen, K.P. Campbell, and J.A. Hirsch, *Neuron*, 2004, **42**, 389-399.

PLEUROMUTILINS: ANTIBIOTIC OPTIMISATION FOR HUMAN THERAPY

Richard Jarvest

GlaxoSmithKline R&D, Medicines Research Centre, Gunnels Wood Road, Stevenage, Herts. SG1 2NY, England. E-mail: Richard.L.Jarvest@gsk.com

1 INTRODUCTION

The growing problem of bacterial resistance has been well documented, to the extent that the topic can produce regular headline stories in the general media. It is worth noting that behind the headlines, reports of learned societies, legislative bodies and NGOs have highlighted wider medical and societal implications. For example, from the UK House of Lords Select Committee on Science and Technology 1998 report on Resistance to Antibiotics and other Antimicrobial Agents: "This enquiry has been an alarming experience, which leaves us convinced that resistance to antibiotics and other anti-infective agents constitutes a major threat to public health, and ought to be recognised as such more widely than it is at present". The 2001 WHO Global Strategy for the Containment of Antimicrobial Resistance comments: "Resistance costs money, livelihoods and lives and threatens to undermine the effectiveness of health delivery programmes. It has recently been described as a threat to global stability and national security." Finally, a 2008 Royal Society Policy Document notes: "We are in danger of returning to a pre-antibiotic era".

The development and spread of resistance mechanisms to established antibiotics drives the need for new antibiotic classes that act by novel mechanisms of action and are thus not affected by existing clinical resistance patterns. In its "Call to Action for the Medical Community" the Infectious Diseases Society of America has highlighted three key medical needs for new antibiotics:[1]

- global pandemic of methicillin-resistant *Staphylococcus aureus* (MRSA) infections;
- global spread of drug resistance among common respiratory pathogens, including *Streptococcus pneumoniae;*
- epidemic increases in multidrug-resistant (and increasingly, truly pan-resistant) Gram-negative bacilli.

The GSK Infectious Diseases Centre of Excellence in Drug Discovery has programs underway to address all three of these needs. However, this presentation will focus on the application of the pleuromutilin scaffold to the first and second of these pathogen profiles.

To develop a new antibiotic, we need to translate a list of target pathogenic bacteria into a target disease profile that a new agent would be aiming to treat. For the pleuromutilin drug discovery endeavour we are focusing on two separate disease indications – skin and soft tissue infections (SSTI) and respiratory tract infections (RTI). The key pathogens for the SSTI profile are the Gram positive organisms *S. aureus* (including MRSA) and *Streptococcus pyogenes*, an organism that causes a range of skin infections from the very mild to the "flesh-eating" necrotising fasciitis. The RTI profile has to encompass a wider diversity of organisms including *S. aureus*, *S. pneumoniae*, and the Gram negative organisms *Haemophilus influenzae* and *Moraxella catarrhalis*. The broader phylogenetic diversity of the target and the challenges of Gram negative cellular penetration make this a harder profile to achieve. For a successful RTI agent, the antibiotic also needs to be orally bioavailable, whereas an SSTI agent would be of utility as either an oral or topical therapy.

1 PLEUROMUTILIN OPTIMISATION

2.1 Activity and Mechanism of Action

Pleuromutilin (Figure 1) was originally isolated from the fungus *Pleurotus mutilis* (now *Clitopilus scyphoides*). It has a compact fused tricyclic core with a glycolate ester at the 14 position. A semi-synthetic derivative of pleuromutilin, tiamulin, had already been developed by Novartis as a veterinary medicine with excellent activity against mycoplasmal infections.

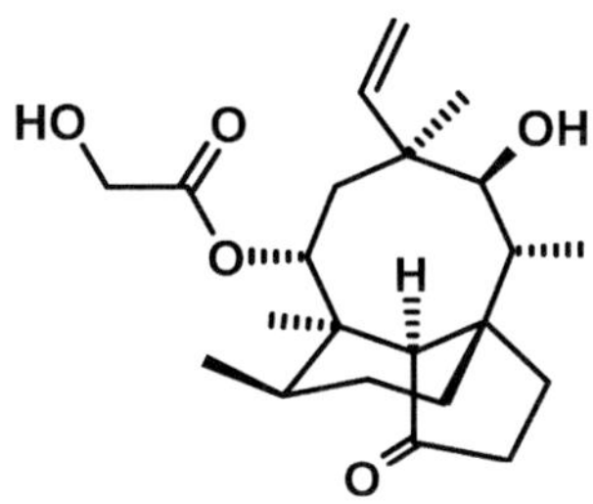

Figure 1 *Pleuromutilin chemical structure.*

Pleuromutilin was tested against panels of clinical isolates of human pathogenic bacteria at GSK to determine the minimum inhibitory concentration required to block the growth of 90% of the isolates (MIC_{90}). Against *S. pneumoniae* it had an MIC_{90} of 16 ug/ml and against *S. aureus* an MIC_{90} of 4 ug/ml. In both cases this is substantially above our target for further progression of an MIC_{90} of 1 ug/ml for Gram positive organisms. Clearly the intrinsic antibacterial activity would need to be enhanced in an antibiotic derivative for human therapy.

Pleuromutilin was also profiled in developability assays to assess its strengths and weaknesses as a scaffold from which to design new human antibiotics. It was confirmed that the pleuromutilin nucleus is very susceptible to oxidative metabolism. For example, it is rapidly degraded when incubated *in vitro* with preparations of human or animal liver microsomes. Assays for metabolic liability would have to be used as part of our strategy to optimise the whole molecule properties for a potential antibiotic medicine.

The antibacterial activity of pleuromutilin derivatives is mediated through interactions with the ribosome, the RNA-protein complex that performs cellular protein synthesis by translation of messenger RNA. Pleuromutilins bind in the P-site region of the ribosomal 50S subunit *via* a unique interaction. For example, pleuromutilins can bind concurrently with the macrolides, the major clinical class of antibiotics acting by ribosomal inhibition, indicating that the binding sites of these two classes do not overlap. As a consequence of the unique interaction mode of pleuromutilins, target-based cross resistance with other antibiotics is rare.

Our studies with pleuromutilin analogues confirmed the binding site by photoaffinity labelling[2] and also footprinting studies.[3] In fact, during the course of our studies a crystal structure of an extremophile ribosome-tiamulin complex was published, clearly defining the binding site.[4] We showed that pleuromutilin compounds inhibit the catalytic peptidyl transfer step of the ribosome, unlike macrolides which do not specifically inhibit the peptidyl transfer step. In further studies it was shown that pleuromutilins inhibit three essential processes associated with the ribosome: inhibition of the catalytic peptidyl transferase step of protein synthesis; inhibition of translation initiation by blocking the P-site and inhibition of ribosomal assembly.[3,5-7]

We wanted to be able to understand pleuromutilin structure-activity relationships (SAR) at the ribosomal target unimpeded by variables such as differences in bacterial penetration of molecules. The peptidyl transfer assay is not ideal for this purpose and so we designed a pleuromutilin binding assay. A fluorescent pleuromutilin derivative was prepared and used to develop a competitive binding assay based on a fluorescence polarisation read-out.[2,3]

2.2 Targeting an RTI Antibiotic

Challenges for designing a pleuromutilin analogue to meet the RTI profile include the following:

- enhance antibacterial activity, especially improved activity against *S. pneumoniae* and bringing the Gram negative *H. influenzae* into the spectrum.
- reduce the rate of metabolic clearance and enhance aqueous solubility. These could be addressed by increasing molecule polarity with polar or basic substituents.
- design synthetic strategies that allow a significant throughput of new analogues with a meaningful diversity of polar functionality
- achieve the multiple developability criteria for a safe, orally bioavailable candidate compound

The medicinal chemistry strategy focused on the 14-positon of the pleuromutilin scaffold as shown in Figure 2.

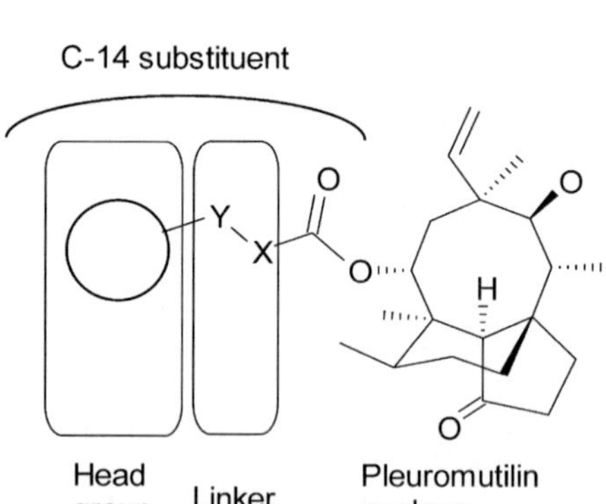

Figure 2 *Pleuromutilin analogue strategy.*

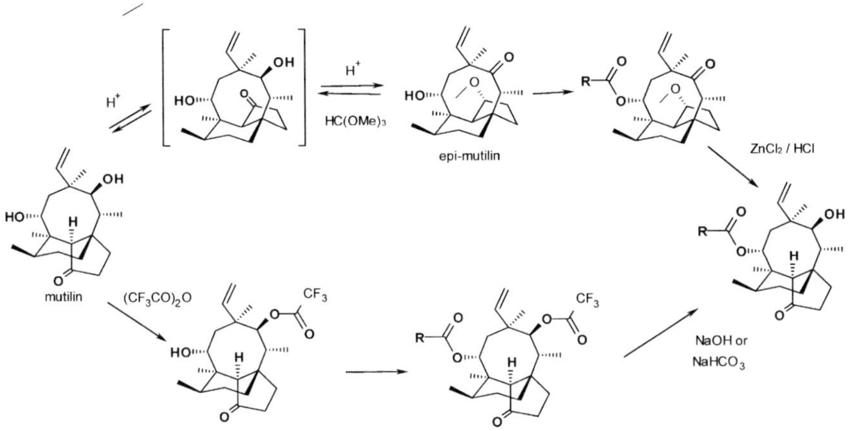

Figure 3 *Synthesis scheme for mutilin 14-O derivatives.*

The bulk of the pleuromutilin nucleus is left constant – manipulations of the vinyl group and 5-membered ring are possible, but they tend not to afford major advantages in antibacterial potency. The 14-carbonyl group is essential for good ribosomal binding and antibacterial activity.

Changes in the C-14 substituent are used to adjust the biological and physicochemical properties of the molecule in a beneficial manner. Polarity in the substituent increases metabolic stability whilst both the head and linker groups modulate antibacterial activity, stability and solubility. For the RTI indication the focus is on the acyl-carbamate functionality for the linker group, as this moiety imparts favourable properties both in terms of good antibacterial activity and increasing polarity.[8]

The prime synthetic routes for preparing C-14 derivatives are shown in Figure 3. The glycolate ester is removed from pleuromutilin to give the mutilin core. Treatment of mutilin with acid and orthoformate results in an epimerisation followed by a 1,5-hydride shift to give the so-called epi-mutilin. Epi-mutilin only has one free hydroxyl group so can be acylated and then acid or Lewis acid catalysed rearrangement back to the pleuromutilin scaffold delivers the 14-acylated derivative. Alternatively the 11-position can be selectively protected with a labile ester such as trifluoracetate or dichloroacetate. Subsequent 14-acylation following by mild deprotection again affords the C-14 derivative.

One major strategy was to focus on preparing key late stage pleuromutilin intermediates that could be transformed into a wide variety of polar derivatives. This is illustrated in Figure 4. On the left hand side, three acylcarbamate intermediates with transformable functionality are shown. The three intermediates can be used with different sources of chemical diversity to achieve Michael addition products, amide, urea and sulfonamide derivatives and sulfoxide and sulfone compounds respectively. The R- groups illustrate the diversity achieved in addition to the polar functionality involved in the transformation.

In primary antibacterial assays, pleuromutilin analogues from all these acylcarbamate series were able to achieve target levels of antibacterial activity (Gram positive MIC ≤ 1 ug/ml, Gram negative MIC ≤ 2 ug/ml). Examples derived from each of the intermediates in Figure 4 are shown in Figure 5. As can be seen, *M. catarrhalis* is generally very sensitive to pleuromutilin analogues. The compounds all showed potent binding affinity to the

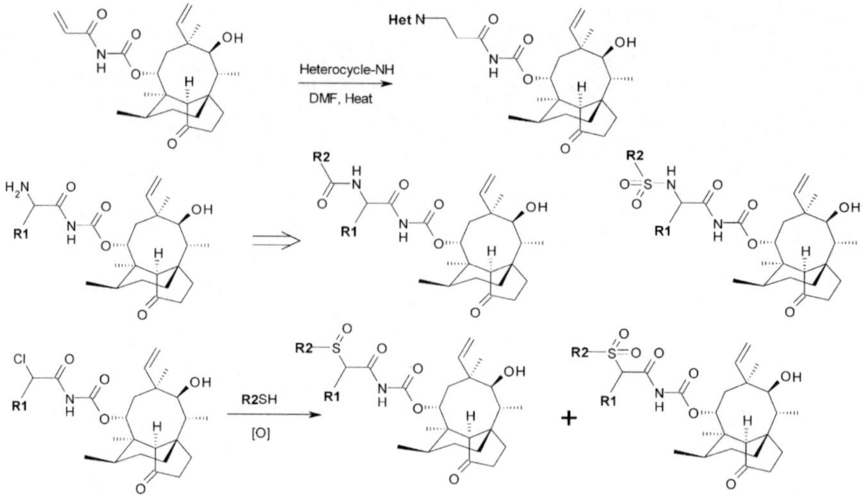

Figure 4 *Late stage approaches for rapid introduction of chemical diversity.*

ribosome, the K_d values under 10 nM are probably at the limit of the assay due to the concentration of ribosome in the assay.

The analogues also showed useful SAR in developability assays. Selected data for five heterocyclic series analogues that pass primary antibacterial profiling are shown in Figure 6. The compound on the left hand side is the least polar of the set and has a high rate of metabolic clearance in human microsomes *in vitro*, though less than that of pleuromutilin.

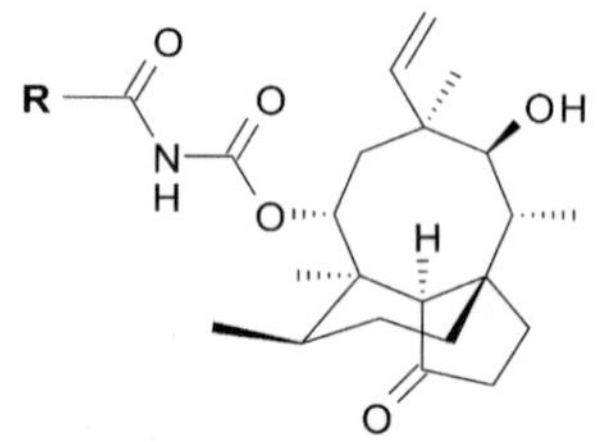

R	pyrazol-$(CH_2)_2$	oxamyl-NHCHMe-	$MeSO_2NH$-CH_2	$HO(CH_2)_2SO_2$-CH_2
	Minimum Inhibitory Concentration (ug/ml)			
S. aureus	< 0.06	0.25	0.25	0.5
S. pneumoniae	0.125	0.25	0.5	1
H. influenzae	1	1	1	1
M. catarrhalis	< 0.06	< 0.06	< 0.06	< 0.06
Kd (nM)				
	~ 5	25	~7	19

Figure 5 *Primary antibacterial activity of examples from acylcarbamate series against* S. aureus *Oxford,* S. pneumoniae *1629,* H. influenzae *Q1 and* M. catarrhalis *1502.*

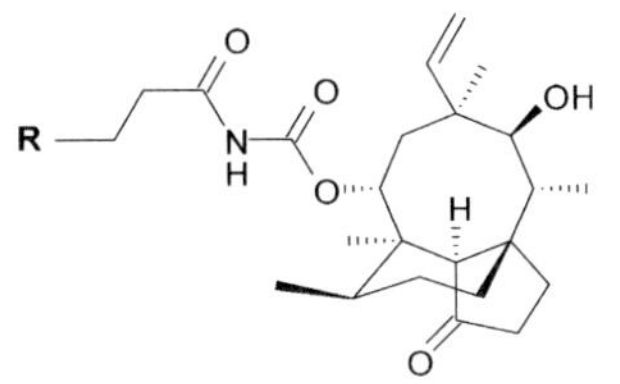

R	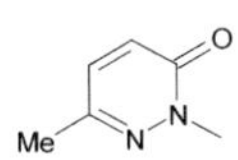	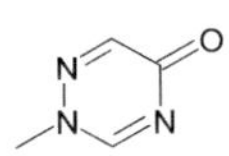	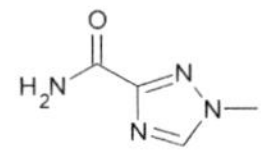	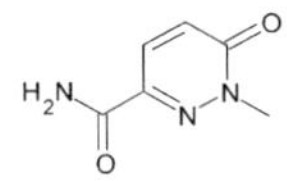	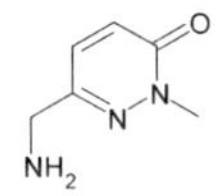
	Minimum Inhibitory Concentration (ug/ml)				
S.pneumoniae	0.125	0.25	0.125	< 0.06	0.125
H. influenzae	2	2	1	1	1
Metabolism	High	Medium	Low	Medium	Low
3A4 IC_{50} uM	3 - 10	> 30	10 - 30	3 - 10	3 - 10
Aq. solubility	Medium	Not detected	Low	Medium	High

Figure 6 *Activity against* S. pneumoniae *N1387 and* H. Influenza *Q1, metabolism in human microsomes, inhibition of cytochrome P450 3A4 and aqueous solubility of a selection of mutilin derivatives.*

By contrast, the most polar analogues have attained the target low clearance level. A modulation in affinity for the cytochrome P450 3A4 enzyme is also seen across the series, as is also the case for aqueous solubility. High aqueous solubility is achieved with a basic group in the side chain as in the compound on the right.

Overall for the RTI profile, the pleuromutilin acylcarbamate derivatives show a very encouraging profile. These compounds can achieve MIC_{90} values against the target pathogens ≤ 1 ug/ml, have up to 2-orders of magnitude greater metabolic stability, reduced interactions with cytochrome P450 3A4, acceptable oral bioavailability in rats and good efficacy in infection models. The development of an oral pleuromutilin derivative is an ongoing challenge but I want to move on now to the other therapeutic profile I mentioned in the introduction: a topical antibiotic with a spectrum encompassing MRSA.

3 DEVELOPMENT OF A TOPICAL ANTIBIOTIC

3.1 Selection of Retapamulin

The need for new topical antibiotics, as for other indications, is driven by bacterial resistance – for example, globally 10% of *S. aureus* isolates are resistant to the topical agent mupirocin. The topical route of administration also has some specific therapeutic advantages:

- a higher antibiotic concentration applied directly to the infection site
- spares the use of oral antibiotics, potentially leading to less development of resistance in enteric commensal bacteria
- less systemic exposure and side effects

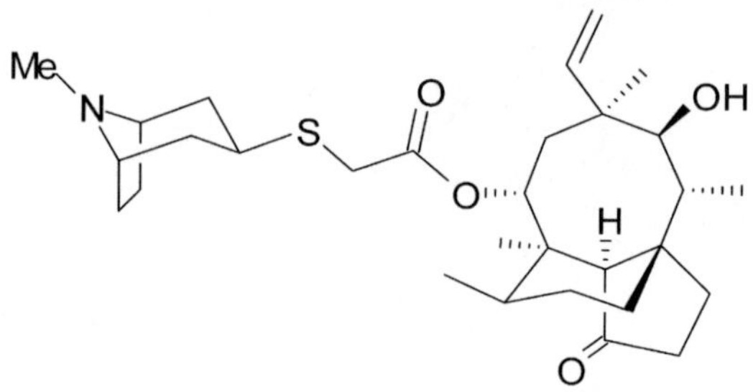

Figure 7 *Retapamulin chemical structure.*

In addition to a novel mechanism of action and potent activity against resistant skin pathogens, the target profile for a topical agent has some special features. In terms of physicochemical properties the agent must be amenable to a topical formulation but deliver only low systemic exposure from topical administration. Furthermore, it was desirable to have a shorter and more convenient course of therapy than current topical antibiotics – a target was set of twice daily application over a period of one week or less (in comparison to three times daily for 7-10 days required for the agents mupirocin or fusidic acid).

To select the topical candidate, analogues of the structure in Figure 2 were evaluated with 28 monocyclic or bicyclic amines at the head group and a variety of linkers. In the linker position the sulfanyl acetate group stood out as having an exceptional combination of antibacterial potency and stability. The molecule that most successfully passed the selection criteria is shown in Figure 7. This compound was later given the name retapamulin. Retapamulin had the best balance of key properties – antibacterial activity, solubility, stability in solution, limited systemic exposure, topical efficacy, and synthetic accessibility – and furthermore was not irritant to skin when applied at efficacious concentrations.

The potent antimicrobial activity of retapamulin against key classes of skin pathogens is evidenced by the MIC data obtained for candidate selection. Remarkably the MIC_{100} for *S. aureus* isolates (n = 51) including MRSA was just 0.06 ug/ml, and for *S. pyogenes* (n = 24) it was 0.03 ug/ml. In comparison with the topical antibiotic mupirocin, retapamulin was between 2-fold and 32-fold more potent against mupirocin-sensitive strains and was fully active against mupirocin-resistant strains.

In studies of the metabolism of retapamulin, a high rate of metabolic clearance was observed in *in vitro* experiments with rat, monkey and human liver microsomes. Clearance occurred by multiple oxidative pathways and there were no human-specific metabolites. The rapid clearance by multiple pathways is ideal for a topical agent where minimal systemic exposure is desired.

3.2 Synthesis and Comparative Activity of Retapamulin

A synthesis of retapamulin is shown in Figure 8. Selective mesylation of the primary hydroxyl group of pleuromutilin affords an activated intermediate whilst the other starting material, tropine alcohol, is converted to a thiol intermediate. Reaction of these two intermediates followed by work-up and purification affords retapamulin in a short sequence without the need for protecting groups.[9] Retapamulin is conveniently obtained and used as the anhydrous, crystalline, free base form.

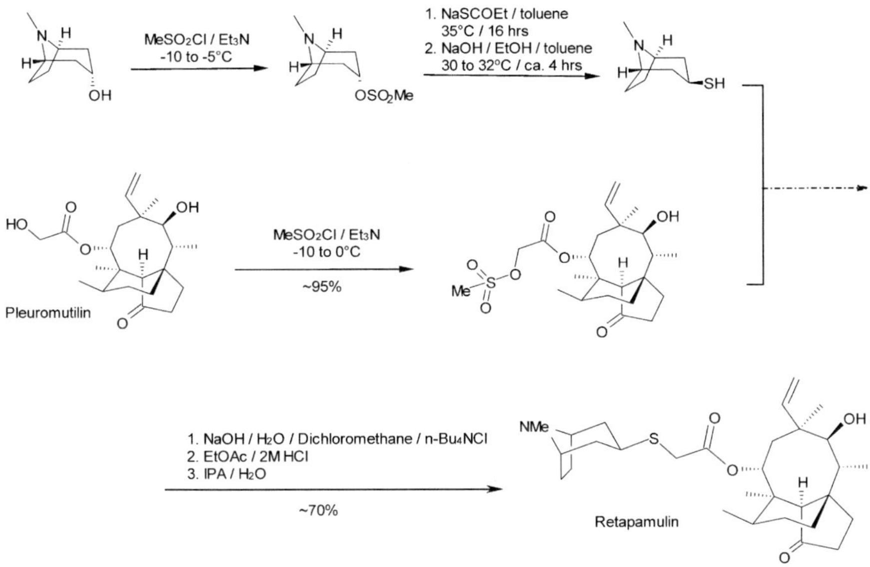

Figure 8 *Synthesis of retapamulin from pleuromutilin*

Efficacy of retapamulin was demonstrated by topical treatment in an *S. aureus* wound infection model. In studies in this model retapamulin as a 1% w/w formulation was effective against all *S. aureus* isolates tested, including a variety of resistance phenotypes. This formulation with twice daily application consistently gave efficacy equivalent or superior to three times daily treatment with mupirocin (2%) or fusidic acid (2%).[10]

Organism / Phenotype	N	MIC (μg/mL) MIC_{50}	MIC_{90}	Range
Staphylococcus aureus	1,975	0.06	0.12	0.004–0.5
methicillin-susceptible	1,326	0.06	0.12	0.015–0.5
methicillin-resistant	649	0.06	0.12	0.004–0.5
macrolide-resistant	956	0.06	0.12	0.015–0.5
mupirocin-resistant	194	0.06	0.12	0.03–0.5
fusidic acid-resistant	134	0.06	0.12	0.004–0.5
Coagulase-negative Staphylococci	975	0.06	0.06	0.002–0.5
methicillin-susceptible	397	0.06	0.06	0.002–0.5
methicillin-resistant	578	0.06	0.12	0.015–0.5
mupirocin-resistant	341	0.06	0.12	0.015–0.5
All Staphylococci isolates	2,950	0.06	0.12	0.002–0.5

Retapamulin Global Surveillance Study (2004 –2005)
Stevens et al. and Bouchillon et al. ICAAC 2005; Data on file, GlaxoSmithKline

Figure 9 *Minimum inhibitory concentration of* retapamulin *against 2950 staphylococcal SSTI isolates.*

Organism / Phenotype	N	MIC_{50}	MIC_{90}	Range
		MIC (µg/mL)		
Streptoococcuspyogenes	1918	0.03	0.03	0.004–0.25
macrolide-susceptible	1498	0.03	0.03	0.004–0.12
macrolide-intermediate	55	0.03	0.06	0.008–0.12
macrolide-resistant	365	0.03	0.06	0.008–0.25
Viridans Streptococci	930	0.06	0.25	?0.002–0.5
Streptococcusagalactiae	941	0.03	0.06	0.002–0.5
Other beta hemolytic Streptococci	8	-	-	0.008–0.12
All Streptococci isolates	3797	0.06	0.12	? 0.002–0.5

Retapamulin Global Surveillance Study (2004 – 20005)
Stevens et al. and Bouchillon et al. ICAAC 2005; Data on file, GlaxoSmithKline.

Figure 10 *Minimum inhibitory concentrations of* retapamulin *against 3797 streptococcal SSTI isolates.*

Post candidate selection, retapamulin was profiled against much larger panels of staphylococcal and streptococcal pathogens isolated from skin and soft tissue infections.[11-13] The results were extremely encouraging and examples of early studies are shown in Figures 9-11. Profiling nearly 3000 staphylococcal clinical isolates with a range of resistance phenotypes gave an MIC_{90} of 0.12 ug/ml (Figure 9). Likewise from nearly 3,800 streptococcal isolates the MIC_{90} was also 0.12 ug/ml (Figure 10). In both cases the highest MIC observed was just 0.5 ug/ml, indicating that all strains in these large clinical panels were susceptible to retapamulin.

Antimicrobial	MIC $_{90}$(µg/mL)				
	S. *aureus* (n = 1182)	Coag -ve staphylococci (n = 442)	*S. pyogenes* (n = 1164)	*S. agalactiae* (n = 462)	*Viridans streptococci* (n = 471)
Retapamulin	0.12	0.06	0.06	0.06	0.25
Mupirocin	4	> 256	0.5	1	2
Fusidic acid	1	8	8	16	32
Bacitracin	> 128	> 128	64	128	128
Gentamicin	2	16	16	32	16
Neomycin	64	16	64	> 64	> 64
Clindamycin	> 32	> 32	0.25	32	8
Tetracycline	16	> 32	16	> 32	> 32
Erythromycin	> 32	> 32	2	32	> 32
Cephalothin	8	4	0.25	0.5	2
Penicillin	32	16	0.06	0.12	0.5

Johnson B et al. Poster presented at the 45 th ICAAC, December 16 -19, 2005; Washington, DC

Figure 11 *Retapamulin: Comparative activity against 3721 North American SSTI isolates.*

The MIC profile of retapamulin was compared with that of other antibiotics in a panel of over 3,700 North American skin and soft tissue isolates of *Staphylococci* and *Streptococci* (Figure 11). Against this panel the MIC_{90} values of retapamulin were between 0.06 and 0.25 ug/ml for different organisms. This stands in marked contrast to the profiles for the other antibiotics where the high MIC_{90} values again evidence the prevalence of bacterial resistance.

Retapamulin was compared to fusidic acid and mupirocin in studies examining the rate of development of resistance. A panel of *S. aureus* isolates was passaged in the presence of each antibiotic and susceptibility to that antibiotic was measured after one, five, ten, fifteen and twenty passages. The development of resistance was defined as a > 4-fold increase in MIC. It can be seen from Figure 12 that significant resistance to fusidic acid and mupirocin was observed at five and ten passages, at which points no retapamulin resistance was observed. Of course, resistance to retapamulin eventually develops and at 20 passages all strains are resistant to the antibiotic present. Whilst these studies may not provide an exact model for the development of resistance in the clinic, it does suggest that retapamulin has a lower intrinsic propensity to development of resistance by spontaneous mutation than these other topical antibiotics.

The mechanism of action studies described earlier were repeated with retapamulin, profiling it against other relevant clinical antibiotic agents – the topical agents fusidic acid and mupirocin, and the more recent macrolides azithromycin and telithromycin. All these agents target inhibition of various stages of bacterial protein synthesis. Retapamulin was found to bind at a single unique site on the ribosome and was the only compound to show binding in the pleuromutilin binding assay. It retained the multiple modes of action of the class, inhibiting the catalytic peptidyl transfer step and blocking P-site binding. This data fulfils the desired criteria for a new topical agent by having a different mode of action to other antibiotics in clinical use and thus very rare target-based resistance.[14]

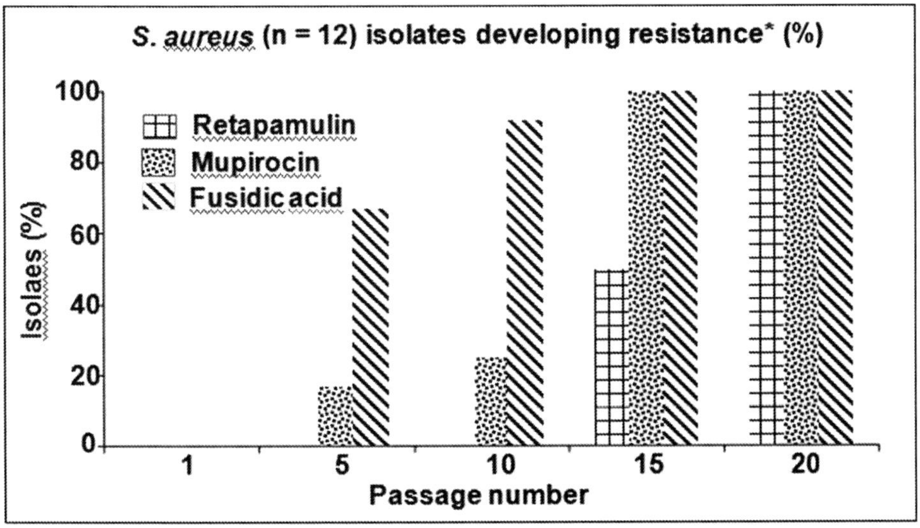

Figure 12 *Retapamulin demonstrates a low propensity for the development of resistance (*defined as > 4 increase in MIC) in* S. aureus.

3.2 A New Clinical Antibiotic

Following candidate selection of retapamulin in March 2002, the development process to a clinical medicine was prosecuted in an unusually short time frame. The first dose in humans occurred in December of 2002, followed by a commit-to-medicine development decision in October of the following year. This decision leads to large scale clinical trials of the efficacy of the agent in bacterial infections. The NDA submission for a new antibiotic medicine was made in November of 2005 and the first regulatory approval occurred in April 2007.

Retapmulin was launched as a topical antibiotic as a 1% ointment under the names Altabax™ in the USA and Altargo™ in Europe. The dosing protocol of twice daily for a period of 5 days met the criteria for a more convenient dosing regimen that had been set earlier in the programme. Retapamulin represents the first new class of topical antibacterial agent in clinical usage for nearly two decades.

Acknowledgements

I acknowledge and thank the many scientists and others in GSK who have contributed to the discovery and development of pleuromutilin antibiotics and especially of retapamulin.

References

1 B. Spellberg, R. Guidos, D. Gilbert, J. Bradley, H.W. Boucher, W.M. Scheld, J.G. Bartlett and J. Edwards, Jr., *Clin. Infect. Dis.*, 2008, **46**, 155.
2 E. Hunt, *Drugs of the Future,* 2000, **25**, 1163.
3 K. Yan, L. Madden, A.E. Choudhry, C.S. Voigt, R.A. Copeland and R.R. Gontarek, *Antimicrob. Agents Chemother.*, 2006, **50**, 3875.
4 F. Schlünzen, E. Pyetan, P. Fucini, A. Yonath and J.M. Harms, *Mol. Microbiol.*, 2004, **54**, 1287.
5 S.M. Poulsen, M. Karlsson, L.B. Johansson and B. Vester, *Mol. Microbiol.*, 2001, **41**, 1091.
6 K.S. Long, L.H. Hansen, L. Jakobsen, and B. Vester, *Antimicrob. Agents and Chemother.,* 2006, **50**, 1458.
7 W.S. Champney, *Infectious Disorders – Drug Targets*, 2006, **6**, 377.
8 G. Brooks, W. Burgess, D. Colthurst, J.D. Hinks, E. Hunt, M.J. Pearson, B. Shea, A.K. Takle, J.M. Wilson and G. Woodnutt, *Bioorg. Med. Chem.*, 2001, **9**, 1221.
9 B. Boyd and J. Castañer, *Drugs of the Future*, 2006, **31**, 107.
10 S. Rittenhouse, C. Singley, J. Hoover, R. Page and D. Payne, *Antimicrob. Agents Chemother.*, 2006, **50**, 3886.
11 R.N. Jones, T.R. Fritsche, H.S. Sader and J.E. Ross, *Antimicrob. Agents Chemother.*, 2006, 50, 2583.
12 S. Rittenhouse, S. Biswas, J. Broskey, L. McCloskey, T. Moore, S. Vasey, J. West, M. Zalacain, R. Zonis and D. Payne, *Antimicrob. Agents Chemother.*, 2006, **50**, 3882.
13 N. Woodford, M. Afzal-Shah, M. Warner and D.M. Livermore, *J. Antimicrob. Chemother.*, 2008, **62**, 766.
14 Subsequent studies have identified rare resistance mechanisms.[15,16]
15 K.S. Long, J. Poehlsgaard, C. Kehrenberg, S. Schwarz and B. Vester, *Antimicrob. Agents Chemother.*, 2006, **50**, 2500.
16 D.R. Gentry, L. McCloskey, M.N. Gwynn, S.F. Rittenhouse, N. Scangarella, R. Shawar and D.J. Holmes, *Antimicrob. Agents Chemother.*, 2008, **52**, 4507.

OXAZOLOMYCINS AND EQUISETINS: NOVEL ANTIBIOTIC LEAD STRUCTURES

Muhammad Anwar, Claire Bagwell, Yongchul Jeong, Mark Moloney and Muhammad Yaqoob.

University of Oxford, Chemistry Research Laboratory, Mansfield Road, Oxford OX1 3TA
E-mail: mark.moloney@chem.ox.ac.uk

1. INTRODUCTION

Sessile bacterial colonies, called biofilms, form a distinct part of the bacterial life cycle, differentiated from the more commonly recognized planktonic state and arising as a result of sophisticated small molecule mediated cell signaling. Such quorum sensing is choreographed by acyl homoserine lactone derivatives, but it has recently been recognized that these compounds rearrange to give tetramic acid derivatives which exhibit antibacterial activity.

There is, therefore, a switch from a positive feedback signal encouraging the growth of a biofilm, to one in which further bacterial growth is inhibited. In fact, tetramic acids and their derivatives are a particularly important sub-group of natural products and are already well known for their potent antibiotic, antiviral and antifungal as well as cytotoxic activities. Those which contain structural units derived from N-methylserine or N-methylthreonine include the equisetins, the oxazolomycins, the salinosporamides and the cinnabarimides (Figure 1). The synthesis of this key structural motif[1-3] continues to attract interest and our contribution in this area stems from the finding that suitable serine and threonine-derived oxazolidine templates can be used for highly chemo- and diastereo-selective ring closure reactions leading to tetramic acid and pyroglutamate derivatives.[4] We report here on further development of this work which has led to the identification of novel antibiotics based upon the natural products oxazolomycin, reutericyclin and equisetin.

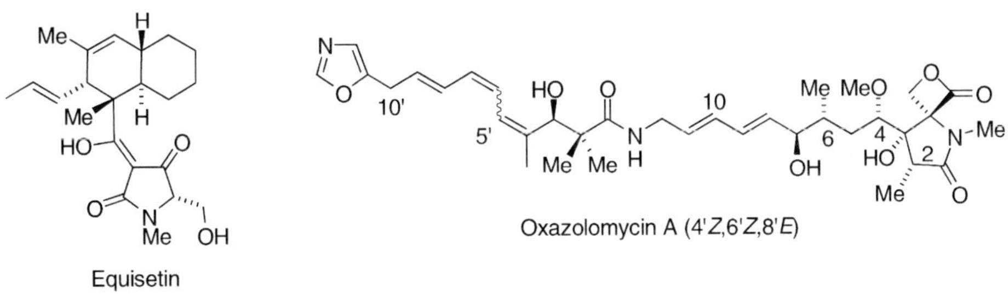

Figure 1 *Natural products containing a common tetramic acid derived core*

1.1 Production and Biological Activities of Tetramic Acid Derivatives

Equisetin was isolated from filamentous fungi of the *Fusarium* genus where it is produced in a typical titre of 5g/kg of the growth substrate.[5] It contains a β-acyltetramic acid core, with a highly substituted decalin side chain and has activity in the range of 0.5 - 4.0 μg/ml against both Gram-positive bacteria such as *Bacillus subtilis*, *Mycobacterium phlei* and MRSA and Gram-negative bacteria such as *Neisseria perflava*. Low mammalian toxicity (LD_{50} 63 mg/kg in mice) has encouraged the consideration of equisetin as an antibiotic, although it also inhibits the integrase of HIV-1 in infected cells at 25 μM.

Oxazolomycins A, B and C, which vary only in their stereochemistry, were isolated from *Streptomyces* spp., which secrete in the order of 2 μg/ml of the compounds into fermentation broths.[6] The structures are unusual in that they contain β-lactone and oxazole units linked by a triene and diene containing spacer, although they are closely related to neooxazolomycin, curromycin, phthoxazolin A and inthomycin. The oxazolomycins exhibit antiviral, antibacterial and cytotoxic activities and these have been attributed to their protonophoric properties which act across the endosomal membrane. Like equisetin, they display low mammalian toxicity (LD_{50} 11 mg/kg in mice). The intention of our current work is to use these compounds as lead structures to identify simpler compounds which have similar or improved bioactivities.

2 EQUISETIN ANALOGUES

2.1 Synthesis and Activity of Equisetin Analogues

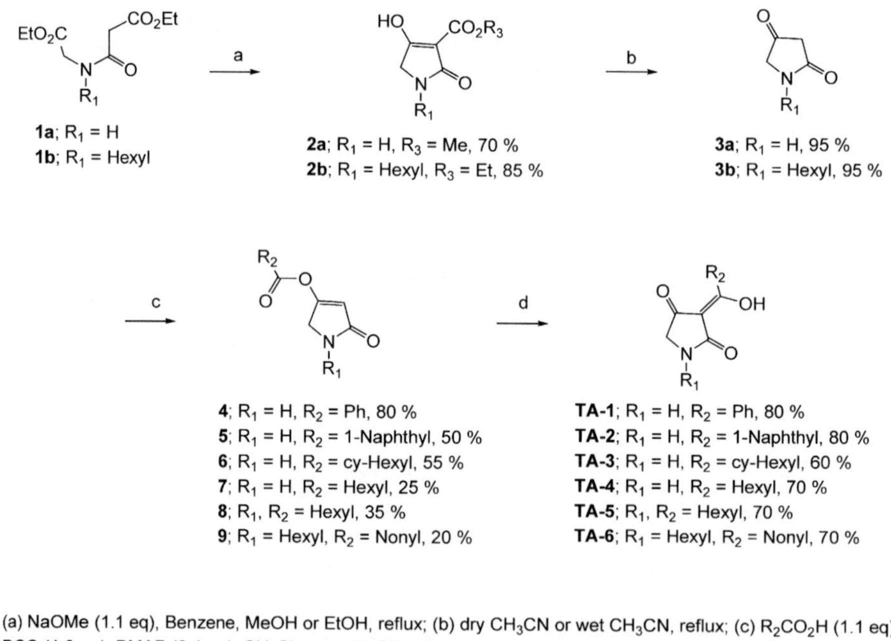

(a) NaOMe (1.1 eq), Benzene, MeOH or EtOH, reflux; (b) dry CH_3CN or wet CH_3CN, reflux; (c) R_2CO_2H (1.1 eq), DCC (1.0 eq), DMAP (0.1 eq), CH_2Cl_2, r.t.; (d) $(CH_3)_2C(OH)CN$ (0.5 eq), Et_3N (2.0 eq), CH_3CN, r.t..

Scheme 1 *The synthesis of analogues of equisetin.*

Table 1 *Antibiotic activity of the equisetin analogues at 1 mg/ml and, in parentheses, at 4 mg/ml. The minimum inhibitory concentratiom against* S. aureus *is shown on the right.*

Compound	Diameter of inhibition zone (cm)		MIC *vs S. aureus* (μg/ml)
	S. aureus	*E. coli*	
Cephalosporin C	1.1	2.3	200
TA-1	(NA)	NA	-
TA-2	NA (0.6)	NA	-
TA-3	NA (0.5)	0.2	-
TA-4	0.8 (2.0)	0.9	400
TA-5	3	NA	2
TA-6	-	NA	0.4

Dieckmann cyclisation of malonylglycinamide diester derivatives **1** and decarboxylation provides the cyclic ketolactams **3** which are coupled with chemically diverse acids using DCC and rearranged to give the completed five membered lactam (Scheme 1).

The activities of these analogues against *Staphylococcus aureus* and *Escherichia coli* were determined by the disc diffusion assay on agar plates and are shown in Table 1. Encouraging activities against both these common pathogens were apparent in several of the analogues and they compare favourably with those of previously reported tetramic acid derivatives[7-9] made by routes which are less well suited to the introduction of diverse substituents and do not have the facility of the current synthesis.

3 OXAZOLOMYCIN ANALOGUES

3.1 Synthesis and Activity of the Lactam Mimics

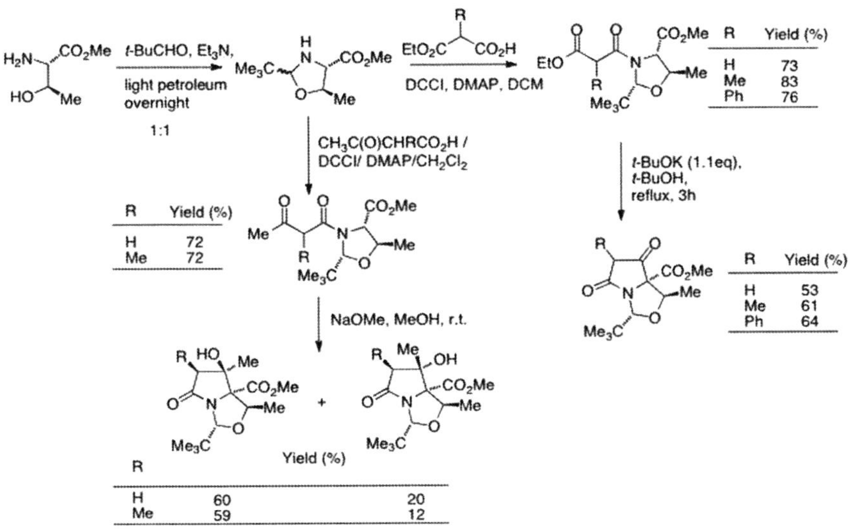

Scheme 2 *The synthesis of the threonine derived lactam mimic.*

Threonine methyl ester is protected by the formation of an oxazolidine ring (Scheme 2) so that the exposed amine can be condensed with substituted malonate half esters or β-keto esters which, after base catalysed cyclisation, provide the lactam mimics in the protected form as single separable enantiomers of each diastereoisomer. The biological activity of a selection of the compounds synthesized using this chemistry is presented in Table 2

Table 2 *The antibiotic activity of protected and deprotected lactam mimics.*

OXAZOLIDINE	Bioactivity: *S. aureus*	Bioactivity: *B. Subtillis*	DEPROTECTED	Bioactivity: *S. aureus*	Bioactivity: *B. Subtillis*
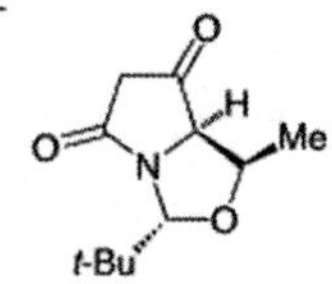	Active at 2mg/ml	Active at 2mg/ml	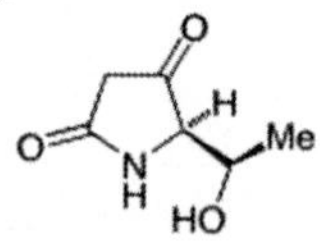	Active at 2mg/ml	X
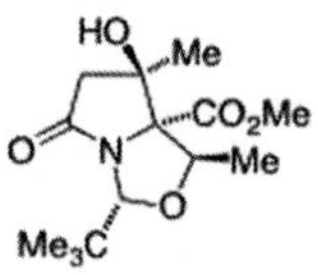	Active at 2mg/ml	X	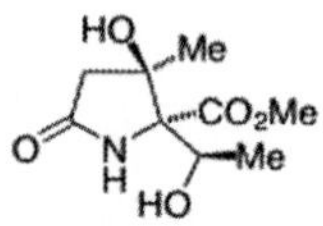	X	Active at 4mg/ml
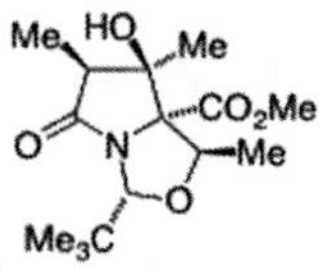	Active at 2mg/ml	Active at 2mg/ml	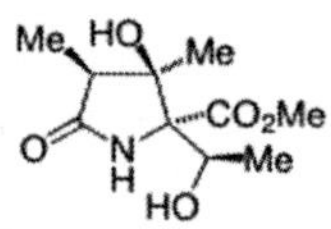	X	Active at 5mg/ml

Diverse substituents may also be introduced at the ketone moiety of the fused bicyclic ring system in good yields (Scheme 3) by the use of Grignard reagents, but in the event, one such example, the fused bicyclic alkyne (R = Ph) showed only weak activity against *E. coli*.

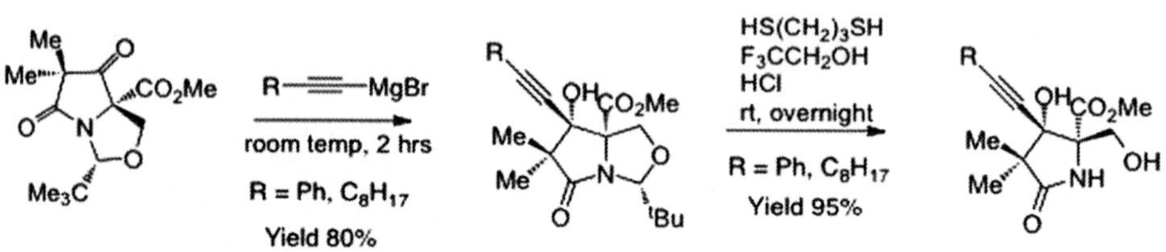

Scheme 3 *Attachment of the linker unit to the lactam analogue.*

Since delicate functionality would not be expected to survive the relatively harsh conditions of deprotection, the more challenging β-lactone ring of oxazolomycin A can be accessed directly, though in moderate yield, from the unprotected γ–lactam skeleton (Scheme 4). The mixed sulphonic anhydride is generated by the addition of *p*-toluenesulphonyl chloride to the unprotected γ–lactam and the anhydride may then be displaced by either the adjacent primary or secondary alcohol to form a β-lactone ring either fused or spiro to the γ–lactam.

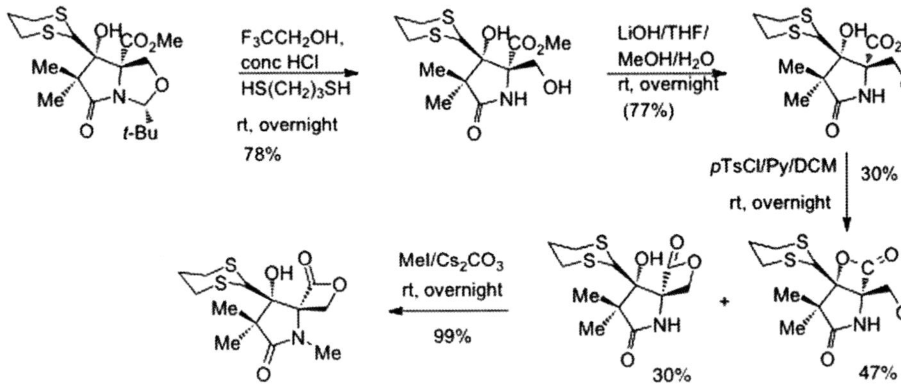

Scheme 4 *Formation of the β-lactone ring.*

3.2 The Central Section of Oxazolomycin

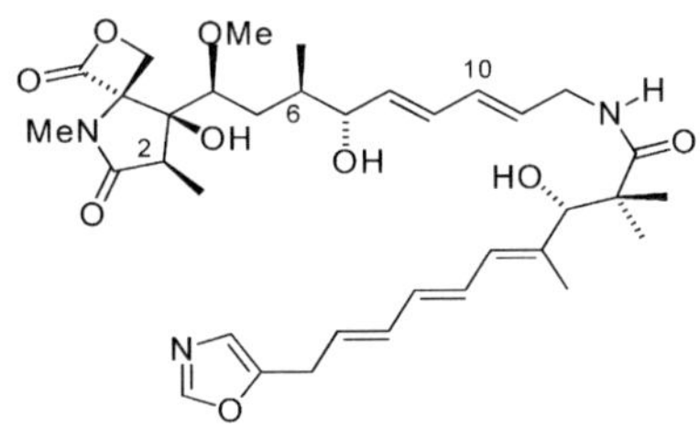

Figure 2 *The folding of oxazolomycin B.*

Although cursory inspection of the central section of oxazolomycins suggests an unstructured chain, closer assessment indicates the folding shown in Figure 2. X-ray diffraction reveals the presence of a hydrogen bond of 2.29 ångström between the secondary alcohol and the amide NH in the centre of the chain (illustrated by a model compound in Figure 3) and this forms an effective six membered ring favouring the superposition of the oxazole and lactam units thus promoting the formation of hydrophobic interactions between the two unsaturated sections of the linker. This section of the molecule can be represented by its primary components: the amide, an adjacent gem dimethyl and the alcohol. The chemistry we have developed to synthesise analogues of this moiety[10,11] is shown in Scheme 5 and their antibiotic activity in Table 3.

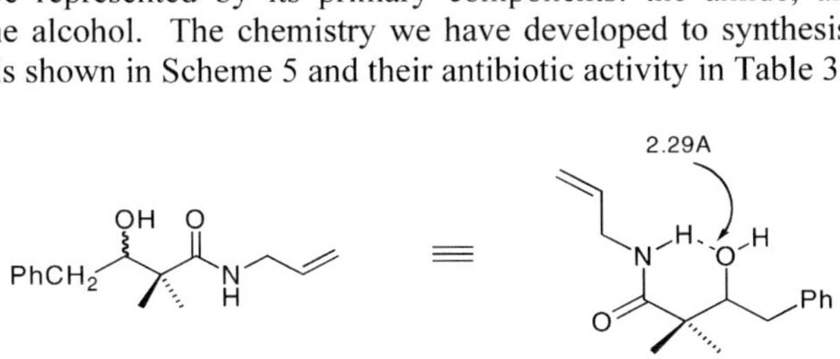

Figure 3 *Hydrogen bonding in the central section of the oxazolomycin A chain.*

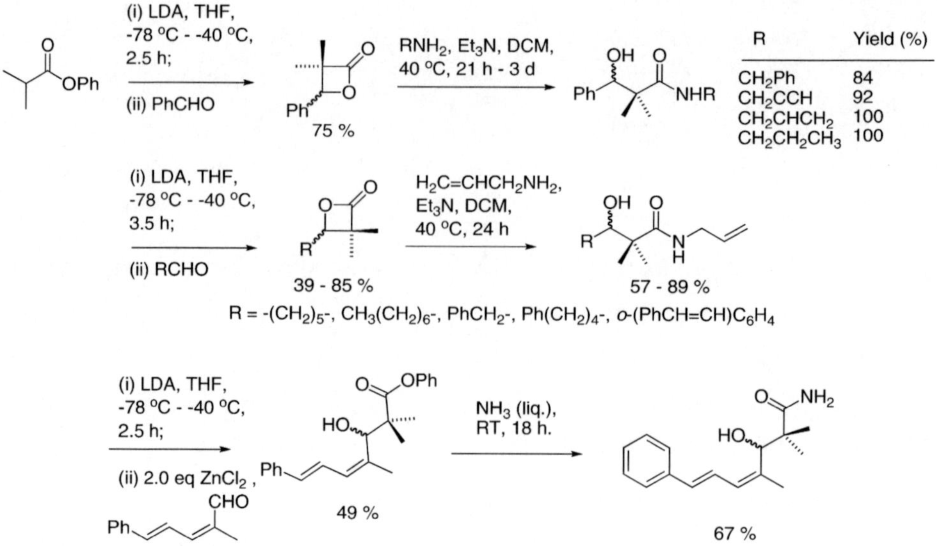

Scheme 5 *Synthesis of analogues of the central section of oxazolomycin A.*

Table 3 *Antibiotic activity of analogues of the central section of oxazolomycin A.*

	Bioactivity	
	S. aureus	*E. coli*
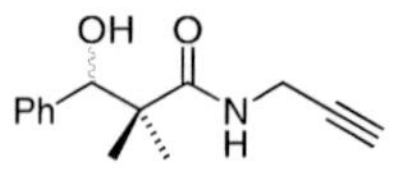	X	Weakly active at 4mg/ml
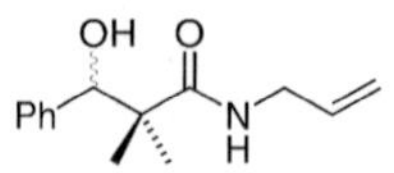	X	X
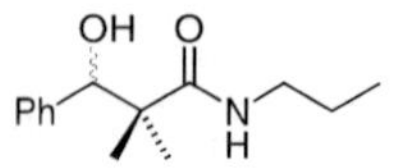	X	X
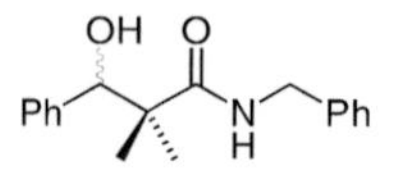	X	Very active at 4mg/ml
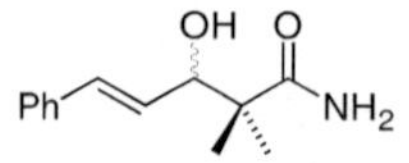	X	X
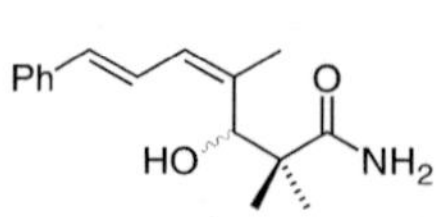	Weakly active at 4mg/ml	Very active at 4mg/ml

	Bioactivity	
	S. aureus	*E. Coli*
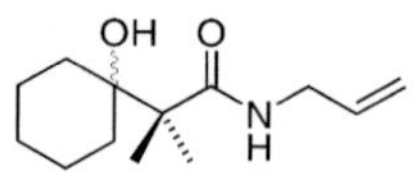	X	X
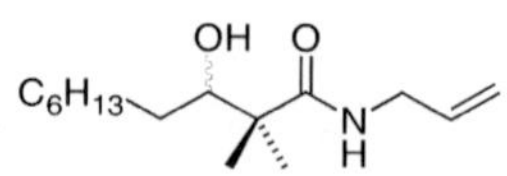	X	Very active at 4 mg/ml
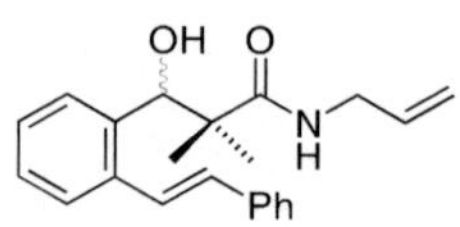	Weakly active at 4mg/ml	Active at 4mg/ml
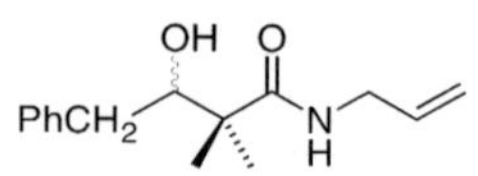	X	X
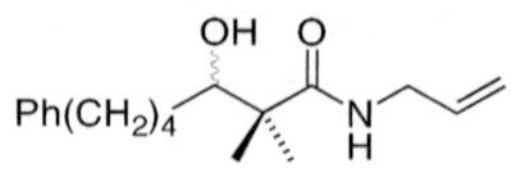	Active at 4mg/ml	Very active at 4mg/ml

4 CONCLUSION

Our work suggests that there are contributions to the antibiotic activity of the tetramic acids from several portions of the antibiotic molecules and we have developed robust, stereospecific chemistry to access analogues of these fragments so as to develop simpler compounds for use as antibacterial agents.

References

1 M.D. Andrews, A.G. Brewster, K.M. Crapnell, A.J. Ibbett, T. Jones, M.G. Moloney, C.K. Prout and D. Watkin, *J. Chem. Soc., Perkin Trans. 1*, 1998, 223.

2 M.G. Moloney and M. Yaqoob, *Synlett.*, 2004, 1631.

3 M. Anwar and M.G. Moloney, *Tetrahedron Lett.*, 2007, **48**, 7259.

4 M.D. Andrews, A.G. Brewster and M.G. Moloney, *Synlett.*, 1996, 612.

5 H.R. Burmeister, G.A. Bennett, R.F. Vesonder and C.W. Hesseltine, *Antimicrob. Agents Chemother.*, 1974, **5**, 634.

6 M.G. Moloney, P.C. Trippier, M. Yaqoob, Z. Wang, *Curr. Drug Discovery Technol.*, 2004, **3,** 181.

7 T. Rosen, P.B. Fernandes, M.A. Marovich, L.Shen, J. Mao, A.G. Pernet, *J. Med. Chem.*, 1989, **32**, 1062.

8 K. Matsuo, M. Kimura, T. Kinuta, N. Takai and K. Tanaka, *Chem. Pharm. Bull.*, 1984, **32** ,4197.

9 K. Matsuo, I. Kitaguchi, Y. Takata and K. Tanaka, *Chem. Pharm. Bull.,* 1980, 28, 2494

10 Z.Y. Wang and M.G. Moloney, *Tetrahedron Lett.,* 2002, **43**, 9629.

11 P.G. Bulger, M.G. Moloney and P.C. Trippier, *Org. Biomol. Chem.*, 2003, **1**, 3726.

ERUCIN, AN ISOTHIOCYANATE FROM ROCKET SALAD SPECIES, ENHANCES THE BIOLOGICAL ACTIVITY OF TRANSFORMING GROWTH FACTOR-β

A. Melchini,[1] M. Traka,[2] R. Mithen,[2] M.F. Taviano,[3] N. Miceli,[3] S. Catania,[4] C. Costa[4]

[1]Foundation "Prof. Antonio Imbesi", Messina, Italy.
[2]Institute of Food Research , Norwich Research Park, Norwich, United Kingdom.
[3]Pharmaco-Biological Department, School of Pharmacy, University of Messina, Italy.
[4]Interdepartmental Centre of Experimental, Environmental and Occupational Toxicology, University of Messina, Italy.
[1]E-mail: amelchini@pharma.unime.it

1 INTRODUCTION

Erucin is a volatile dietary isothiocyanate present in cruciferous vegetables, such as rocket salads (*Eruca sativa* Mill., *Diplotaxis* spp.), that has recently been considered a promising cancer chemopreventive phytochemical (Figure 1). It is produced by enzymatic hydrolysis of glucoerucin or by the reduction *in vivo* of the corresponding sulphoxide, sulforaphane, that is prevalent in broccoli (*Brassica oleracea* L. ssp. *italica*) (Figure 1).[1-3]

Literature data suggest that isothiocyanates can be considered natural chemotherapeutic agents in cancer treatment.[4] Several studies have showed that isothiocyanates can retard or prevent the process of carcinogenesis by multiple mechanisms perturbing the three major steps, initiation, promotion and progression, as well as the later stages, angiogenesis and metastasis.[5]

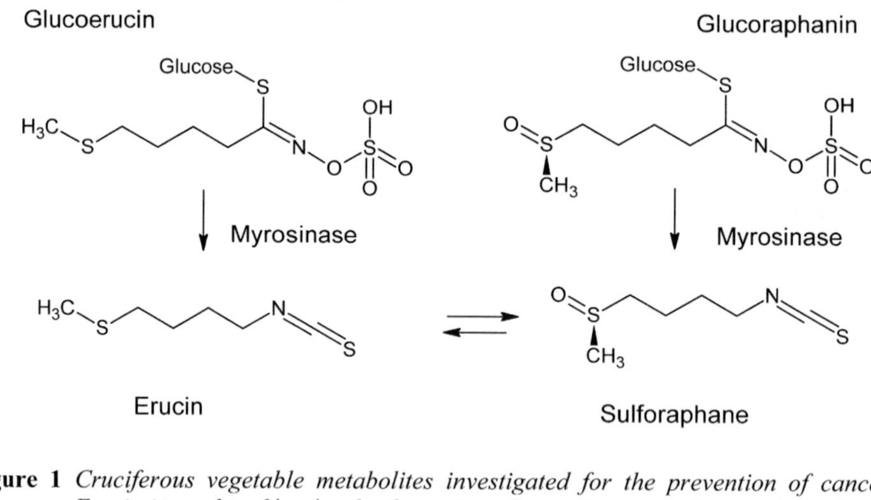

Figure 1 *Cruciferous vegetable metabolites investigated for the prevention of cancer. Erucin is produced* in vivo *by the action of myrosinase on glucoerucin or by the reduction of sulforaphane.*

Transforming growth factor-β (TGF-β) is a multifunctional cytokine that plays an important role in restraining mammalian cell growth by cytostatic and apoptotic actions under physiological conditions.[6] The TGF-β superfamily of ligands bind to membrane receptor serine and threonine kinases and promote the activation of Smad proteins that form high-affinity transcriptional complexes to control several hundreds of target genes.[7] Loss of TGF-β control is involved in the development and progression of cancer.[8]

The *in vivo* interconversion of the dietary isothiocyanates erucin and sulforaphane and their structural similarity has suggested similar biological activity. We have already demonstrated that sulforaphane can perturb the TGF-β signalling pathway and enhance TGF-β1/Smad-mediated transcription.[9] In the present study we investigate the *in vitro* effects of erucin on TGF-β signalling, analyzing its secretion and its effect on Smad-mediated transcription in human lung adenocarcinoma A549 cells.

2 THE EFFECT OF ERUCIN ON ONCOGENIC SIGNALLING PATHWAYS

2.1 Erucin has No Effect on TGF-β1 Secretion

The human A459 cell line was used as an *in vitro* system to investigate the lung cancer chemoprevention by erucin and provide additional mechanistic information in relation to its effects on the TGF-β signalling pathway. A549 human adenocarcinoma lung epithelial cells were obtained from the European Collection of Cell Cultures (ECACC 86012804) and were cultured as monolayers in Dulbecco's Modification of Eagle's medium which was supplemented with 10% foetal calf serum and incubated at 37° C in a humidified atmosphere containing 5% CO_2.

The A549 cells then were incubated for one day in serum-free media with a range of concentrations of erucin from 0 to 10 μM and aliquots of the supernatant were withdrawn and acidified with 1M HCl for ten minutes to convert latent TGF-β1 to the immunoreactive form. Quantitative determinations of the activated human TGF-β1 concentration in the cell culture supernatants were performed using the Quantikine® system and the optical density of the solutions was determined in a spectrophotometer at 450 nm. The limit of detection of the immunoassay is 4.61 pg/ml of TGF-β1.

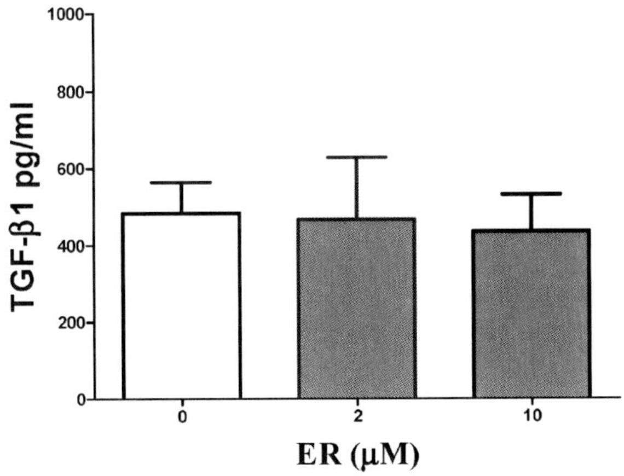

Figure 2 *Effects of erucin (ER) on the secretion of TGF-β1 by A549 cells.*

The results obtained from the assay of culture supernatants are given in graphical form in Figure 2 and show that there are no significant differences between the secretion of TGF-β1 by A549 cells in basal conditions (~500 pg/ml) and after treatment with low concentrations (2 and 10 μM) of erucin (~ 450 pg/ml).

2.2 Enhancement of TGF-β1/Smad Mediated Transcription by Erucin

Having established that erucin does not affect the secretion of TGF-β1, we sought other mechanisms by which erucin might modulate the transcription of Smad proteins. We designed a luciferase reporter gene assay which would reveal the extent of Smad transcription by using the following conditions. Mouse embryonic fibroblast cell NIH 3T3 cells stably transfected with a CAGA12-luc plasmid, which responds to Smad activation, were cultured in Dulbecco's Modification of Eagle's medium supplemented with 10% foetal calf serum, 1% penicillin, 1% streptomycin, 1% L-glutamine and 0.4 mg/ml geneticin. The cells were treated with TGF-β1 (2 ng/ml) alone, TGF-β1 (2 ng/ml) and 10 mM dithiothreitol (DTT), which disrupts the active TGF-β1 dimer, or TGF-β1 (2 ng/ml) and erucin (2 and 10 μM). All samples were pre-incubated for 30 minutes at 37° C. Control cells received no treatment or treatment with erucin alone. Chemiluminescence was measured with a commercially available kit and normalised to the protein concentration of each sample.

The results of these experiments are shown in Figure 3. Exposure of cells to TGF-β1 induced luciferase activity demonstrating Smad-mediated transcription as expected and the addition of dithiothreitol inhibited this response. But when cells that had been pre-incubated with erucin (2 and 10 μM) were exposed to TGF-β1, there was an increase in Smad-mediated transcription compared to exposure to TGF-β1 alone. Exposure of cells to erucin alone did not result in enhanced transcription (data not shown) suggesting that erucin induces Smad activation indirectly.

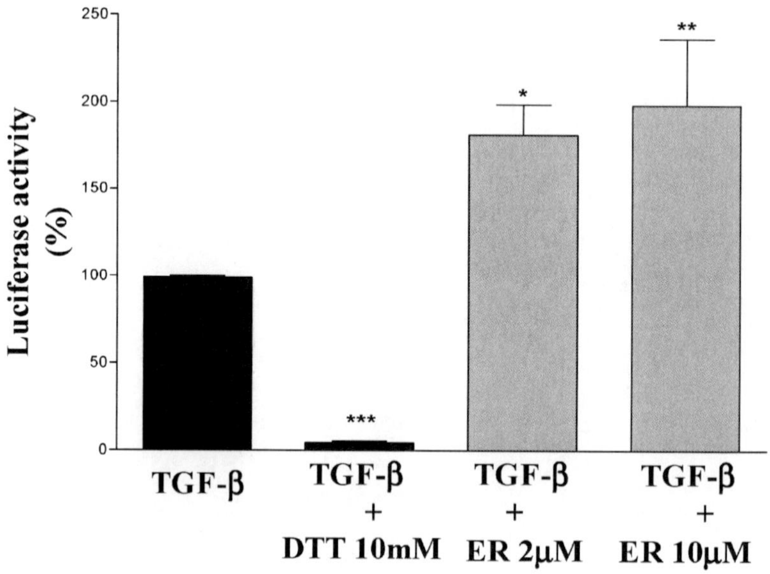

Figure 3 *Enhanced activation of TGF-β1/Smad mediated transcription by erucin (ER).*

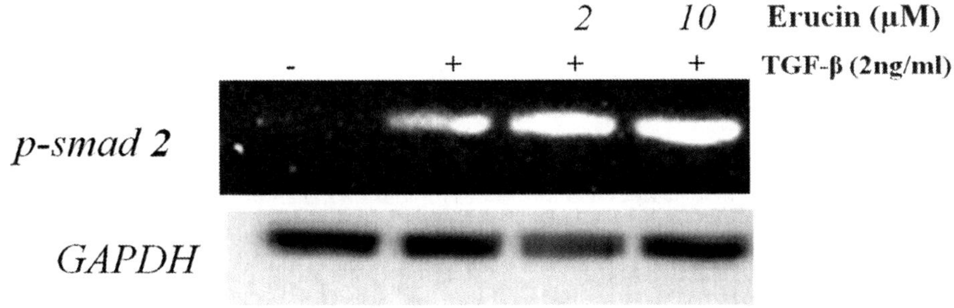

Figure 4 *Modulation of TGF-β1 induced phospho-Smad 2 protein expression by erucin.*

2.3 Modulation of Phospho-Smad 2 Protein Expression by Erucin

As phospho-Smad2 proteins have been reported to play an important role in TGF-β1 transcription, Western blotting analysis was performed to measure phospho-Smad2 protein levels in A549 cells treated with TGF-β1 (2 ng/ml) and erucin in serum-free media for 1 hour. Anti-phospho-Smad2 antibody was purchased from Cell Sig. Tech and GAPDH was used as loading control

Our results (Figure 4) show the transcription of Smad2 in response to TGF-β1 addition, but a significant increase in phospho-Smad2 protein expression when the sample was incubated with both TGF-β1 and erucin. The enhancement in the transcription of Smad proteins by erucin does not appear to be dose-dependent at the concentrations tested.

3 CONCLUSION

A growing body of clinical evidences indicates that several growth factors play an important role in restraining mammalian cell growth under physiological conditions.[10] Loss of the physiological control exerted by these growth factors which results from alterations to their signalling pathway cascade is strongly associated with cancer development and progression.[11]

Recent observations have provided consistent evidence that consumption of cruciferous vegetables has a protective role against lung cancer.[12] It is widely believed that the health benefits of diets high in cruciferous vegetables are associated with the presence of isothiocyanates, that exert their chemopreventive effect by several mechanism of actions.[13]

Results obtained in our study indicate that erucin does not affect TGF-β1 secretion by human lung adenocarcinoma A549 cells at the concentrations tested. However, we observed a significant increase in Smad-mediated transcription when NIH 3T3 cells were pre-incubated with erucin and exposed to TGF-β1 compared to those exposed to TGF-β1 alone. Data obtained from Western blotting analyses confirmed an enhancement in TGF-β mediated pSmad2 protein expression by erucin.

On the basis of the fundamental role of TGF-β1 as a growth-inhibitory cytokine in the earlier stages of cancer progression, these results suggest that the enhancement of

TGFβ1/Smad-mediated transcription by dietary erucin could be one mechanism involved in the prevention of cancer by isothiocyanates.

Acknowledgements

This work was supported by Biotechnology and Biological Sciences Research Council (BBSRC) and Foundation "Prof. Antonio Imbesi" (Messina, Italy).

References

1 R. Gmelin, M. Schluter, *Arch Pharm Ber Dtsch Pharm Ges*, 1970**, 303**, 330.
2 O. Leoni, R. Iori, S. Palmieri, E. Esposito, E. Menegatti, R. Cortesi, C. Nastruzzi, *Bioorg Med Chem*, 1997, **5**, 1799.
3 K. Kassahun, M. Davis, P. Hu, B. Martin, T. Baillie, *Chem Res Toxicol* 1997, **10**, 1228.
4 F. Bianchini, H. Vainio, *Drug Metab* 2004, **36**, 655.
5 J.D. Clarke, R.H. Dashwood, E. Ho, *Cancer Lett*, 2008, **269**, 291.
6 P.M. Siegel, J. Massagué, *Nature Review Cancer*, 2003, **3**, 807.
7 B. Schmierer, C.S. Hill, *Nature Review Molecular Cell Biology*, 2007, **8**, 970.
8 J. Massagué, *Cell*, 2008, **134**, 215.
9 M. Traka, A.V. Gasper, A. Melchini, J.R. Bacon, P.W. Needs, V. Frost, A. Chantry, A.M. Jones, C.A. Ortori, D.A. Barrett, R.Y. Ball, R.D. Mills, R.F. Mithen, *PLoS ONE*, 2008, **3**, e2568.
10 V. Ellenrieder, *Anticancer Res*, 2008, **28**, 1531.
11 K.J. Gordon, G.C. Blobe, *Biochim Biophys Acta*, 2008**, 1782**, 197.
12 A. Sapone, A. Affatato, D. Canistro, L. Pozzetti, M. Broccoli, J. Barillari, R. Iori, M. Paolini, *Mutat Res*, 2006, **635**, 146.
13 Y.S. Keum, W.S. Jeong, A.N. Kong, 2005, **18**, 445.

ADVANCES IN PHARMACEUTICAL APPLICATIONS OF IMINOSUGARS

Robert J. Nash

Phytoquest Limited, Plas Gogerddan, Aberystwyth, Ceredigion, SY23 3EB, UK

1 INTRODUCTION

The iminosugars are a widespread group of plant and microbial compounds that have been attracting interest due to their ability to interact with human glycosidases, other proteins and sugar receptors. In their simplest form they resemble furanose and pyranose monosaccharides with a nitrogen replacing the oxygen in the ring (Figure 1). There are five N-containing ring structures that are most common in nature: pyrrolidine, piperidine, pyrrolizidine, indolizidine and *nor*-tropane (Figure 2). These compounds have been given several names in the past in the literature, including aza-sugars, glycosidase inhibitors and sugar analogues and the variety and distribution of natural compounds have been reviewed.[1-3] There are now around 200 reported as natural products.[4] However, very few have been widely available for drug discovery and most research has concentrated on the first compounds discovered, i.e. deoxynojirimycin (DNJ), castanospermine, swainsonine, fagomine, 1,4-dideoxy-1,4-imino-D-arabinitol (DAB) and derivatives and variants of these compounds.

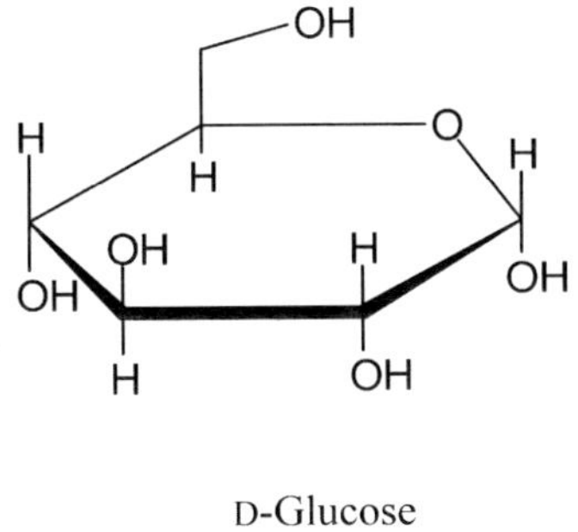

D-Glucose

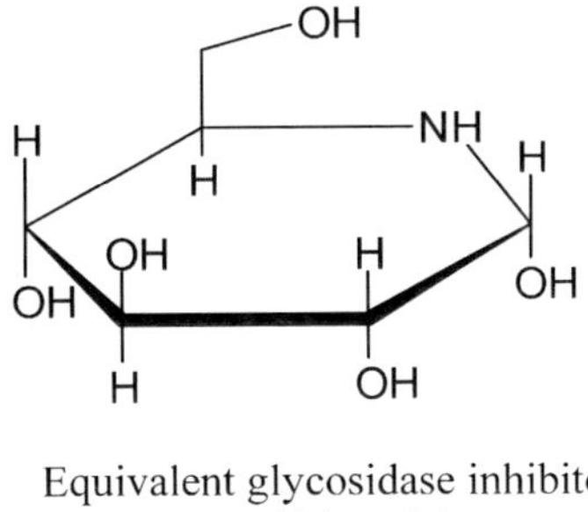

Equivalent glycosidase inhibitor (Nojirimycin)

Figure 1 *The basic structure of iminosugars.*

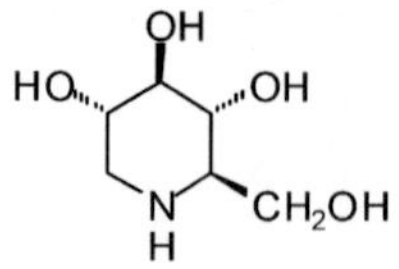

Piperidine
e.g. DNJ

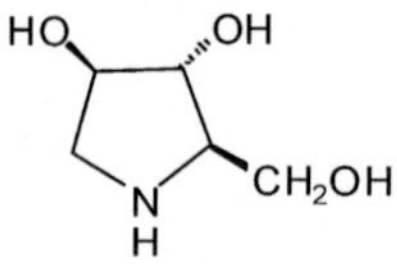

Pyrrolidine
e.g., DAB

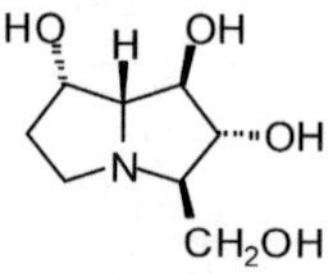

Pyrrolizidine
e.g., australine

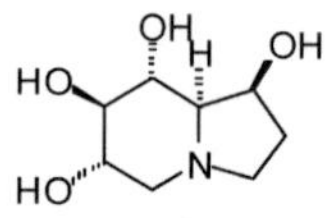

Indolizidine
e.g., castanospermine

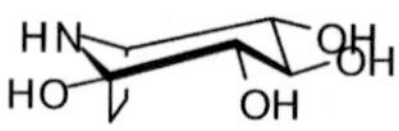

Nor-tropane
e.g., calystegine B_2

Figure 2 *The most common natural structural classes of the iminosugars.*

There has been a long-standing reticence within mainstream Pharmaceutical companies to engage with carbohydrate-like hits and leads, which are often deemed to be too polar for delivery and bioavailability purposes. Glycobiology has had a reputation for being technically difficult but it offers great potential in pharmaceutical development since glycosylation is crucial to how the human body functions. With the productivity of the drug discovery process having declined significantly over the last decades and the number of new chemical entities entering the clinic declining in contrast to ever increasing costs, there is a need and an opportunity, to bring to the table less conventional approaches to the discovery and development of new small molecules with pharmacological activity. In essence, the iminosugars represent a privileged scaffold and the timing is perfect because of an increased understanding of the structures and functions of carbohydrates. The challenge is to tune this bioactive template for specific therapeutic applications.

2 NEW GENERATION IMINOSUGARS WITH ENHANCED EFFICACY AND SPECIFICITY

Iminosugars display a wide range of biomedically relevant activities: anti-viral properties [against HIV-1, herpes simplex virus (HSV), bovine viral diarrhoea virus (BVDV) and hepatitis C virus (HCV)],[5] anti-diabetic properties (Miglitol, Glyset has been in the clinic for some time),[6] for use in lysosomal storage disorders, such as Gaucher's and Niemann-Pick type C disease (*N*-butyl deoxynojirimycin, Miglustat, marketed as Zavesca)[7,8] and immune modulation and anti-cancer activity.[3,9] Due to the great importance of carbohydrates in mammalian biochemistry and disease progression, it is clear that imino sugars have potential as small molecule therapeutics in a wide range of disease areas in addition to their current successes. However, the lack of specificity of the agents that have been tested in the clinic so far and the difficulties involved in isolating new iminosugars, mean that new approaches are required to exploit their potential.

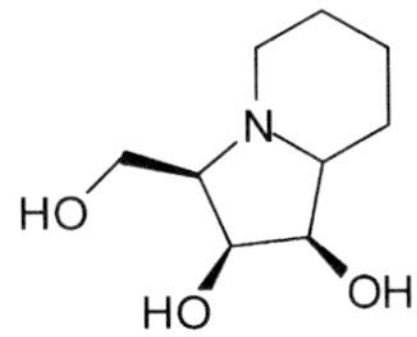

Figure 3 *Steviamine.*

The identification of novel natural iminosugars and the synthesis of new active structures have been slow up to now, even with the exciting therapeutic effects seen with DNJ, castanospermine and the other early iminosugar swainsonine. There are a number of reasons for the slow development which include their high water solubility and the lack of a chromophore, which increases the technical challenges of working with these compounds. In plants and microbes, iminosugars are also masked by high concentrations of sugars and amino acids. A great step forward has been made recently with the realization that glycosidase inhibition is not necessarily required for their therapeutic activity.[9-11] Many of the iminosugars isolated or synthesized and disregarded due to lack of glycosidase inhibition, consequently deserve further evaluation. Indeed, lack of glycosidase inhibition may reflect the limited range of assays readily available and used; for example, one of the more interesting iminosugars isolated recently, steviamine (Figure 3),[12] is very specific in inhibiting an α-N-acetylgalactosaminidase that is rarely assayed but has important roles in the body.[13]

3 HISTORY OF NATURAL IMINOSUGARS AS THERAPEUTIC AGENTS

The first natural iminosugar, nojirimycin, was discovered in Japan in 1966 as a novel antibiotic from a *Streptomyces* isolate.[3] The closely related compound, 1-deoxynojirimycin (DNJ), was isolated from mulberry in 1970 by workers studying the anti-diabetic activity of the plant. DNJ was found to be a potent inhibitor of the enzyme family of α- and β-glucosidases.

Medicinal chemistry was applied to the DNJ sugar template to produce *N*-hydroxyethyl-DNJ (Miglitol) (Figure 4), which was successfully developed as an anti-diabetic (Glyset®) by Bayer and was the first iminosugar drug to reach the market. Since 1996 it has been approved for non-insulin-dependent diabetes mellitus where it reduces the rate of complex carbohydrate digestion, thereby controlling the absorption of glucose after meals and preventing hyperglycaemia.

DNJ was also adapted by Oxford University and Oxford Glycosciences, to enhance cell uptake, to produce *N*-butyl-DNJ (Zavesca®) (Figure 4), which is used for the treatment of type I Gaucher's disease and Niemann-Pick type C disease, lysosomal storage disorders. There was no prior treatment for Niemann-Pick type C disease and prior to Zavesca® patient management was restricted to symptom relief. Zavesca® functions by inhibiting an enzyme involved in the glucosylation of many sphingolipids. This decreases the excessive cellular storage of glycolipids in neural tissue. Despite its efficacy, Zavesca® has also been associated with side effects, which has had an impact on the product's development.[3,4] *N*-Butyl-DNJ was also found to be effective at inhibiting HIV infectivity *in vitro* by blocking viral envelope glycoprotein trimming. This means that the viral particles produced in the

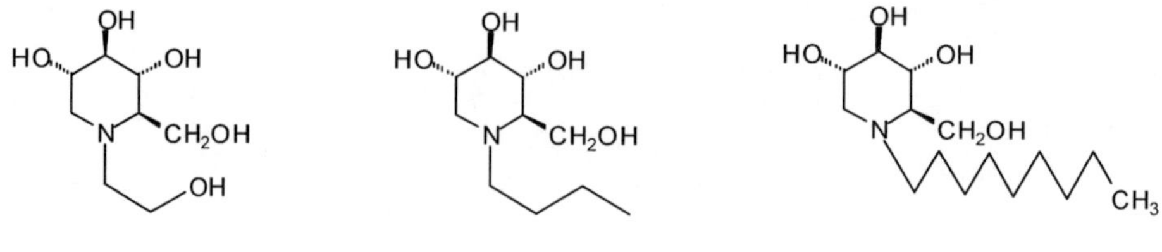

Figure 4 *Derivatives of deoxynojirimycin.*

presence of the imino sugar are non-infectious as the conformation of their envelope protein has been altered and they can no longer bind to their respective cell receptor. *N*-Butyl-DNJ was evaluated by Searle-Monsanto in a Phase II clinical trial and although antiviral effects were observed, there were side effect issues again – this time a high prevalence of GI tract complications, attributable to the non-specific activity of the drug. In addition, high serum levels of the drug were required for efficacy.

Castanospermine (Figure 2), isolated from the Australian tree *Castanospermum australe*, was developed by Merrell Dow who, in order to increase its cell uptake, modified the sugar by the addition of a butanoyl group to form the pro-drug known as Celgosivir (6-*O*-butanoyl-castanospermine). Celgosivir was shown to have anti-HIV activity and was evaluated in a Phase II clinical trial. However, similarly to *N*-butyl-DNJ, the clinical development was problematic, due to the compound's toxicity profile and competition from other less toxic anti-HIV drugs.

Celgosivir was also shown to have antiviral activity against other viruses including CMV, influenza and the hepatitis C virus (HCV) surrogate BVDV, *in vitro*.[3,5] Despite advances in antiviral therapeutics, hepatitis C virus (HCV) infection continues to be a major worldwide health concern. In the search for newer agents with novel mechanisms of action, such as compounds which target virus-specific enzymes, inhibition of alpha-glucosidase I is considered an attractive anti-HCV strategy since this enzyme is involved in the biosynthesis of glycoproteins that, when expressed on the viral surface, are essential for virus-host interactions. Castanospermine is an α-glucosidase I inhibitor with marked antiviral activity against a number of viruses. Unfortunately, the agent also inhibits intestinal sucrases and causes osmotic diarrhea. In contrast, celgosivir is a relatively inactive inhibitor of intestinal sucrases and appears to be non-toxic to the gastrointestinal tract. It possesses antiviral activity that is 30-fold greater than the parent compound, its active metabolite. Celgosivir has displayed potent antiviral activity *in vitro* and *in vivo* against several viruses, including HIV-1, herpes simplex virus (HSV), bovine viral diarrhea virus (BVDV) and HCV, and the agent was chosen for further development as a treatment for HCV infection. The antiviral efficacy and safety of celgosivir were demonstrated in clinical trials in HIV-1-infected patients.[5]

N-nonyl-deoxynojirimycin (N,N-DNJ) has an additional anti-viral mode of action against HCV by inhibiting formation of the p7 ion channel. HCV p7 is capable of forming

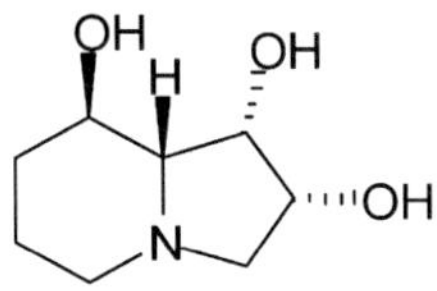

Figure 5 *Swainsonine.*

cation selective ion channels in planar lipid bilayers. *In vitro*, HCV p7 ion channel activity can be inhibited by a range of compounds including NN-DNJ.[14] *N*-(*n*-Nonyl)-deoxygalactonojirimycin (N,N-DGJ) can also reduce the amount of HBV DNA produced in culture and has been under consideration for development as a human therapeutic. N,N-DGJ does not appear to inhibit HBV DNA polymerase activity or envelope antigen production.[15]

Swainsonine (Figure 5) has been isolated from both plants and microorganisms.[3] It is a potent inhibitor of α-mannosidases and inhibits α-mannosidase II activity in the *N*-glycan biosynthesis pathway, blocking production of complex-type oligosaccharides. Swainsonine has been of great use in the study of *N*-glycan functions with many important results published since its discovery. The anti-tumour activity of swainsonine has also been previously examined. Swainsonine exhibits not only cytotoxicity, but inhibits cancer cell metastasis, decreases the toxicity of chemotherapeutic drugs and works as immunomodulator. Despite its side effects (mainly vacuolation caused by inhibition of lysosomal mannosidases), clinical studies on patients have shown that swainsonine is of some benefit as a chemotherapeutic drug, suggesting that it might have further applications in this field.[3,16]

The pyrrolidine iminosugar DAB (Figure 2), isolated initially from the African legume species *Angylocalyx braunii*, is an inhibitor of glycogen phosphorylase but is probably not an ideal therapeutic agent itself due to potent inhibition of α-glucosidases.[17,18] DAB was shown to be nearly completely orally available in rats and dogs and it can reduce glucagon-induced and spontaneous hyperglycaemia. Inhibition of hepatic glycogen phosphorylase may benefit glycaemic control in patients with type 2 diabetes. It is of interest that iminosugars have recently started to arouse wider interest as potential anti-diabetic and weight control agents through activities not related to inhibition of glucosidases.[19]

Some of the first generation imino sugars have been shown to be potent agonists of the human glucose sensor, sodium/glucose cotransporter type 3 (hSGLT3).[20] Potent hSGLT3 activation was shown by the imino sugars (DNJ), *N*-hydroxylethyl-DNJ (miglitol), *N*-butyl-DNJ, *N*-ethyl-DNJ and DNJ-1-sulfonic acid, with $K_{0.5}$ values of 0.5 to 9 μM. 1-Deoxygalactonojirimycin (DGJ) activated hSGLT3 with a $K_{0.5}$ value of 11 mM, a 3000-fold less potent interaction than was observed for DNJ (4 μM).

4 A NEW GENERATION OF IMINOSUGAR THERAPEUTIC AGENTS

Iminosugars have an excellent drug profile – they are small and simple in structure, stable and are orally available. Of those studied to date, *in vivo* they are not metabolised or incorporated and are excreted unchanged primarily in urine. There is no doubt that many of the natural iminosugars will eventually find therapeutic applications once they are evaluated in suitable assays. Casuarine is a good example. It is a pentahydroxylated pyrrolizidine alkaloid with the unusual hydroxymethyl moiety at C-3 of the pyrrolizidine iminosugars (Figure 6).

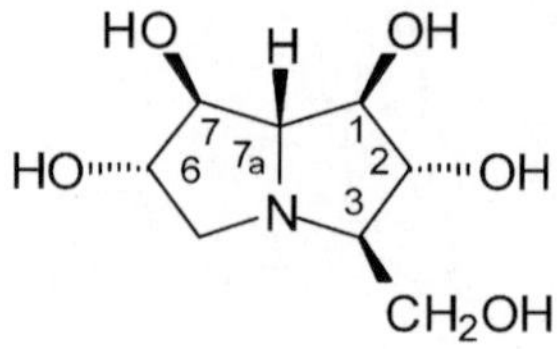

Figure 6 *Casuarine.*

It was first found as the major alkaloid of *Casuarina equisetifolia* bark in 1994.[21,22] It is a good glucosidase inhibitor but subsequently was also found to be able to prime the immune system for an increased type-1 immune response, increasing levels of cytokines such as IL-2, IL-12, and IFN-γ, and it has potent activity in the B16/F10 melanoma in mice given a single oral dose of 100μg/ml. What was particularly interesting was that glucosidase inhibition was not necessary for the immunological activity.[9,23]

In addition to the possibilities of increasing or correcting a deficient immune response to cancer cells using immune modulators such as casuarine, the iminosugars can be selective competitive and non-competitive inhibitors of carbohydrate manipulation enzymes, such as glycosidases, which are involved in tumour cell invasion and migration.[24] Such enzymes are also responsible for the attachment of oligosaccharides to the cell surface of tumour cells, displayed as glycoproteins, glycolipids, and proteoglycans, which play an important role in malignant phenotype and tumour growth. Furthermore, cancer cells show an extremely active lysosomal system which is reflected by enhancement of glycoprotein turnover. Iminosugars interact with glycosyl hydrolases responsible for this kind of action in cancer cells and thus introduce a new compound class into the research field of finding new anti-cancer activities. The specificity of enzyme inhibition that is becoming apparent with natural and synthetic variants makes it likely that the exciting activities of compounds such as swainsonine can be utilized without the off-target activities.

Pharmacological chaperones act on misfolded, unstable mutant proteins that exhibit residual biological activity. The ability of iminosugars to act as pharmacological chaperones appears to be opening up new therapeutic possibilities. The compounds can bind to the catalytic sites of the deficient enzymes and lead to improved trafficking in the endoplasmic reticulum, resulting in improved enzyme function. Fabry disease is an X-linked lysosomal storage disorder caused by mutations in the gene encoding alpha-galactosidase A (alpha-Gal A), with consequent accumulation of its major glycosphingolipid substrate, globotriaosylceramide (GL-3). Over 500 Fabry mutations have been reported; approximately 60% are missense. The iminosugar 1-deoxygalactonojirimycin (DGJ, migalastat hydrochloride, AT1001) is a pharmacological chaperone that selectively binds alpha-Gal A, increasing physical stability, lysosomal trafficking, and cellular activity.[25] Importantly, the chaperoning activity of DGJ is observed at concentrations which do not inhibit the alpha-Gal A enzyme or other galactosidases. DGJ sometimes occurs alongside DNJ in plants and is found as various C-1 derivatives in the Chinese medicinal plant *Adenophora triphylla* (Campanulaceae), e.g. β-1-*C*-Butyl-DGJ and β-1-*C*-butenyl-1-DGJ (Figure 7).[26] *N*-butyl-DNJ also acts as a pharmacological chaperone for the deficient lysosomal α-glucosidase causing Pompe's disease.[27]

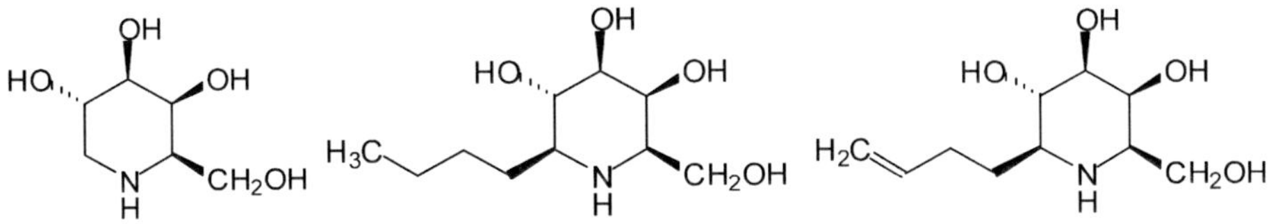

Figure 7 *DGJ, β-1-C-Butyl-DGJ and β-1-C-butenyl-1-DGJ.*

Other pharmacological chaperones have been designed based on natural iminosugar templates. An example is isofagomine (Plicera) for treatment of Gaucher's disease (Figure 8). The natural product fagomine occurs in Mulberry (Morus species) alongside DNJ, is a weak inhibitor of glucosidases, and potentiates insulin release.[3] Isofagomine binds selectively to N370S glucocerebrosidase (deficient in Gaucher's disease) and increases activity three-fold.[28] One potential problem with isofagomine is that it also inhibits glycogen phosphorylase.

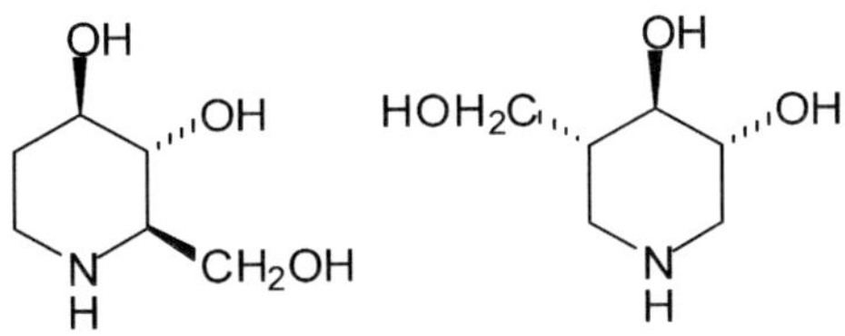

Figure 8 *Fagomine and isofagomine.*

The competitive glucosidase inhibitor DAB (Figure 2) and the non-competitive glucosidase inhibitor enantiomer LAB[29] have been used as templates to produce inhibitors of β-N-acetylglucosaminidases.[30] 2-Acetamido-1,4-imino-1,2,4-trideoxy-**L**-arabinitol (LABNAc) was synthesised along with the *N*-benzyl and *N*-butyl analogues and they were found to be potent inhibitors. The enantiomers DABNAc (Figure 9) and *N*-Bn-DABNAc were also synthesized. The **L**-iminosugar LABNAc and its derivatives were found to be potent noncompetitive inhibitors of some β-*N*-acetylhexosaminidases, while the **D**-iminosugar DABNAc and its derivatives were found to be weaker competitive inhibitors. These results support the previous studies postulating that **D**-iminosugar mimics inhibit **D**-glycohydrolases competitively, and that their corresponding and usually un-natural **L**-enantiomers show noncompetitive inhibition of these enzymes. Molecular modelling studies confirmed that the spatial organisation in enantiomeric inhibitors leads to a different overlay with the monosaccharide substrate.[30]

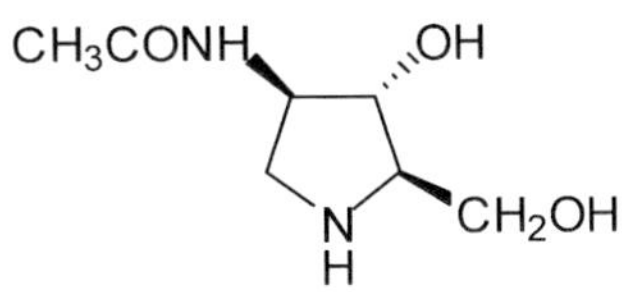

Figure 9 *DABNAc.*

N-acetylhexosaminidases are of considerable importance in mammals and are involved in various significant biological processes. In humans, deficiencies of these enzymes in the lysosome, resulting from inherited genetic defects, cause the glycolipid storage disorders Tay-Sachs and Sandhoff diseases. One promising therapy for these diseases involves the use of β-*N*-acetylhexosaminidase inhibitors as chemical chaperones to enhance the enzyme activity above sub-critical levels. Initial cell-based studies suggest that *N*-Bn-LABNAc can act as a chaperone to enhance the deficient enzyme's activity to levels that may cause a positive pharmacological effect. LABNAc, *N*-Bn-LABNAc, and *N*-Bu-LABNAc are potent and selective inhibitors of β-*N*-acetylhexosaminidase and as such may be useful as therapeutic agents for treating adult Tay-Sachs and Sandhoff diseases.[30]

While iminosugar pharmacological chaperones that bind to catalytic sites hold potential as treatment of disorders arising from aberrant protein folding, including in particular lysosomal storage diseases, there are now indications that some iminosugars can have chaperoning activity *via* binding to other sites on these enzymes. These compounds could have the advantage of being free of concerns over possible inhibition of non-target glycosidase enzymes.[11] IsoDAB and isoLAB (isomers of DAB and LAB) (Figure 10) have been synthesized.[10] IsoDAB was a potent and specific inhibitor of a number of α-glucosidases. In contrast, isoLAB showed no significant inhibition of any glycosidase. DAB is a good inhibitor of glycogen phosphorylase[18] and a moderate inhibitor of glycoprotein processing glucosidases;[3] neither isoDAB or isoLAB showed any inhibition of these enzymes. The combination of potency and specificity of isoDAB as an α-glucosidase inhibitor may provide a useful agent in the study of diabetes.

The cystic fibrosis transmembrane conductance regulator (CFTR) is a protein involved in the transport of chloride ions across cell membranes which is mutated in cystic fibrosis. *N*-butyl-DNJ partially rescues the defective F508del-CFTR function in CF-KM4 cells and thus may have potential for the chemotherapeutic treatment of cystic fibrosis (CF);[31] calnexin may be a therapeutic target for miglustat in CF. Although isoLAB showed no significant inhibition of any glycosidase, in preliminary experiments isoLAB also rescued CFTR function and thus is likely to be of value in the investigation of CF. This may be another example of an imino sugar that apparently acts as a chaperone but is not a glycosidase inhibitor.

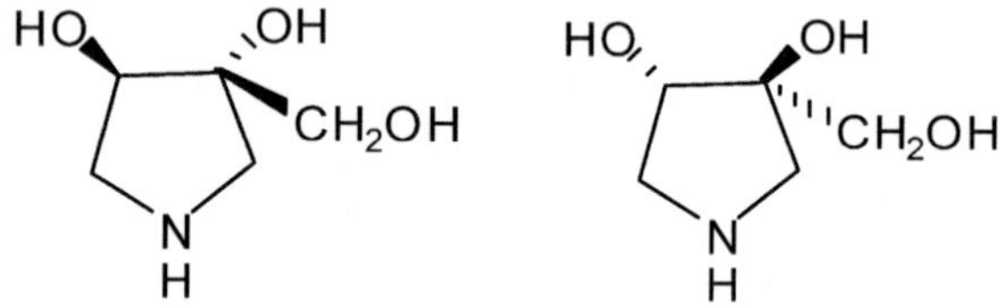

isoDAB isoLAB

Figure 10 *IsoDAB and isoLAB.*

5 IMINOSUGARS IN BIOACTIVE PLANTS AND MICROBIAL CULTURES AND FUTURE DRUG DISCOVERY

One of the big gaps in herbal product knowledge is in their water soluble components, as these have been the most difficult to analyse. This is a pity because most herbs were traditionally prepared in water. Very few commercial herbal products are prepared in water and water- soluble components are rarely identified. It is also true that drug discovery programmes using plants have typically used solvents such as methanol, chloroform or hexane for extraction. In the case of microbial fermentations, the high concentrations of water-soluble media components often make it difficult to isolate water-soluble secondary metabolites. It is becoming clear that iminosugars are common components of plants and some have been reported from bacteria and fungi.[3] It is probable that many more microorganisms produce them than are currently documented.

It seems likely that plants produce iminosugars to protect their carbohydrate products produced *via* photosynthesis. Microorganisms probably produce them to reduce competition from other microorganisms by inhibiting their glycosidases. The glycosidase inhibitors may also help to regulate endogenous glycosidase activity. However, many of the natural products isolated in recent years are not inhibitors of any glycosidase tested as yet. This suggests that these compounds can have other functions and this is perhaps supported by the discovery of chaperoning activity and the ability of some agents to modulate the immune response of mammals directly and indirectly.[23]

One of the most surprising facts about herbal medicines is that we still do not know that the active components of most of them work nor how exactly they work. The same is true of most fruit and vegetables, as the main nutrients are known but many health promoting properties seem to be due to unidentified components. Great importance is given to different plant components at different times as knowledge increases; anti-oxidants, for example, have aroused great interest, but there is almost no clinical evidence supporting their health benefits. Carbohydrates are of great importance in mammalian biochemistry. They help to determine the three-dimensional structure of proteins which is inherently linked to their function and efficacy. Iminosugars can act by modifying carbohydrate processing, as chaperones to produce otherwise defective glycoproteins, by inhibiting glycosidases elevated by tumour cells, controlling cancer cell glycosylation, by specifically altering viral and bacterial cell walls (reducing infectivity, the number of glycoforms and altering immune responses) and by binding to carbohydrate receptors. Due to their several mechanisms of action they provide great flexibility to the class and it may well be that the elusive active components of many herbal products could be iminosugars.

The iminosugars have potential for great stereoselectivity in biological activity. Such selectivity is due to the chemical and biological diversity in terms of structural information of small sugars. For instance, there are 2^9 stereoisomers in sucrose alone, and even glucose has $2x2^5$ possible isomers in the pyranose and furanose forms. This amazing diversity in such small molecules displays a remarkable economy in structural information in nature, completely surpassing in molecular weight terms anything achieved by amino acids. Most plants producing iminosugars produce complex mixtures of related compounds but with diverse and often very specific activites. To realise the potential of iminosugars from plants and microorganisms the individual compounds need to be isolated and identified for bioassays. Modification of the natural products by synthesis and semi-synthesis has shown great success. The increased knowledge of carbohydrate biochemistry and improved technology should mean that this field of drug discovery will now move fast.

References

1. R.J. Nash, A.A. Watson, A.L. Winters, G.W.J. Fleet, M.R. Wormald, S. Dealler, E. Lees, N. Asano, H. Kizu, in *Phytochemical Diversity: A Source of New Industrial Products*, ed. S. Wrigley, M. Hayes, R. Thomas and E. Chrystal. Royal Society of Chemistry, Cambridge, 1997, p 106.
2. A.A. Watson, R.J. Nash, in *Biodiversity: New Leads for the Pharmaceutical and Agrochemical Industries,* ed. S.K. Wrigley, M.A. Hayes, R. Thomas, E.J.T. Chrystal and N. Nicholson. Royal Society of Chemistry, Cambridge, 2000, p 169.
3. A.A. Watson, G.W.J. Fleet, N. Asano, R.J. Molyneux, R.J. Nash, 2001, *Phytochem.,* **56,** 265.
4. N. Asano, *Cell. Mol. Life Sci.,* 2009, **66**, 1479.
5. L.A. Sorbera, J. Castaner, L. Garcia-Capdevila, 2005, *Drugs of the Future*, **30**, 545.
6. L.J. Scott, 2000, *Drugs*, **59**, 521.
7. T.D. Butters, 2007, *Curr. Opin. Chem. Biol.*, **11**, 412.
8. Medical News Today, 2009, July 21st.
9. A.A. Watson, R.J. Nash, E. Evinson, 2004, PCT/GB2004/000198, Immunomodulatory compositions (WO 04/064715).
10. D. Best, S.F. Jenkinson, A.W. Saville, D.S. Alonzi, M.R. Wormald, T.D. Butters, C. Norez, F. Becq, Y. Bleriot, I. Adachi, A. Kato, G.W.J. Fleet, *Tetrahedron Lett.*, 2010, **51**, 4170.
11. A. Kawamura, A. Roach, A.F. Wilson, J. Tinsley, R. Storer, R.J. Nash, 2008, PCT/GB2008/003885, Treatment of Protein Folding Disorders (WO/2009/066069).
12. A.L. Thompson, A. Michalik, R.J. Nash, F.X. Wilson, R. van Well, P. Johnson, G. W.J. Fleet, C.-Y. Yu, X.-G. Hu, R.I. Cooper and D.J. Watkin, *Acta Cryst.,* 2009, ***E65***, o2904
13. X.-G. Hu, B. Bartholomew, R.J. Nash, F.X. Wilson, G.W.J. Fleet, S. Nakagawa, A. Kato, Y.-M. Jia, R. van Well, and C.-Y. Yu, *Org. Lett.* 2010, ***12***, doi. 10.1021/ol1007718.
14. D. Pavlovic, D.C. Neville, O. Argaud, B. Blumberg, R.A. Dwek, *et al.*, *Proc. Natl. Acad. Sci. USA.*, 2003, 100, 6104.
15. X. Lu, T. Tran, E. Simsek, T.M. Block, *J. Virol.*, 2003, **77**, 11933.
16. J. Hamaguchi, H. Nakagawa, M. Takahashi, T. Kudo, N. Kamiyama, B. Sun, T. Oshima, Y. Sato, K. Deguchi, S. Todo, S-I. Nishimura, *Mol. Cancer*, 2007, **6**, 58.
17. G.W.J. Fleet, S.J. Nicholas, P.W. Smith, S.V. Evans, L.E. Fellows and R.J. Nash, *Tetrahedron Lett.*, 1985, **26**, 3127.
18. P. Mackay,L. Ynddal, J.V. Anderson, J.G. McCormack, *Diabetes Obes. Metab.*, 2003, **5**, 397.
19. W-H. Kong, S-H. Oh, Y-R. Ahn, K-W Kim, J-H. Kim, J-H., S-W. Sec, *J. Agric. Food Chem.* 2008, **56**, 2613-2619.
20. A.A. Voss, A. Diez-Sampedro, B.A. Hirayama, D.D.F. Loo, E.M. Wright, *Mol. Pharmacol.*, 2007, **71**, 628.
21. R.J. Nash, P.I. Thomas, R.D. Waigh, G.W.J. Fleet, M.R. Wormald, P.M. de Q. Lilley, D.J. Watkin, *Tetrahedron Lett.,* 1994, **35**, 7849.
22. M.R. Wormald, R.J. Nash, A.A. Watson, B.K. Bhadoria, R. Langford, R.M. Sims, G.W.J. Fleet, *Carbohydr. Lett.*, 1996, **2**, 169.
23. R.J. Nash, 2008, patent WO/2008/009894 Use of imino sugars in immunotherapy (GB2007/002597).

24. T.M. Wrodnigg, A.J. Steiner, B.J. Ueberbacher, *Anti-Cancer Agents Med. Chem.*, 2008, **8**, 77.
25. E.R. Benjamin, J.J. Flanagan, A. Schilling, L. Agarwal, E. Katz, X. Wu, C. Pine, B. Wustman, R.J. Desnick, D.J. Lockhart, K.J. Valenzano, *J. Inherit. Metab. Dis.*, 2009, **32**, 424.
26. A. Kato, Y. Yamashita, S. Nakagawa, Y. Koike, I. Adachi, J. Hollinshead, R.J. Nash, K. Ikeda, N. Asano, *Bio-org. Med. Chem.*, 2010, **18**, 3790.
27. C. Porto, M. Cardone, F. Fontana, B. Rossi, M. R. Tuzzi, A. Tarallo, M. V. Barone, G. Andria and G. Parenti, *Mol. Therapy*, 2009, **17**, 964 .
28. R.A. Street, S. Chung, B. Wustman, A. Powe, H. Do, S.A. Kornfeld, *Proc. Natl. Acad. Sci. USA.*, 2006, **103**, 13813.
29. A.M. Scofield, L.E. Fellows, R.J. Nash and G.W.J. Fleet, *Life Sci.,* 1986, **39** 645.
30. J.S.S. Rountree, T.D. Butters, M.R. Wormald, R.A. Dwek, N. Asano, K. Ikeda, E.L. Evinson, R.J. Nash, G.W.J. Fleet, *ChemMedChem*, 2009, **4**, 378.
31. C. Norez, S. Noel, M. Wilke, M. Bijvelds, H. Jorna, P. Melin, H. DeJonge, F. Becq, *FEBS Letters*, 2006, **580**, 2081.

DEOXYSCHIZANDRIN AND γ-SCHIZANDRIN RESTORE THE CYTOTOXIC ACTION OF DOXORUBICIN IN MULTI-DRUG RESISTANT LUNG CANCER CELLS COR-L23/R

Jiří Slanina, Lenka Adámková, Ludmila Koubíková, Jindřiška Hammerová and Iva Slaninová
Department of Biochemistry and Department of Biology, Faculty of Medicine, Masaryk University, Kamenice 5, 625 00 Brno, Czech Republic.
E-mail: jrslanina@med.muni.cz

1 INTRODUCTION

Dibenzo[a,c]cyclooctadiene lignans are natural products occurring in *Schisandra chinensis* (Schisandraceae), a well-known medicinal plant in traditional Chinese medicine. The fruits and seeds have been used for centuries as a tonic and antitussive and many studies have indicated that the active ingredients are lignans possessing an unusual structure derived from dibenzo[a,c]cyclooctadiene (Figure 1). These lignans have been shown to exhibit a broad range of biological effects, including hepatoprotective and antiviral properties. Recently, dibenzocyclooctadiene lignans have been discussed as compounds that are able to overcome multidrug resistance.[1,2]

A panel of nine dibenzo[a,c]cyclooctadiene lignans isolated from seeds of *S. chinensis* was examined for their effect on multidrug resistance, as well as their anti-proliferative and pro-apoptotic activities. COR-L23/R, a multidrug-resistant sub-line over-expressing multidrug resistance-associated protein 1 (MRP1) together with its parent cell line COR-L23 (human lung cell carcinoma) and HL60 (human promyelocytic leukemia) cell lines were used. The results concerning COR-L23/R and COR-L23 cells were published by us previously.[2]

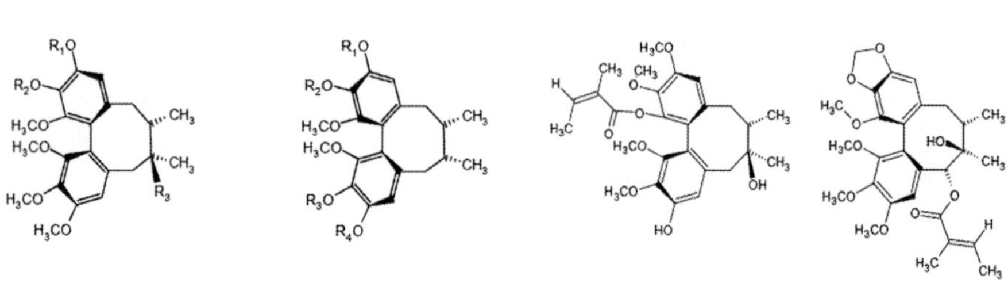

	R_1	R_2	R_3
(+)-schizandrin	$-CH_3$	$-CH_3$	-OH
(+)-gomisin A	$-CH_2-$		-OH
(+)-deoxyschizandin	$-CH_3$	$-CH_3$	-H
(+)-γ-schizandrin	$-CH_2-$		-H

	R_1	R_2	R_3	R_4
(-)-γ-schizandrin	$-CH_3$	$-CH_3$	$-CH_2-$	
(-)-gomisin N	$-CH_2-$		$-CH_3$	$-CH_3$
(-)-wuweizisu C	$-CH_2-$		$-CH_2-$	
(-)-gomisin J	-H	$-CH_3$	$-CH_3$	-H

(+)-angeloylgomisin H

(-)-tigloylgomisin P

Figure 1 *The structures of the lignans investigated for activity against multidrug resistance mechanisms. R-(+) or S-(-)-biaryl configurations are shown.*[2]

2 THE BIOLOGICAL ACTIVITIES OF THE LIGNANS

2.1 Cytotoxicity of the Lignans to Cancer Cells

The lignans, *R*-(+)-deoxyschizandrin, *R,S*-(±)-γ-schizandrin and *S*-(-)-gomisin N showed only weak cytotoxicity against the COR-L23 cell line when compared with the positive control, doxorubicin (Table 1). In the doxorubicin resistant cell line, COR-L23/R, the IC_{50} value for doxorubicin is increased one hundred fold, but the three lignans gave similar IC_{50} values for the paternal COR-L23 cells and the resistant strain. This suggests that the lignans are not substrates for MRP1 since their export would be expected to reduce the toxicity to the resistant cells. The remaining lignans show no toxicity and verapamil, which is an inhibitor of MRP1, shows a slight toxicity for both cell lines.

Table 1 *IC_{50} values (μg/ml) of the lignans, doxorubicin and verapamil against COR-L23 sensitive and multidrug resistant carcinoma cell lines and HL60 leukemia cell line.*

Compound	COR-L23	COR-L23/R	HL60
deoxyschizandrin	20	20	20
γ-schizandrin	12	20	25
gomisin N	13	18	22
doxorubicin	0.01	1.25	-
gomisin J	58	70	-
gomisin A	80	100	40
wuveizisu C	> 100	> 100	35
schizandrin	> 100	> 100	85
angeloylgomisin H	> 100	> 100	-
tigloylgomisin P	> 100	> 100	-
verapamil	115	70	-

2.2 Inhibition of Drug Export Mechanisms

The inhibition of export mechanisms would be expected to potentiate the toxicity of doxorubicin in the resistant, COR-L23/R cells and the percentage survival of these cells after treatment with verapamil or lignans in the presence of doxorubicin is shown in Table 2. Increasing doses of doxorubicin alone make little difference to the number of resistant cells surviving (first entry) due to its export from the cell. However, the addition of verapamil, which inhibits export mechanisms, restores the cytotoxic action of doxorubicin in a manner dependent on the dose of doxorubicin (second entry).

The lignans, deoxyschizandrin and γ-schizandrin give a dose dependent effect similar to verapamil suggesting that they might act by the same mechanism of negating the resistance of the COR-L23/R cells by inhibiting the export of doxorubicin. However, lignans possessing the *S*-biaryl configuration such as gomisin N and wuweizisu C or which have a hydroxyl group at C-8 as is the case in schizandrin and gomisin A, gave weak or negligible potentiation of the activity of doxorubicin in resistant cells. This spectrum of activity correlates with further properties of the lignans described in subsequent sections. First we sought to demonstrate the accumulation of doxorubicin in resistant cells which had been treated with lignans.

Table 2 *Percentages of viable COR-L23/R cells after treatment with doxorubicin in combination with lignan modulators, compared to controls. Significant differences at *$p < 0.05$ and **$p < 0.01$ relative to doxorubicin treated cells were noted (Dunnett's test).*[2]

MODULATOR	DOXORUBICIN			
10 mg/l	0 mg/l	0.01 mg/l	0.1 mg/l	0.3 mg/l
NONE	94 ± 7	89 ± 11	81 ± 11	75 ± 7
verapamil	68 ± 6	67 ± 11	36 ± 17 **	28 ± 12 **
deoxyschizandrin	74 ± 11	65 ± 4 *	59 ± 16 *	36 ± 8 **
γ-schizandrin	76 ± 16	66 ± 12 *	66 ± 14	49 ± 7 *
gomisin N	68 ± 3	81 ± 15	71 ± 9	72 ± 14
gomisin J	69 ± 7	77 ± 2	59 ± 4	58 ± 9
gomisin A	94 ± 1	86 ± 1	88 ± 8	75 ± 1
wuweizisu C	107 ± 4	93 ± 9	83 ± 9	57 ± 3
schizandrin	88 ± 7	83 ± 6	79 ± 8	57 ± 7
angeloylgomisin H	68 ± 4	83 ± 14	70 ± 11	72 ± 15
tigloylgomisin P	72 ± 10	86 ± 5	75 ± 5	66 ± 7

2.3 Lignans Increase the Accumulation of Doxorubicin in COR-L23/R cells

We tested our hypothesis that the lignans restore the toxicity of doxorubicin in the resistant cell line by inhibiting export mechanisms and causing the antibiotic to accumulate inside the cells. COR-L23/R cells were incubated with doxorubicin (10 μM) in the presence or absence of the lignans (10 μg/ml) or of verapamil using a flow cytometry assay (FL2 channel, emission at 575 nm). The fluorescence without modulator was 100 %.

As may be seen in Figure 2, lignans and the known MRP1 inhibitor verapamil increased the accumulation of doxorubicin inside the doxorubicin resistant COR-L23/R

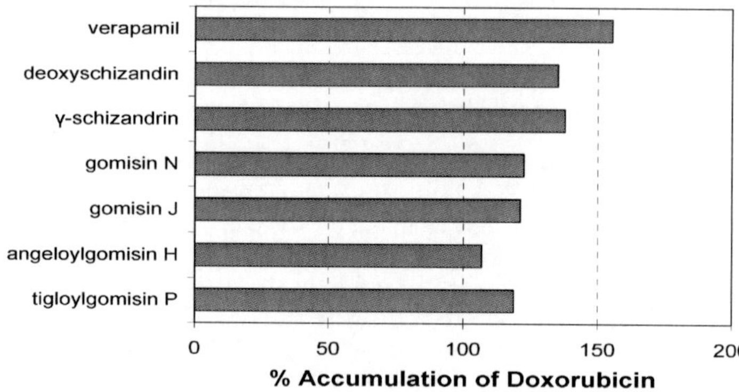

Figure 2 *The effect of lignans (10 mg/l) on intracellular doxorubicin (10 μM) accumulation determined by flow cytometry. Results are expressed as the percentage ratio of the doxorubicin fluorescence in the presence or absence of the lignans or verapamil. Verapamil and the lignans deoxyschizandrin and γ-schizandrin cause the greatest accumulation of intracellular doxorubicin.*

cells. Thus, the lignans deoxyschizandrin and γ-schizandrin, which were most effective in restoring the toxicity of doxorubicin to resistant cell lines were also the most effective in causing the accumulation of the antibiotic inside the resistant cell lines.

2.4 Lignans Act Synergistically with Doxorubicin to Induce Cell Cycle Arrest

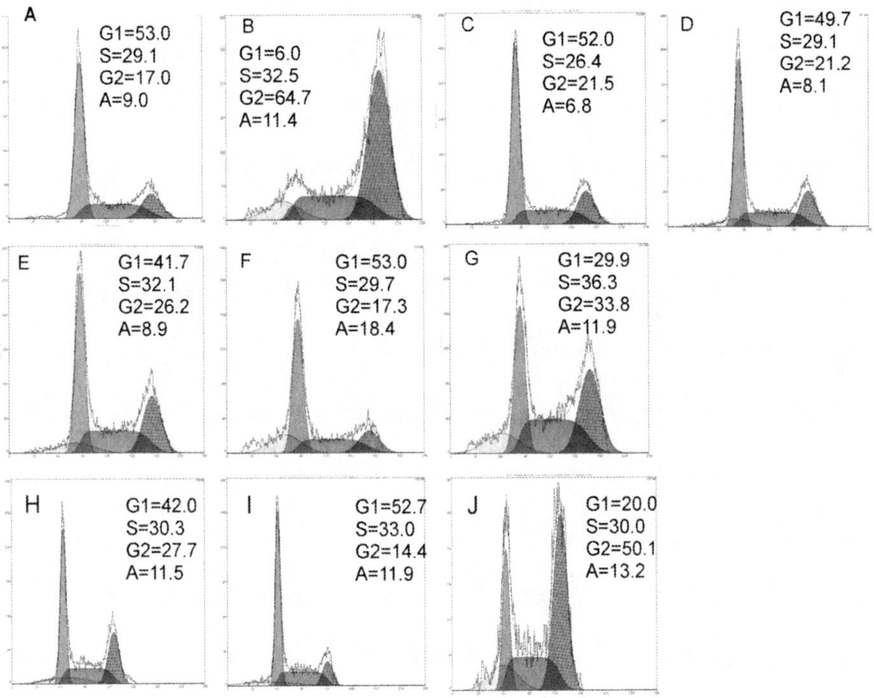

Figure 3 *The cell cycle analysis of propidium iodide stained* COR-L23/R *cells treated for 48 h. The DNA content analysis included the determination of the percentage of cells in G_1, G_2/M (G_2), S-phase and sub-G_1 (A, apoptotic) fractions. A-G first experiment (A) negative control (DMSO alone); (B) doxorubicin 1 μg/ml; (C) doxorubicin 0.1 μg/ml; (D) γ-schizandrin 10 μg/ml; (E) doxorubicin 0.1 μg/ml in combination with γ-schizandrin 10 μg/ml; (F) γ-schizandrin 50 μg/ml; (G) doxorubicin 0.1 μg/ml in combination with γ-schizandrin 50 μg/ml. H-J second experiment: (H) doxorubicin 0.1 μg/ml; (I) deoxyschizandrin 10 μg/ml; (J) doxorubicin 0.1 μg/ml in combination with deoxyschizandrin 10 μg/ml. Lignans γ-schizandrin and deoxyschizandrin alone, at the concentrations shown and doxorubicin at sub IC_{50} concentrations (1.25 μg/ml) had no effect on the cell cycle. Deoxyschizandrin or γ-schizandrin in combination with doxorubicin (0.1 μg/ml) increased the proportion of cells in the G2/M phase.*

Low concentrations of deoxyschizandrin (< 10 μg/ml) or of γ-schizandrin (< 50 μg/ml) did not affect the cell cycle of COR-L23/R cells (Figure 3). A sub-toxic dose of doxorubicin (0.1 μg/ml) had no discernible effect and similarly caused no changes in the cell cycle, but the combination of sub-toxic doses of doxorubicin and low concentrations of either lignin increased the proportion of cells in the G2 phase by between 5 and 25%. Arrest in the G2/M phase is typical of the behaviour of cells when treated with doxorubicin at toxic doses and this result is therefore consistent with toxicity due to an increased concentration of doxorubicin inside the cells.

2.5 The Expression of Multidrug Resistance Genes in COR-L23 Cells

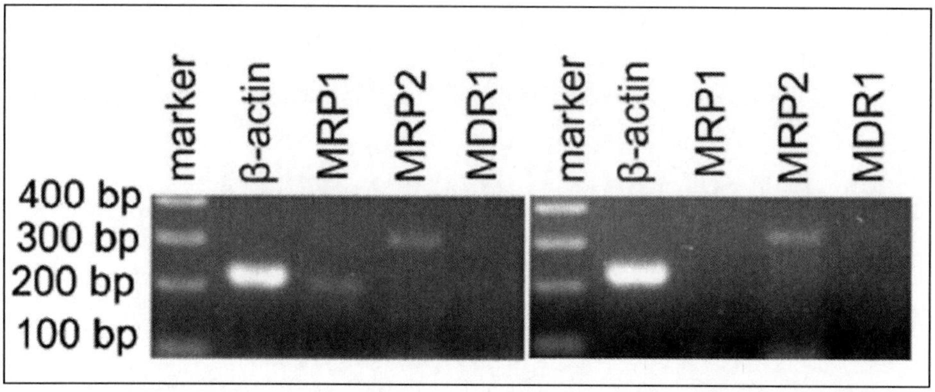

Figure 4 *Expression of MRP1 gene only in the resistant COR-L23/R cells (left-hand frame), whereas MRP2 was expressed and P-glycoprotein (MDR1) was absent in both paternal (right-hand frame) and resistant cells.*[2]

In order to analyze the expression of MRP1, MRP2 and MDR1 genes in COR-L23 and COR-L23/R cells, reverse transcriptase-polymerase chain reaction (RT-PCR) was carried out using the β-actin gene as a positive control (Figure 4). We detected high β-actin expression in both cell lines and expression of MRP1 and MRP2 in COR-L23/R cells. However, we were unable to detect any expression of MRP1 and weak expression of MRP2 in COR-L23 suggesting that these proteins are associated with the doxorubicin resistance of the COR-L23/R cell line. MDR1 expression was not detected in any cells.

3 CONCLUSION

We found that the lignans *R*-(+)-deoxyschizandrin and *R,S*-(±)-γ-schizandrin restored the cytotoxic action of the MRP1 substrate doxorubicin to COR-L23/R cells at relatively non-toxic concentrations of the lignans. Deoxyschizandrin and γ-schizandrin also showed the greatest propensity to enhance the accumulation of doxorubicin in drug resistant cells.

These lignans alone had no effect on the cell cycle; however, in combination with sub-toxic doses of doxorubicin, cells were arrested in the G2/M phase, which is typical of the effect of doxorubicin at toxic doses. The common structural features of the active lignans are the *R*-biaryl configuration and the absence of a hydroxyl group at C-8.

Acknowedgements

This work was supported by the Czech Science Foundation (522/07/0995) and the Ministry of Education of the Czech Republic (VZ MSM0021622415 and LC06077).

References

1 Q.R. Pan, Q.H. Lu, K. Zhang, X. Hu, *Cancer Chemoth. Pharm.*, 2006, **58**, 99.
2 I. Slaninová, L. Březinová, L. Koubíková, J. Slanina, *Toxicol. In Vitro*, 2009, **23**, 1047.

4 Biosynthesis and Genetics

ELUCIDATION OF PATHWAYS FOR ANTIBIOTIC BIOSYNTHESIS IN *STREPTOMYCES COELICOLOR* A3(2)

Stuart W. Haynes, Paulina K. Sydor, Christophe Corre, Gregory L. Challis

Department of Chemistry, University of Warwick, Coventry, CV4 7AL, UK

1 INTRODUCTION

1.1 *Streptomyces coelicolor* A3(2)

Natural sources have proved to be invaluable pools for the discovery of clinically-useful compounds for more than 50 years and represent over two thirds of new antibiotics approved for clinical use in the last 25 years.[1] A wide variety of these naturally-derived compounds are produced by the bacterial genus *Streptomyces* and related actinobacteria, which are well known for their ability to produce a large number of structurally diverse secondary metabolites.[2] *Streptomyces* species have been extensively studied in an attempt to understand the biosynthetic processes involved in the formation of these secondary metabolites. This knowledge has already been applied in numerous cases to produce analogues of compounds with promising pharmaceutical properties by combinatorial biosynthesis or mutasynthesis approaches.[3]

S. coelicolor is one of the most well studied streptomycetes and is a known producer of several structurally and medicinally interesting antibiotics. It is known to produce prodiginine and methylenomycin antibiotics, the biosynthesis of which will be subject of this article, and also calcium-dependent antibiotics (CDAs) and actinorhodin. The complete sequence of the gene clusters responsible for directing the biosynthesis of these secondary metabolites are known as a result of the complete sequencing of the *Streptomyces coelicolor* genome along with the sequencing of the giant linear plasmid SCP1, present in 'wild-type' *S. coelicolor*.[4,5]

1.2 Prodiginines

Prodiginines are a group of red-pigmented secondary metabolites that are all based around the conjugated tripyrrole ring system first observed in progidiosin.[6] Since the discovery of prodigiosin several prodiginine analogues have been isolated including those produced by *S. coelicolor*, undecylprodiginine and its cyclic derivative streptorubin B.[7] These structurally intriguing metabolites have been demonstrated to display immunosuppressant activity at non-toxic doses,[8,9,10,11,12] and the cyclic undecylprodiginine derivatives, most notably metacycloprodigiosin (streptorubin A), have been shown to exhibit potent anti-malarial activity.[13] Furthermore a synthetic prodiginine analogue, obatoclax, is currently in

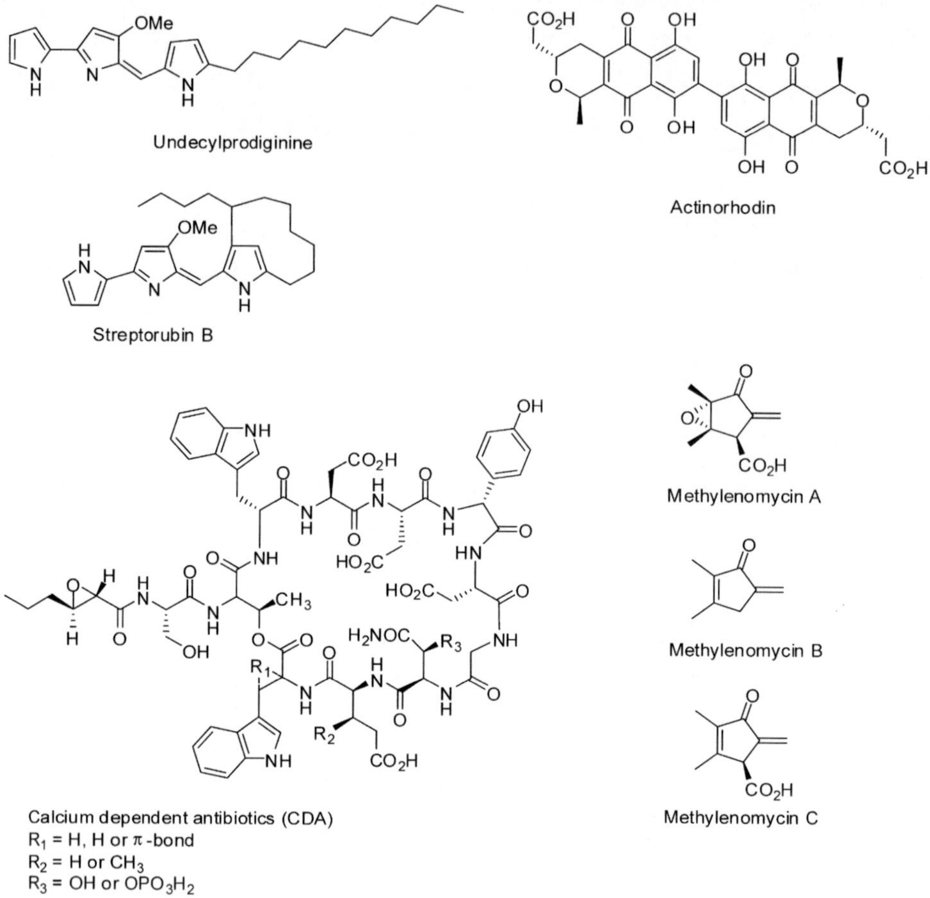

Figure 1 *Structures of several characterised antibiotics produced by* S. coelicolor

phase 1/2 clinical trials for a variety of solid tumours and multiple myeloma.[14] The activity of obatoclax appears to derive from its interaction with the Bcl-2 family of proteins, which are involved in mediating apoptotic processes within the cell. Anti-apoptotic Bcl-2 proteins (Bcl-x_L, Mcl-1, and Bcl-2 itself) have been shown to be overexpressed in numerous cancer cell lines, making them promising targets for oncology treatments. The Bcl-2 family of proteins are known to bind to pro-apoptotic proteins within the cell (including Bad, Bid and Bim), which are involved in initiating the signalling cascade that triggers cytochrome c release that in turn activates caspases responsible for dismantling the cell during apoptosis.[15] Obatoclax has been shown to disrupt the binding of the anti-apoptotic Bcl-2 proteins with pro-apoptotic proteins and hence restore normal cell death processes within immortal cell lines overexpressing Bcl-2.[16,17]

1.3 Methylenomycins

Methylenomycin A and B were first isolated in 1974 from *S. violaceoruber*,[18] although methylenomycin B has since been shown to result from an acid catalysed degradation of methylenomycin A.[22] Methylenomycin A and B were subsequently shown to possess a cyclopentanone core common to several metabolites of *Streptomyces* species including sarkomycin, which has been clinically-used in Japan for the treatment of a variety of carcinomas and sarcomas.[19] Methylenomycin A displays a wide spectrum of antibiotic activity against Gram-positive bacteria and some Gram-negative bacteria particularly *Proteus* species.[20] However its mode of action still remains unclear. *Streptomyces coelicolor* was subsequently shown to produce methylenomycin A and the desepoxy-4,5-dehydro derivative, methylenomycin C.[21,22,47]

1.4 Actinorhodin

Actinorhodin was initially identified as a pigmented antibiotic fragment, before its structure was fully determined.[23] It belongs to a class of aromatic antibiotics known as the benzoisochromanequinones some of which possess anticancer, antibacterial, anticoccidial, or platelet aggregation inhibitory activities.[24] The structure of actinorhodin consists of a naphthazarin dimer. Its characteristic pH-dependent colour change, resulting from this naphthazarin dimer, has greatly-facilitated genetic studies of its biosynthesis.[25]

1.5 Calcium-dependent Antibiotics (CDAs)

A variety of calcium-dependent antibiotics (CDAs) have been isolated from *S. coelicolor*, which vary in functional groups but all share the same nonribosomally-biosynthesised peptide sequence.[26] They can be classified more broadly into a group of acidic cyclic lipopeptide antimicrobial agents including friulimicins, amphomycin and daptomycin.[27,28] All of these lipopeptides possess a decapeptide lactone or lactam core made up of several nonproteinogenic residues and acidic residues essential for biological activity, which bind Ca^{2+}.[29] Daptomycin has recently been approved for clinical treatment of skin and soft tissue infections caused by Gram-positive pathogens including *Staphylococcus aureus* strains resistant to methicillin and vancomycin. The proposed mode of action believed to be common to CDAs and daptomycin, involves disruption of Gram-positive bacterial cell membranes by aggregates of the decapeptide-Ca^{2+} complex. This disruption subsequently results in depolarization of the membrane and ultimately cell death.[30] This has resulted in considerable interest in the biosynthesis of CDAs and daptomycin with the aim of generating analogues of these compounds *via* combinatorial biosynthesis approaches.[31,32]

2 ELUCIDATION OF PRODIGININE BIOSYNTHESIS IN *STREPTOMYCES COELICOLOR*

2.1 Incorporation of Labelled Precursors and Identification of the Biosynthetic Gene Cluster

Prodiginine biosynthesis has been investigated by incorporation experiments using labelled precursors.[33,34,35] The results led to the suggestion that undecylprodiginine is made by the condensation of the two putative intermediates 4-methoxy-2,2'-bipyrrole-5-carboxaldehyde (MBC) and 2-undecylpyrrole (2-UP) (Scheme 1). MBC is biosynthesised from one unit of acetate, proline and serine (with loss of the carboxyl group) and the 2-UP is biosynthesised from seven units of acetate and a unit of glycine (again with loss of the carboxyl group). Butyl-*meta*-cycloheptylprodiginine (streptorubin B) is proposed to be formed from undecylprodiginine by an oxidative cyclisation.[36]

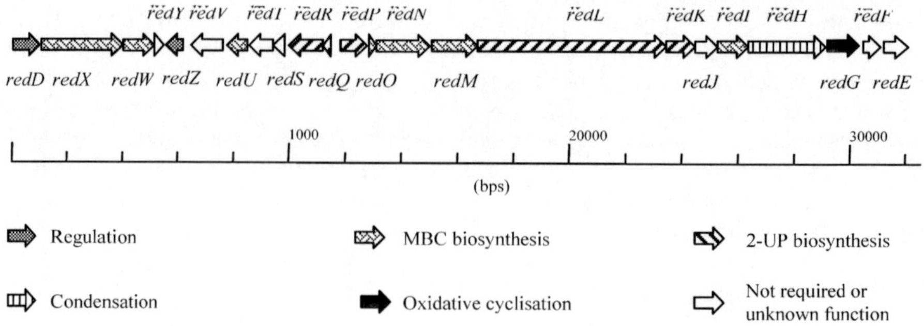

Protein	Proposed function	Protein	Proposed function
RedE	Unknown (not required)	RedQ	ACP
RedF	Oxidoreductase (not required)	RedR	β-ketoacyl-ACP synthase II
RedG	Cyclase	RedS	Unknown
RedH	Condensing enzyme	RedT	Unknown
RedI	O-methyltransferase	RedU	Phosphopantetheinyl transferase
RedJ	Thioesterase (not required)	RedV	Oxidoreductase (not required)
RedK	Reductase	RedZ	Transcriptional regulator
RedL	Hybrid PKS/NRPS (A-ACP-KS-AT-ACP-OAS)	RedY	Unknown (not required)
RedM	Prolyl-PCP synthetase	RedW	Prolyl-PCP-oxidase/desaturase
RedN	PKS (ACP-ACP-OAS)	RedX	PKS (KS-KS)
RedO	PCP	RedD	Transcriptional regulator
RedP	β-ketoacyl-ACP synthase III		

Figure 2 *Organisation of the* S. coelicolor *prodiginine biosynthetic gene cluster and putative functions of encoded enzymes*

The *red* cluster, identified in 2001, which directs prodiginine biosynthesis in *S. coelicolor*, contains 23 genes. They are arranged in 4 transcription units and functions of the products of most of the genes were suggested on the basis of sequence comparisons (Figure 2). These putative functions led to a proposed biosynthetic pathway,[36] which has subsequently been revised in light of experimental evidence.[37,38,39] Two genes (*redD* and *redZ*) encode pathway specific regulators.[40,41,42] Seven genes are known to be required for the biosynthesis of MBC,[37] and five genes are required for the biosynthesis of 2-UP.[38,43] An analogous set of MBC biosynthetic genes resides within the *pig* gene cluster responsible for prodigiosin biosynthesis in *Serratia* species.[44] The *redH* gene has been shown to be responsible for the condensation of MBC and 2-UP to give undecylprodiginine[39] and the *redG* gene is proposed to be required for the oxidative cyclisation of undecylprodiginine to form streptorubin B. Five genes (*redE*, *redF*, *redJ*, *redV* and *redY*) appear not to be required for prodiginine biosynthesis. Two genes (*redT* and *redS*) are of unknown function, but the latter appears to encode a truncated, non functional protein (Figure 2).[36]

2.2 4-Methoxy-2-2'-bipyrrole-5-carboxyaldehyde (MBC) Biosynthesis

4-methoxy-2,2'-bipyrrole-5-carboxaldehyde (MBC) has been shown to be an intermediate in prodiginine biosynthesis in *Streptomyces* species as well as in prodigiosin biosynthesis in *Serratia* species.[44] The mechanism of MBC biosynthesis, which appears to be similar in both species, was proposed by Cerdeño *et al.* and Walsh *et al.* and then later revised by Williamson *et al.* and Stanley *et al.*[36,37,44,45,46] L-proline is adenylated by RedM thus activating it for loading onto the phosphopantetheine arm of the peptidyl carrier protein (PCP) RedO. L-prolyl-RedO is then dehydrogenated by RedW in an FAD-dependent reaction to give pyrrole-2-carboxyl-RedO.[44,36] The pyrrole-2-carboxyl group attached to RedO is then transferred to the C-terminal ketosynthase (KS) domain of RedX and condensed with concomitant decarboxylation with a malonyl group attached to one of the two acyl carrier protein (ACP) domains in RedN to give a β-keto-ACP-thioester. In the next step, the C-terminal α-oxoamine synthase (OAS) domain of RedN catalyses decarboxylative condensation of L-serine with the RedN ACP-thioester, resulting in cleavage from RedN. A subsequent spontaneous cyclisation and dehydration would give 4-hydroxy-2-2′-bipyrrole-4-methanol (HBM). SAM-dependent methylation of the pyrrole hydroxyl group catalysed by RedI and finally oxidation of the remaining hydroxyl group by a currently undefined enzyme would give 4-methoxy-2,2'-bipyrrole-5-carboxaldehyde (MBC) (Scheme 1).[36]

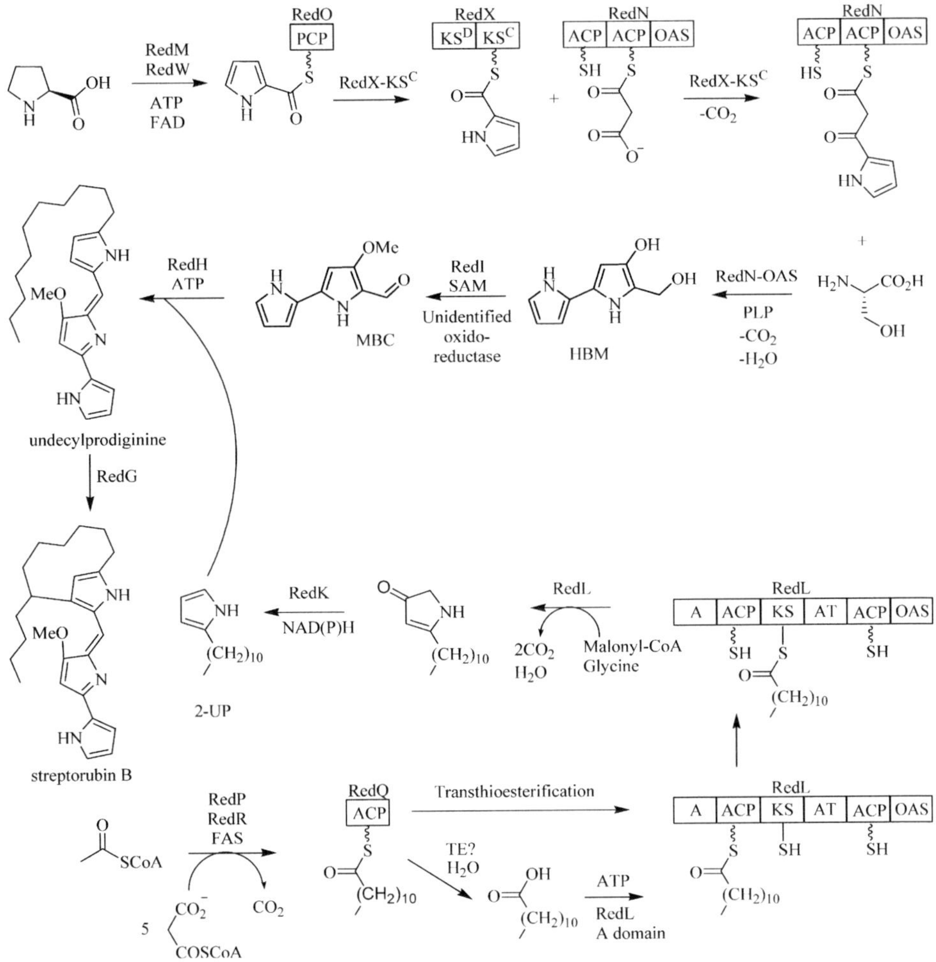

Scheme 1 Proposed biosynthetic pathway to the undecylprodiginine and streptorubin B

2.3 2-Undecylpyrrole (2-UP) Biosynthesis

It was originally proposed by Cerdeño *el al.* that RedP, RedQ, and RedR are involved in the assembly of dodecanoyl thioester attached to the RedQ ACP, which is elaborated to 2-UP by RedN and RedX.[36] More recent work has shown that RedN and RedX are involved in MBC biosynthesis,[35,36] and that RedL is responsible for assembly of 2-UP from dodecanoyl RedQ.[37]

It is proposed that RedP generates an acetoacyl thioester by decarboxylative condensation of an acetyl-CoA starter unit with a malonyl-CoA extender unit attached to RedQ. The resulting thioester intermediate would then be fully reduced to butyryl-ACP by the corresponding type II fatty acid synthase (FAS) enzymes. Iterative elongation by the action of RedP and RedR and reduction to the alkyl chain by the corresponding FAS enzymes after each elongation would give dodecanoyl-RedQ.[38,42] In the next step the dodecanoyl group would be transferred to the N-terminal ACP domain of RedL in one of two ways: *via* direct transfer from RedQ, or *via* hydrolysis to the free acid which subsequently could be loaded onto the ACP domain after activation catalysed by the NRPS like adenylation domain of RedL. The RedL ketosynthase (KS) domain then catalyses the reaction of the resulting dodecanoyl thioester with a malonyl thioester attached to the C-terminal ACP domain of RedL to form β-ketomyristoyl-ACP. A subsequent decarboxylative condensation with glycine catalysed by the C-terminal pyridoxal 5'-phosphate (PLP)-dependent OAS domain of RedL would result in chain release. Spontaneous cyclisation and elimination of water would lead to 4-keto-2-undecylpyrroline. Reduction of the keto group, catalysed by the NAD(P)H dependent RedK, and elimination of water yields 2-undecylpyrrole (scheme 1).[38]

2.4 Condensation of Late Stage Intermediates

It is known that the condensation of chemically-synthesised MBC and 2-UP is very facile under acidic conditions and it was originally proposed that no enzyme would be required to perform this step in undecylprodiginine biosynthesis.[35] However analysis of an *S. coelicolor* mutant in which *redH* is deleted showed that RedH is required for the condensation reaction of MBC and 2-UP to give undecylprodiginine (Scheme 1).[39] A similar role was hypothesised for the *RedH* homologue *PigC* in *Serratia marcescens* which was shown to be required for the condensation of MBC and 2-methyl-3-n-amyl-pyrrole (MAP) to give prodigiosin.[44,46]

The *redH* gene of *S. coelicolor* is predicted to encode a phosphotransferase with three functional domains, two of which show sequence similarity to domains of pyruvate phosphate dikinase (PPDK) and phosphoenolpyruvate synthetase (PEPS).[35] The RedH N-terminal domain is similar to the ATP-binding domain of PEPS, the C-terminal domain is homologous to the phosphotransfer domain of PPDK. The central domain of RedH does not show similarity to any proteins or protein domains of known function and is thought to be responsible for binding MBC.

Analogous to the known mechanism of PPDK and PEPS it is proposed that the phosphotransfer domain catalyses phosphorylation of the aldehyde oxygen of MBC, resulting in activation of the aldehyde carbon towards nucleophilic attack by C-5 of 2-UP. Elimination of the aldehyde derived oxygen as phosphate from this intermediate would yield undecylprodiginine.[38]

The last step in *S. coelicolor* prodiginine biosynthesis is hypothesised to be the conversion of undecylprodiginine to streptorubin B *via* an oxidative cyclisation reaction. *RedG* encodes a Rieske non-heme iron-dependent oxygenase-like enzyme, which is proposed to catalyse this reaction.[35] However, no experimental evidence for this novel transformation has been published to date.

3 INVESTIGATION OF METHYLENOMYCIN BIOSYNTHESIS IN *STREPTOMYCES COELICOLOR*

3.1 Precursor Incorporation Experiments

The biosynthesis of the methylenomycin antibiotics has been studied since shortly after their discovery in 1974 and is still a subject of much interest.[47] Such studies have benefited from a variety of approaches which have been developed over many years. The primary metabolic origins of methylenomycin A were first investigated by feeding experiments in *S. coelicolor* with radiolabelled precursors, which suggested that acetate played an important role.[48] The availability of stable-isotope based techniques facilitated further investigations. Feeding of [U-^{13}C]acetate to *S. coelicolor* resulted in the incorporation of two intact acetate units into methylenomycin A.[47]

Despite feeding a wide range of labelled precursors, the origin of five of the nine carbon atoms remained unknown for two decades.[47] After a revision of ^{1}H and ^{13}C NMR assignments for methylenomycin A, the two intact acetate units incorporated were found to be adjacent (C-5/C-9 and C-1/C-6) and acetoacetyl-ACP was proposed as an intermediate in the pathway (Scheme 2).[22] In the same study, [U-^{13}C]glycerol was found to be directly incorporated into C-3, C-4, and C-8.[48] However the origin of C-2 and C-7 remained unclear.

Further feeding experiments were carried out with [U-^{13}C]glycerol used as the main carbon source in a minimal medium. These experiments together with the feeding of [U-^{13}C]ribose proved the incorporation of an intact five carbon unit into methylenomycin A.[22] Taken together these experiments indicate that the methylenomycins derive from two molecules of acetic acid and one pentulose unit derived from an intermediate in the pentose phosphate pathway (Scheme 2). No other antibiotic is known to derive from this unique combination of biosynthetic precursors.

The late stages of methylenomycin biosynthesis have also been investigated. Methylenomycin C (desepoxy-4,5-dehydro-methylenomycin A), which is coproduced with methylenomycin A has been shown to be a biosynthetic precursor of methylenomycin A.[46] When radiolabelled methylenomycin C (generated by feeding of radiolabelled acetate) was purified and fed to a methylenomycin producer radiolabelled, methylenomycin A was obtained. It has also been shown that methylenomycin B results from the spontaneous decarboxylation of methylenomycin C in acidic conditions (Figure 1).[22]

3.2 Sequencing and Analysis of the Methylenomycin Biosynthetic Gene Cluster

Interestingly, in both *S. violaceoruber* SANK95570 and *S. coelicolor* A3(2)[21,49] the genes responsible for the resistance to methylenomycin A, as well as its biosynthesis and the regulation of its production are on a plasmid;[50,51] the circular plasmid pSV1 in *S. violaceoruber* and the linear plasmid SCP1 in *S. coelicolor.*[52,53] Both of these plasmids have been sequenced, analysed and compared.[5,54,55] The entire sequence of the *mmy* gene cluster of *S. coelicolor* (Genbank Accession number: AJ276673) has been analysed in detail allowing putative functional assignments for nearly all of the proteins encoded by genes within the cluster (Figure 3). The *mmr* gene encodes a membrane protein belonging

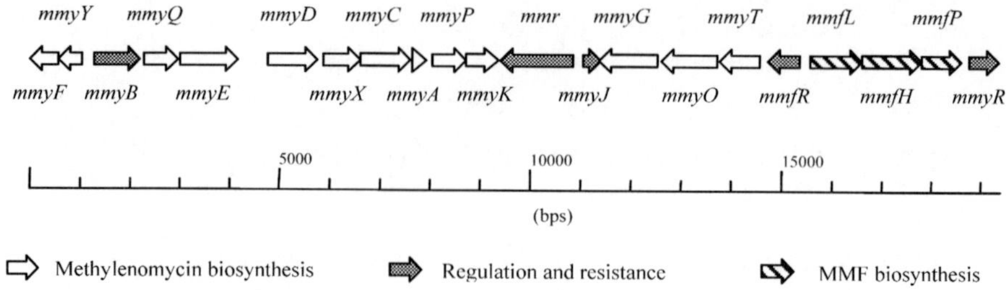

Protein	Proposed function	Protein	Proposed function
MmyA	Acyl carrier protein (ACP)	MmyO	Flavin-dependent monooxygenase
MmyC	Acetoacetyl-ACP synthase	MmyP	Phosphatase
MmyD	Butenolide synthase	MmyQ	Coenzyme F-420-dependent reductase
MmyE	Flavin-dependent enoyl reductase	MmyT	Thioesterase
MmyF	Flavin reductase partner to monooxygenase	MmyX	Kinase
MmyG	NAD(P)-dependent dehydrogenase/ketoreductase	MmyY	Isomerase
MmyK	Kinase		

Figure 3 *Organisation of the* S. coelicolor *methylenomycin biosynthetic gene cluster and putative function of encoded enzymes*

to the major facilitator superfamily that has previously been shown to confer resistance to methylenomycin A. The *mmyR*, *mmfP*, *mmfH*, *mmfL*, *mmfR*, *mmyB* and *mmyJ* genes all encode proteins involved in a complex pathway-specific regulatory network.[56] This network also involves diffusible autoregulatory molecules called methylenomycin furans (MMFs), which are biosynthesised by the products of *mmfP*, *mmfH*, *mmfL* (Section 3.4).[57] The MMFs are thought to be recognised by the DNA-binding receptor protein MmfR that regulates the expression of the methylenomycin biosynthetic genes.[59] The remaining 13 genes encode enzymes believed to be required for biosynthesis of the methylenomycins.

3.3 Proposed Biosynthetic Pathway to Methylenomycin A

Together the data from the feeding studies and sequence analysis suggest plausible early and late steps for methylenomycin A biosynthesis. Thus, it is proposed that methylenomycin biosynthesis is initiated by MmyC-catalysed condensation of malonyl-MmyA (derived from transacylation of malonyl-CoA onto the phosphopantetheinyl arm of holo-MmyA catalysed by malonyl-CoA acyl transferase) with acetyl-CoA to give acetoacetyl-MmyA (Scheme 2). The next step in the pathway is proposed to be catalysed by MmyD, which shows significant similarity to AvrD enzymes involved in the biosynthesis of the *Pseudomonas syringae* syringolide plant defense response elicitors.[58] MmyD could catalyse condensation of acetoacetyl-MmyA with xylulose or ribulose (both derived from pentulose-5-phosphate intermediates of the pentose phosphate pathway) to give the corresponding butenolide intermediate (Scheme 2). Such a biosynthetic pathway involving an unusual condensation of a diketide with a pentose sugar is currently unprecedented.

Several alternative plausible pathways for the elaboration of methylenomycin C from the proposed butenolide intermediate can be envisaged utilising the MmyE, G, K, Q, T, X, and Y enzymes. Previous feeding studies have implicated methylenomycin C as a precursor to methylenomycin A and our analysis of the proteins encoded by the *mmy*

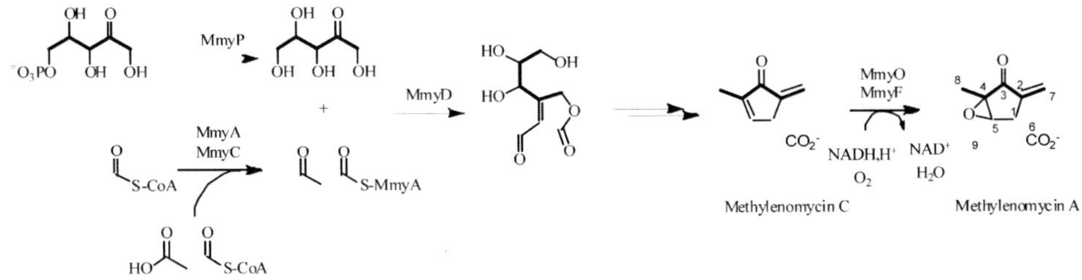

Scheme 2 *Proposed biosynthetic pathway to the methylenomycins*

cluster implicates MmyO and MmyF in the epoxidation of methylenomycin C to form methylenomycin A.

MmyO shows a significant degree of sequence similarity to LimB, which catalyses a $FADH_2$-dependent epoxidation of the 1,2 double bond of limonene using molecular oxygen as a co-substrate. MmyF is similar to NADH-dependent flavin reductases known to be required for supply of $FADH_2$ to flavin-dependent monooxygenases like LimB. Thus, we suggest that MmyO generates a flavin hydroperoxide intermediate (from reduced flavin supplied by MmyF and molecular oxygen), which epoxidises the more sterically hindered electro-deficient 4,5-double bond in methylenomycin C without epoxidising the more electron deficient and less sterically hindered electro-deficient 2,7-double bond, probably via a 1,4-addition – 1,3-elimination mechanism (Scheme 2).

3.4 Role of AHFCAs in Regulating Methylenomycin Biosynthesis

The *mmfLHP* genes were expressed in the SCP1$^-$ *S. coelicolor* M512 strain leading to the discovery of five novel 2-alkyl-4-hydroxymethylfuran-3-carboxylic acids (AHFCAs) named methylenomycin furans (MMFs) 1-5 (figure 4).[59]

These metabolites were shown to specifically induce the production of the methylenomycin antibiotics in *Streptomyces coelicolor*.[59] MMFs were found to derepress the transcription of the methylenomycin biosynthetic genes *via* the transcriptional derepression of the activator protein MmyB.[59]

This regulatory network is thought to be mediated by the interaction of MMFs with the transcriptional repressor MmfR.[59]

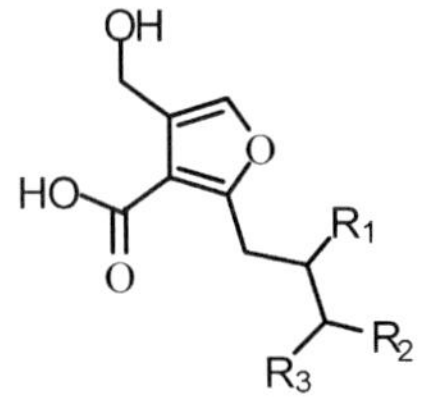

MMF1 (R_1 = Me, R_2 = H, R_3 = H)
MMF2 (R_1 = H, R_2 = H, R_3 = H)
MMF3 (R_1 = H, R_2 = Me, R_3 = Me)
MMF4 (R_1 = H, R_2 = H, R_3 = Me)
MMF5 (R_1 = H, R_2 = H, R_3 = Et)

AHFCAs (2-alkyl-4-hydroxymethylfuran-3-carboxylic acid)

Figure 4 *Structures of the novel AHFCA pathway specific regulators of methylenomycin biosynthesis.*

4 CONCLUSIONS

In recent years, considerable progress has been made in understanding the biosynthetic pathways by which the *S. coelicolor* prodiginine and methylenomycin antibiotics are assembled. However, enzymes that catalyse key transformations in both pathways still remain to be identified. Nevertheless the knowledge already gained has set the stage for developing a detailed understanding of the mechanistic enzymology of several fascinating transformations in both pathways. Ultimately our knowledge of the biosynthesis of these antibiotics will be exploited to generate novel analogues *via* a variety of genetic and chemical approaches and will feed into developing efforts to harness secondary metabolic enzymes for the efficient and environmentally-friendly assembly of completely novel organic structures *via* synthetic biology.

References

1 D.J. Newman, G.M. Cragg and K.M. Snader, *J. Nat. Prod.*, 2003, **66**, 1022.
2 D. Hopwood, *Streptomyces* in Nature and Medicine: *The antibiotic makers*, Oxford University Press, New York, USA, 2007
3 H. Zhou, X. Xie and Y. Tang, *Curr. Opin. in Biotechnol.*, 2008, **19**, 590.
4 S.D. Bentley et al., *Nature*, 2002, **417**, 141.
5 S.D. Bentley et al., *Mol. Microbiol.*, 2004, **51**, 1615.
6 H.H. Wassermann, J. Keggi and F. Bohlmann, W. Luders, *Angew. Chem.*, 1960, **72**, 779.
7 S.W. Tsao, B.A.M. Rudd, X.G. Xe, C.J. Chang and H.G. Floss, *J. Antibiot.*, 1985, **38**, 128.
8 J. Magae, M.W. Miller, K. Nagai and G.M. Shearer, *J. Antibiot.*, 1996, **49**, 86.
9 T. Azuma, N. Watanabe, H. Yagisawa, K. Hirata, M. Iwamura and Y. Kobayashi, *Immunopharmacology*, 2000, **46**, 29.
10 M.H. Lee, T. Kataoka, N. Honjo, J. Magae and K. Nagai, *Immunology*, 2000, **99**, 243.
11 A. Nakamura, J. Magae, R.F. Tsuji, M. Yamasaki and K. Nagai, *Transplantation*, 1989, **47**, 1013.
12 R.F. Tsuji, J. Magae, M. Yamashita, K. Nagai and M. Yamasaki, *J. Antibiot.*, 1992, **45**, 1295.
13 M. Isaka, A. Jaturapat, J. Kramyu, M. Tanticharoen and Y. Thebtaranonth, *Antimicrob. Agents Ch.*, 2002, **46**, 1112.
14 S.T. Trudel, Z.H. Li, J. Rauw, R.E. Tiedemann, X.Y. Wen and A.K. Stewart, *Blood*, 2007, **109**, 5430.
15 N.N. Danial and S.J. Korsmeyer, *Cell*, 2004, **116**, 205.
16 M. Nguyen et al., Proc. *Natl Acad. Sci.* USA, 2007, **104**, 19512.
17 P. Perez-Galan, G. Roue, N. Villamor, E. Campo and D. Colomer, *Blood*, 2007, **109**, 4441.
18 T. Haneishi, A. Terahara, M. Arai, T. Hata and C. Tamura, *J. Antibiot.*, 1974, **27**, 393.
19 G.B. Magill et al., *Cancer Res.*, 1956, **16**, 960.
20 T. Haneishi, N. Kitahara, Y. Takiguchi and M. Arai, *J. Antibiot.*, 1974, **27**, 386.
21 L.F. Wright and D.A. Hopwood, *J. Gen. Microbiol.*, 1976, **95**, 96.
22 C. Corre and G.L. Challis, *ChemBioChem*, 2005, **6**, 2166.
23 H. Brockmann, A, Zeeck, K. Van der Merwe and W. Mueller, *Liebigs Ann. Chem.*, 1966, **698**, 209.
24 H. Brockmann, H. Pini and O. Plotho, *Chem. Ber.*, 1950, **83**, 161.
25 D.A. Hopwood, *Chem. Rev.*, 1997, **97**, 2465.

26 J.H. Lakey, E.J.A. Lea, B.A.M. Rudd, H.M. Wright and D.A. Hopwood, *J. Gen. Microbiol.*, 1983, **129**, 3565.
27 L. Vertesy et al., *J. Antibiot.*, 2000, **53**, 816.
28 M. Debono et al., *J. Antibiot.*, 1988, **41**, 1093.
29 C. Kempter et al., *Angew. Chem. Int. Ed. Engl.*, 1997, **36**, 498.
30 D. Jung, A. Rozek, M. Okon and R.E.W. Hancock, *Chem. Biol.*, 2004, **11**, 949.
31 J. Micklefield, *Pure Appl. Chem.*, 2009, **81**, 1065.
32 K. T. Nguyen, D. Ritz, J.Q. Gu, D. Alexander, M. Chu, V. Miao, P. Brian, R.H. Baltz, *Proc. Natl. Acad. Sci.* USA, 2006, **103**, 17462.
33 H.H. Wasserman, R.J. Sykes, P. Peverada, C.K. Shaw, R.J. Cushley and C.R. Lipsky, *J. Am. Chem. Soc.*, 1973, **95**, 6874.
34 H.H. Wasserman, C. Shaw and R. Sykes, *Tetrahedron Lett.*, 1974, **15**, 2787.
35 N. Gerber, A. McInnes, D. Smith, J. Walter, J. Wright and L. Vining, *Can. J. Chem.*, 1978, **56**, 1155.
36 A.M. Cerdeño, M.J. Bibb and G.L. Challis, *Chem. Biol.*, 2001, **8**, 817
37 A.E. Stanley, L.J. Walton, M. Kourdi-Zerikly, C. Corre and G.L. Challis, *Chem. Commun* , 2006, 3981.
38 S.J. Mo, P.K. Sydor, C. Corre, M.M. Alhamadsheh, A.E. Stanley, S.W. Haynes, L. Song, K.A. Reynolds and G.L. Challis, *Chem. Biol.*, 2008, **15**, 137.
39 S.W. Haynes, P.K. Sydor, A.E. Stanley, L. Song and G.L. Challis, *Chem. Commun.*, 2008, 1865.
40 J. White and M.J. Bibb, *J. Bacteriol*, 1997, **179**, 627.
41 K.E. Narva and J.S. Feitelson, *J. Bacteriol.*, 1990, **172**, 326.
42 E.P. Guthrie, C.S. Flaxman, J. White, D.A. Hodgson, M.J. Bibb and K.F. Chater, *Microbiology*, 1998, **144**, 727.
43 S.J. Mo, B.S. Kim and K.A. Reynolds, *Chem. Biol.*, 2005, **12**, 191.
44 N.R. Williamson, H.T. Simonsen, R.A.A. Ahmed, G. Goldet, H. Slater, L. Woodley, F.J. Leeper and G.P.C. Salmond, *Mol. Microbiol.*, 2005, **56**, 971.
45 M.G. Thomas, M.D. Burkart and C.T. Walsh, *Chem. Biol.*, 2002, **9**, 171.
46 A.K.P. Harris, N.R. Williamson, H. Slater, A. Cox, S. Abbasi, I. Foulds, et al., *Microbiology*, 2004, **150**, 3547.
47 U. Hornemann and D.A. Hopwood, *Tetrahedron Lett.*, 1978, **33**, 2977.
48 U. Hornemann and D.A. Hopwood, Biosynthesis of methylenomycin A: a plasmid-determined antibiotic. In *Antibiotics IV*, Springer-Verlag, Berlin Heidelberg, 1981, pp. 123-131
49 a) T. Haneishi, N. Kitahara, Y. Takiguchi, M. Arai, and S. Sugawara, *J. Antibiot.*, 1974, **27**, 386; b) T. Haneishi, A. Terahara, M. Arai, T. Hata, and C. Tamura, *J. Antibiot.*, 1974, **27**, 393; c) T. Haneishi, A. Terahara, K. Hamano, and M. Arai, *J. Antibiot.*, 1974, **27**, 400.
50 R. Kirby, L.F. Wright, and D.A. Hopwood, *Nature*, 1975, **254**, 265.
51 K.F. Chater and C.J. Bruton, *EMBO J.*, 1985, **4**, 1893.
52 R. Kirby and D.A. Hopwood, *J. Gen. Microbiol.*, 1977, **98**, 239.
53 A. Aguilar and D.A. Hopwood, *J. Gen. Microbiol.*, 1982, **128**, 1893.
54 K. Spatz, H. Kohn, and M. Redenbach, *FEMS Microbiol. Lett.*, 2002, **213**, 87.
55 M. Yamasaki, Y. Ikuto, A. Ohira, K.F. Chater, and H. Kinashi, *Microbiology*, 2003, **149**, 1351.
56 S. O'Rourke, A. Wietzorrek, K. Fowler, C. Corre, G.L. Challis, and K.F. Chater, *Mol. Microbiol.*, 2009, **71**, 763.
57 C. Corre, L. Song, S. O'Rourke, K.F. Chater, and G.L. Challis, *Proc. Natl Acad. Sci. USA*, 2008, **105**, 17510.

58 a) S.L. Midland, et al., J. Org. Chem., 1993, 58, 2940; b) J.P. Henschke and R.W. Rickards, *Tetrahedron Lett.*, 1996, **37**, 3557.
59 C. Corre and G.L. Challis, *Nat. Prod. Rep.*, 2009, **26**, 977.

MODIFYING OXIDATION AND GLYCOSYLATION EVENTS IN BIOSYNTHESES OF NATURAL PRODUCT ANTICANCER DRUGS – CHALLENGES FOR COMBINATORIAL BIOSYNTHESIS

Jürgen Rohr

University of Kentucky, College of Pharmacy, 789 South Limestone Street, Lexington, Kentucky 40536-0596, USA

1 INTRODUCTION

Combinatorial biosynthesis is a relatively new method to manipulate biosynthetic pathways of natural products by modification of existing biosynthetic genes or recombination of given biosynthetic genes with genes from other, usually related pathways of natural products (Figure 1). The goal is either to create new molecules, e.g. with scaffolds unprecedented in nature (usually called unnatural natural products or hybrid natural products, for some examples see Figure 2), or to design optimized analogues of existing natural product drugs through targeted changes of their molecular periphery (often achieved by pathway engineering).[1-4] This book chapter will not review methods and examples for the generation of libraries of unnatural or hybrid natural products, which have been the subjects of recent reviews instead,[4,5] but will focus on analogue generation.

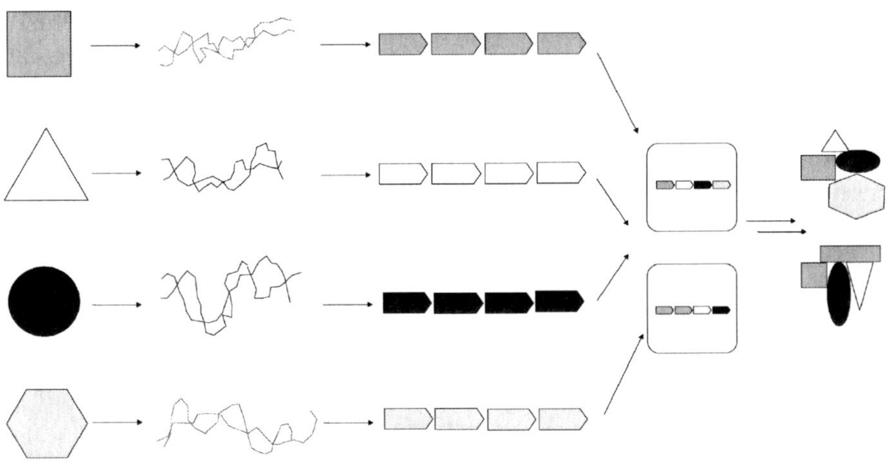

Figure 1 *The principal of combinatorial biosynthesis: isolation and recombination of genes of biosynthetic pathways of natural products, with the goal to generate new natural products.*

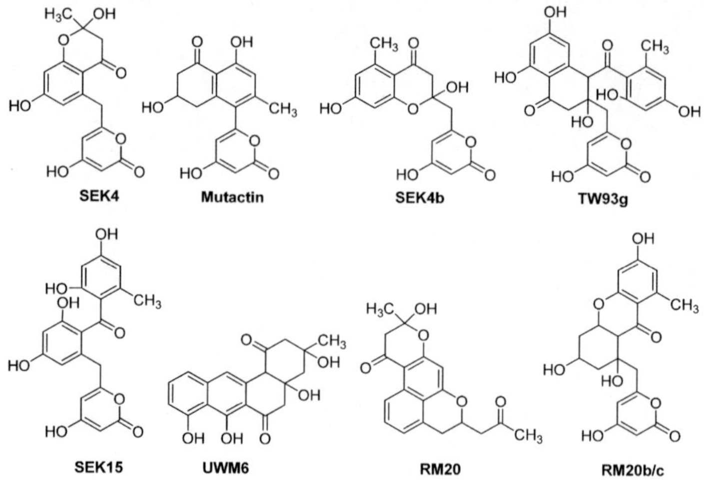

Figure 2 *Examples of unnatural natural/hybrid natural products through recombination of type II polyketide synthase genes.*

As demonstrated by Figure 2, polyketides have been in the centre of the development of combinatorial biosynthetic methods from the beginning onwards, partly because polyketides remain one of the most diverse and most complex groups of natural products with an unusually broad spectrum of biological activities (Figure 3), partly also because polyketide biosyntheses have fascinated natural product chemists since John Norman Collie, i.e. since the end of the 19th century.[6-9]

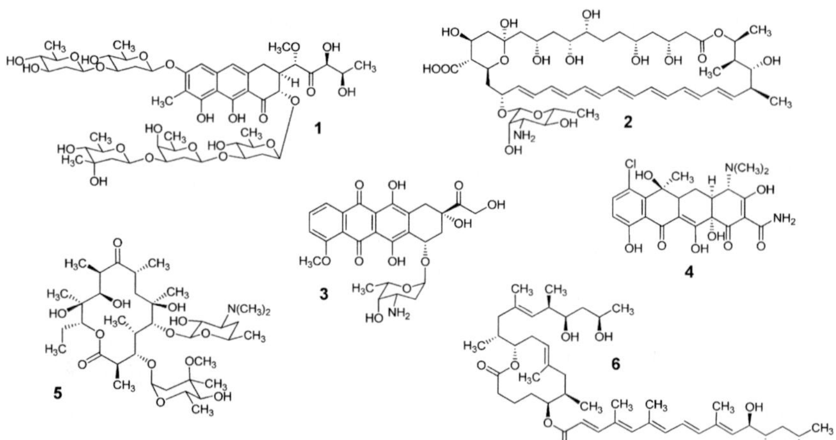

Figure 3 *Clinically relevant polyketides. **1** and **3** are, respectively, the anticancer drugs mithramycin and doxorubicin, **2** the antifungal drug amphotericin B, **4** and **5** the antibiotics chlortetracycline and erythromycin A, and **6** mycolactone, the causative agent responsible for Burulis ulcer skin lesions, a tropical disease.*

As a typical example of the application of combinatorial biosynthetic methods for polyketide drug optimization, the biosynthetic pathway of the aureolic acid class anticancer antibiotic mithramycin (Figure 4) is discussed, and past and current attempts to manipulate its biosynthetic pathway with the goal of creating new analogues with improved bioactivities are reviewed.[10-13]

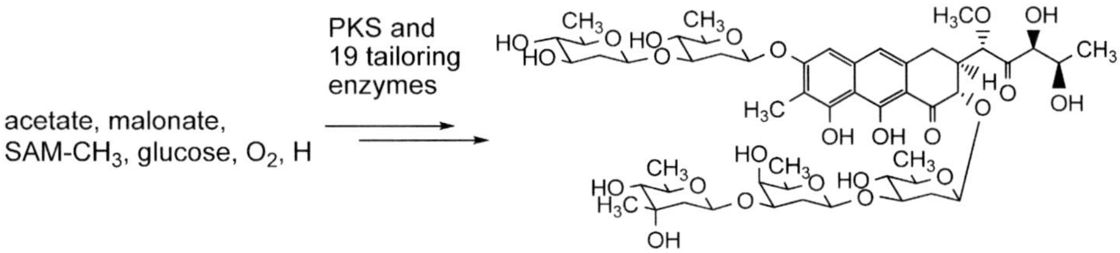

Figure 4 *The polyketide core of mithramycin is biosynthesized from acetate and malonate building blocks, the sugar residues from glucose using the polyketide synthase (PKS) multienzyme complex. Additional post-PKS tailoring oxidation steps (introduction of O-atoms), reduction steps (introduction of H-atoms) and SAM (S-adenosylmethionine)-assisted methyltransfer complete the biosynthesis.*

2 MITHRAMYCIN AND ITS BIOSYNTHETIC PATHWAY

Mithramycin belongs to the small, but distinct family of aureolic acid anticancer drugs, which consists of only five principle members, mithramycin **1**, chromomycin **7**, olivomycin **8**, durhamycin **10** and UCH9 **9** (Figure 5).[10-12,14-17] Mithramycin and chromomycin were clinically applied as anticancer drugs, mithramycin for the treatment of testicular cancers and cancer-related hypercalcemia, and for Paget's bone disease.[18-20] However, severe side effects, such as low white blood counts, anemia, low platelet counts, combined with loss of appetite and bleeding, limited mithramycin's use, so that it is today only considered as a back-up drug. However, recently mithramycin was suggested as a new lead drug for neuroprotection, particularly for the treatment of Huntington's disease, and mithramycin remains a most popular tool to study specificity protein (Sp)-dependent signal transduction pathways.[21-23] Mithramycin's hypocalcemic effect is caused by its ability to inhibit osteoclasts, most likely the c-Src-dependent activation of ruffled borders, since s-Src expression is Sp1-regulated.[24] Mithramycins's mechanism-of-action was investigated by D. Patel and colleagues. It inhibits DNA replication and transcription through cross-linking GC-rich DNA as a Mg^{2+}-coordinated head-to-tail dimer, particularly blocking the DNA-binding of promoter Sp1. In this way, mithramycin blocks the transcription of Sp1-dependent proto-oncogenes involved in tumorigenesis, e.g. *c-myc*, *c-src*, VEGF, SV40 etc.[25-28]

Although some multi-step total syntheses of aureolic acid drugs were achieved, the molecules are far too complex to systematically generate analogues. Thus, the goal of our research was to manipulate the mithramycin biosynthetic pathway by combinatorial biosynthesis, particularly to generate mithramycin analogues with altered saccharide patterns. Therefore, the mithramycin biosynthetic pathway was intensely studied in the author's laboratory in close collaboration with the laboratory of Jose A. Salas and colleagues from the University of Oviedo, Spain.[11,29-35]

Figure 5 shows that the polyketide-derived aglycone moieties of the aureolic acid family members only differ with respect to the 7-alkyl group, which is methyl in mithramycin **1** and chromomycin **7**, isobutyl in durhamycin **10** and UCH9 **9**, and H in olivomycin **8**. However, the aureolic acid drugs differ significantly in their oligosaccharide pattern (mono- to tetrasaccharide chains) and their deoxyhexose building blocks, most of which are D-sugars, occasionally L-sugars (in chromomycin and olivomycin), and a few of these are branched sugars. Some of these sugars are also further decorated, through alkyl residues or methylethers. The fact that there is natural variation in the deoxysugar decoration among the members of the aureolic acid family boded well for combinatorial biosynthetic strategies to further vary these deoxysugar patterns, since nature obviously has used this strategy to modify biological activities, and various combinatorial biosynthetic methods to modify sugar moieties of complex natural products had been already developing since 1995.[36] In addition, Patel's mechanism-of-action studies also revealed that the sugar moieties play an essential role in DNA binding by mithramycin and chromomycin.[25,27]

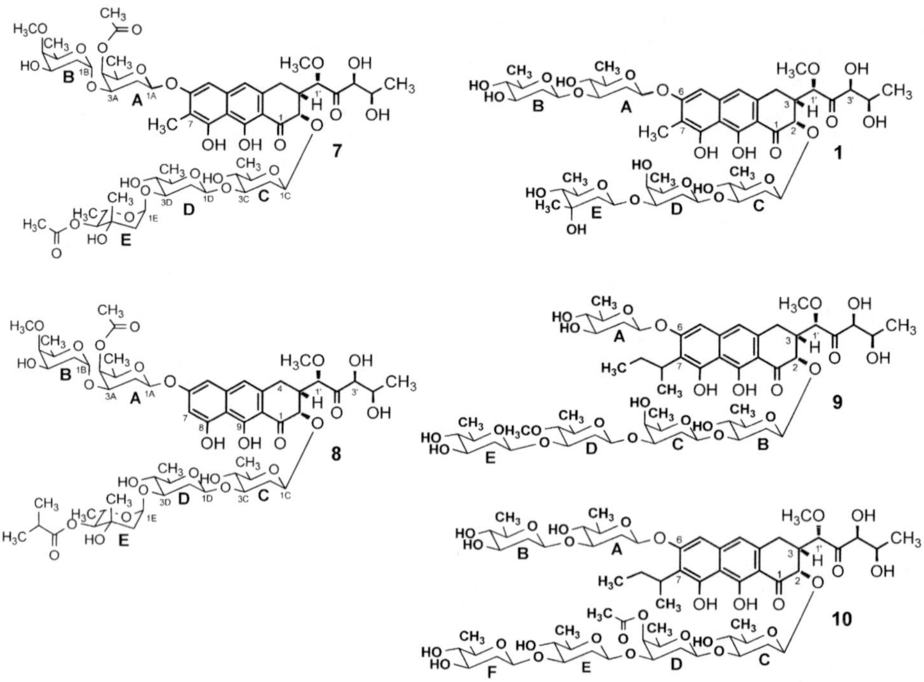

Figure 5 *Aureolic acid family, consisting of five principal molecules, mithramycin **1**, chromomycin A3 **7**, olivomycin **8**, UCH9 **9**, and durhamycin **10**.*

The studies on the mithramcyin biosynthesis included incorporation experiments with various stable isotope-labelled precursors (acetate, glucose) including ^{18}O-labelled acetate and a fermentation in $^{18}O_2$-enriched atmosphere.[5,11] Furthermore, gene inactivation experiments and isolation/structure elucidation of accumulated intermediates/shunt products were applied, after the gene cluster (Figure 6) was cloned and sequenced in the Salas laboratory.[11,29-35] Overall, the picture shown in Figure 7 for the mithramycin biosynthesis emerged. Thus we see that initially the minimal Mtm PKS (encoded by *mtmPKS*) and PKS-associated enzymes, such as cyclases MtmQ, MtmX and MtmY and/or

PKS-associated oxidoreductases Mtm TI, Mtm TII, MtmOI and MtmOII assemble a tetracyclic aglycone-precursor, called demethyl-premithramycinone **11**, which then is *O*- and *C*-methylated as well as successively glycosylated into the fully decorated tetracyclic premithramycin B **18**.

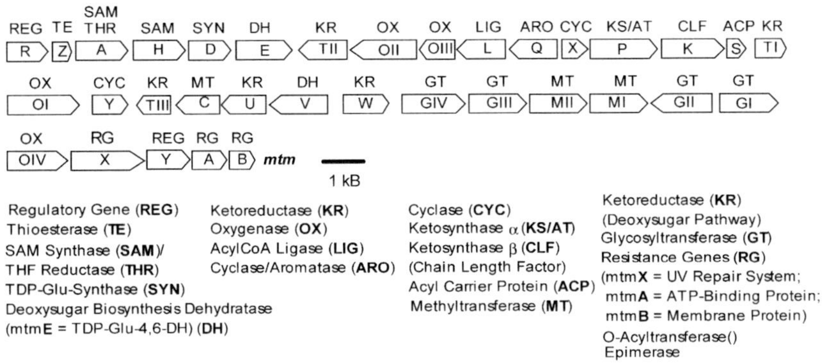

Figure 6 *The mithramycin biosynthetic gene cluster (J.A. Salas et al.[32])*

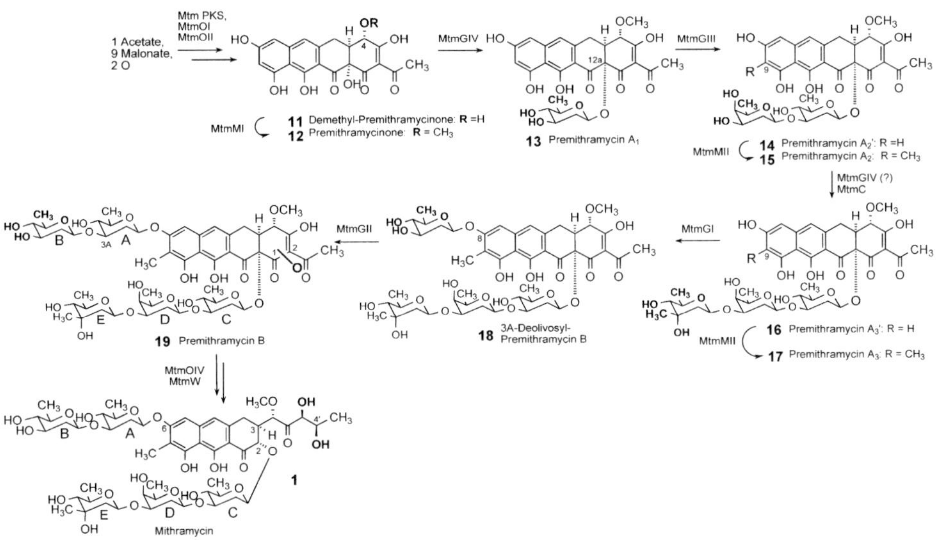

Figure 7 *The biosynthetic pathway of mithramycin. Some ambiguity still remains regarding the double role of glycosyltransferase MtmGIV, which is presumably responsible for the first and third sugar attachment, and the exact timing of the C-methyltransfer at 9-position catalyzed by MtmMII, which occurs either after the second or after the third sugar attachment.*

The two final steps of the mithramycin biosynthesis are the intriguing oxidative C^1-C^2-bond cleavage that establishes the typical tricyclic aglycone of the aureolic acids with its highly functional pentyl side chain in the 3-position, and the reduction of a keto group in this 3-side chain, which then becomes the secondary alcohol (4'-OH) group.[11, 24, 29, 30, 33, 34, 37-40]

3 MITHRAMYCIN ANALOGUES FROM GENE INACTIVATION EXPERIMENTS

Gene inactivation is a commonly used method to elucidate biosynthetic pathways, since ideally stable intermediates accumulate, and can be isolated and analyzed regarding their chemical structures. The method has its limitations, of course, since not all intermediates are stable. They can decompose into shunt products, or downstream enzymes can act on them, so that shunt products accumulate. Finally, the inactivation of multifunctional enzymes or of iteratively acting enzymes (i.e. those that act twice or more) can at best lead to the accumulation of the first substrates of such multi-acting enzymes.

All these complications occurred during the elucidation of the mithramycin biosynthetic pathway, leading to ambiguities. For example, the inactivation of either MtmGI or MtmGII led to the accumulation of premithramycins A_2 **15**, $A_{2'}$ **14**, A_3 **17** and $A_{3'}$ **16** (for structures see Figure 7).[30] Therefore, it remained ambiguous whether the MtmMII-catalyzed *C*-methylation step (at C-9 of the premithramycin core, which becomes C-7 in mithramycin) occurs after the second or after the third sugar attachment. Also, it was unclear which sugar moieties were attached by MtmGI and by MtmGII, until complementation experiments with selected mutants resolved that ambiguity.[35] Often, only experiments with isolated enzymes can resolve such specific questions, which, however, are complicated by enzyme solubility and stability problems and often are hampered due to lack of access of substrates. Nevertheless, such gene inactivation experiments yielded a rough outline of the biosynthetic sequence of events and, more important, also yielded a plethora of new molecules, here all the premithramycins and several mithramycin derivatives.

Since biological activity/DNA binding studies showed that all premithramycins were completely inactive and unable to bind to DNA,[24] the most important derivatives from the gene inactivation experiments for structure-activity-relationships (SAR) and the goal of combinatorial biosynthetic efforts to generate mithramycin analogues were of course the few real mithramycin derivatives, i.e. all those compounds, in which the tricyclic aureolic acid pharmacophore was established. Examples (Figure 8) are 7-demethylmithramycin **20** (accumulated upon inactivation of C-methyltransferase MtmMII), mithramycins SK, SDK and SA (**21**, **22** and **23**, respectively, accumulated upon inactivation of ketoreductase MtmW), 5E-ketomithramycin **24** (accumulated upon inactivation of ketoreductase MtmTIII), and 5C-keto-demycarosylmithramycin **25** (accumulated upon inactivation of C-methyltransferase MtmC and ketoreductase MtmTIII).[24,37,39,41] While the 7-demethyl derivative **20** and mithramycin SA **23** had greatly reduced biological activity (showing the importance of the 7-methyl group for the correct orientation of the lower deoxysugar chain and the importance of the 3-side chain, respectively), mithramycins SK **21** and SDK **22**, both with shorter 3-side chains, showed superior activity as well as reduced toxicity compared to mithramycin.[24,38-40] Both compounds are currently under further investigation and development by the Spanish biotech company EntreChem. Mithramycins SK **21** and SDK **22** result from a β-shift rearrangement from mithramycin DK **26** , the immediate and 'real' product of the MtmW-inactivation (Figure 9), while mithramycin SA **23** is a decomposition product (retro-aldol reaction) occurring in alkaline pH (Figure 9B).[39,40]

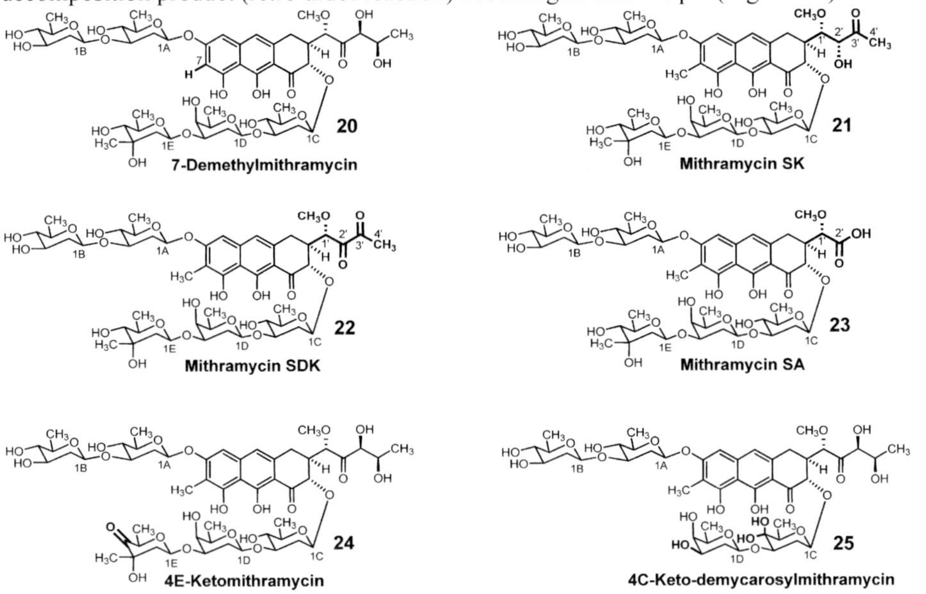

Figure 8 *Examples of mithramycin derivatives from inactivation of selected genes (see text for details). The highlighted (bold) regions indicate structural differences to mithramycin.*

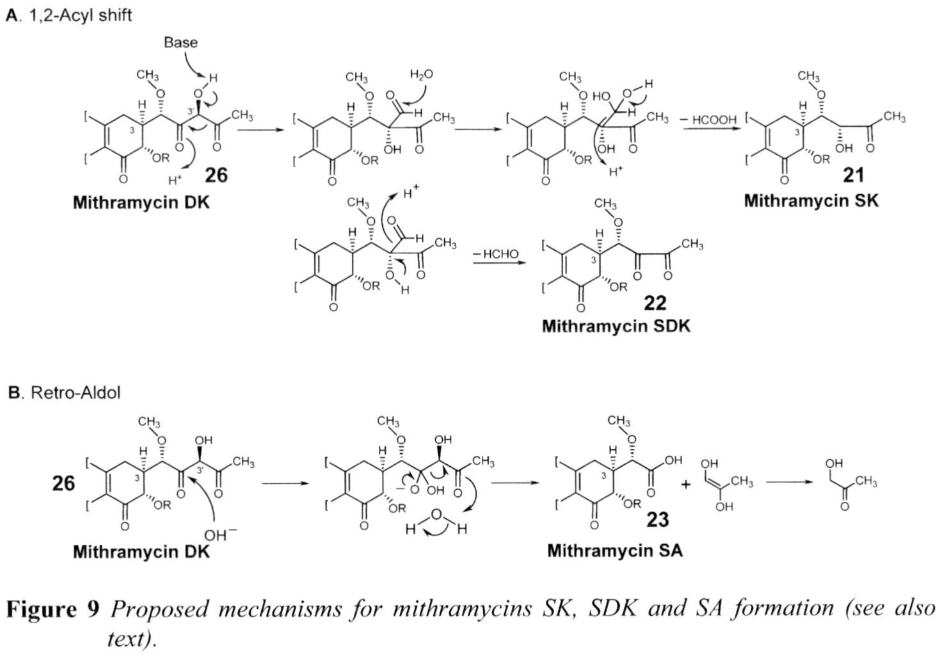

Figure 9 *Proposed mechanisms for mithramycins SK, SDK and SA formation (see also text).*

The two above mentioned ketosugar-containing mithramycins were two of the few mithramycin derivatives with altered saccharide structures resulting from gene inactivation experiments.[37] With regard to SAR, the compounds verified the importance of the E-sugar, and that a 5E-keto group is disadvantageous. For the elucidation of the biosynthetic sequence of events, the MtmC/TIII inactivations suggested that olivose C and mycarose D are presumably transferred as 4-ketosugars and reduced by ketoreductase MtmTIII after attachment to premithramycinone and premithramycin A_2, respectively. Since glycosyltransfer of ketosugars is rare, it was suggested that that the same glycosyltransferase, namely MtmGIV, is responsible for both glycosyl transfer steps (note that the *mtm* gene cluster has only four glycosyltransferase genes, but the mithramycin biosynthesis requires five glycosyltransfer steps!). Of course, the inactivation of MtmGIV yielded only one product, namely premithramycinone **12** (see Figure 7).[29,42]

4 SECOND AND THIRD GENERATION OF MITHRAMYCIN ANALOGUES THROUGH COMPLEMENTATION WITH DEOXYSUGAR GENE CASSETTES

Gene recomplementation experiments in combinatorial biosynthesis aiming at diversification of saccharide structures and patterns has been successfully applied since 1995, when the glycosyltransferase gene *elmGT* along with the genes encoding the polyketide-derived aglycone 12-demethyl-elloramycinone (= 8-demethyltetracenomycin C) from the elloramycin producer *S. olivaceus* Tü2353, all combined into cosmid 16F4, were transformed into the urdamycin producer *S. fradiae* Tü 2717.[36] The urdamycin producer, which among other pathways encodes the biosyntheses of two deoxysugars, D-olivose and L-rhodinose, complemented the permethyl-L-rhamnose biosynthesis genes that were missing from the elloramycin biosynthesis pathway in cosmid 16F4, and the result was 8-D-olivosyl-8-demethyltetracenomycin C, a new elloramycin derivative (Figure 10).

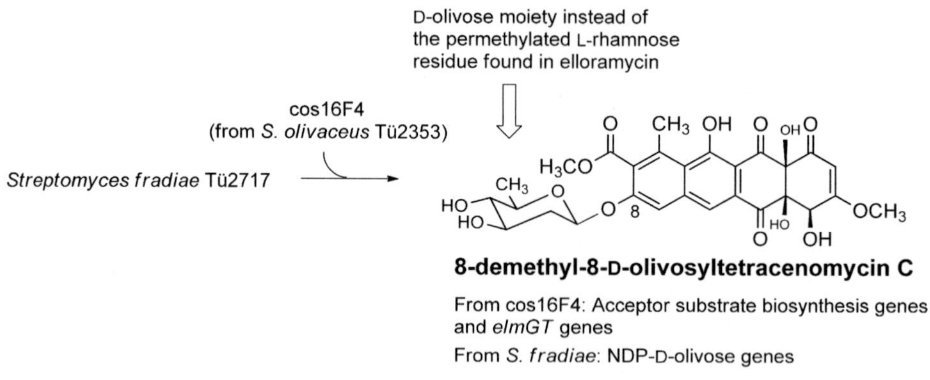

Figure 10 *First hybrid natural product with selectively altered deoxysugar moiety through gene recomplementation.*

The L-rhodinosyl analogue was also found later, and many other deoxysugar pathways leading to various D- or L-nucleoside diphosphate (NDP)-deoxysugars could be exploited to generate a broad spectrum of elloramycin analogues.[43,44] The experiments showed that glycosyltransferase ElmGT is quite promiscuous (can transfer > 10 different sugars!),[45-47] and it was shown later that this is no exception. Glycosyltransferases of secondary metabolite pathways are often quite flexible regarding their NDP-sugar donor substrate. A more elegant variant of this gene recomplementation strategy was later developed in the Salas laboratory, where plasmids (deoxysugar plasmids) encoding entire deoxysugar pathways were assembled from 5 different pathways encoding various D- or L-deoxysugars (some examples of various neutral D- and L-sugar encoding plasmids are shown in Figure 11).[48-50] These deoxysugar plasmids can be used to complement various existing natural product pathways, whenever sugar donor-substrate flexible glycosyltransferases are present. Using one of these recomplementation strategies, numerous analogues of elloramycin, steffimycin, aranciamycin, staurosporine and rebeccamycin were generated.[5] In this context it should be noted that L-rhamnosyltransferases seem to possess particularly broad sugar donor substrate specificity.[5]

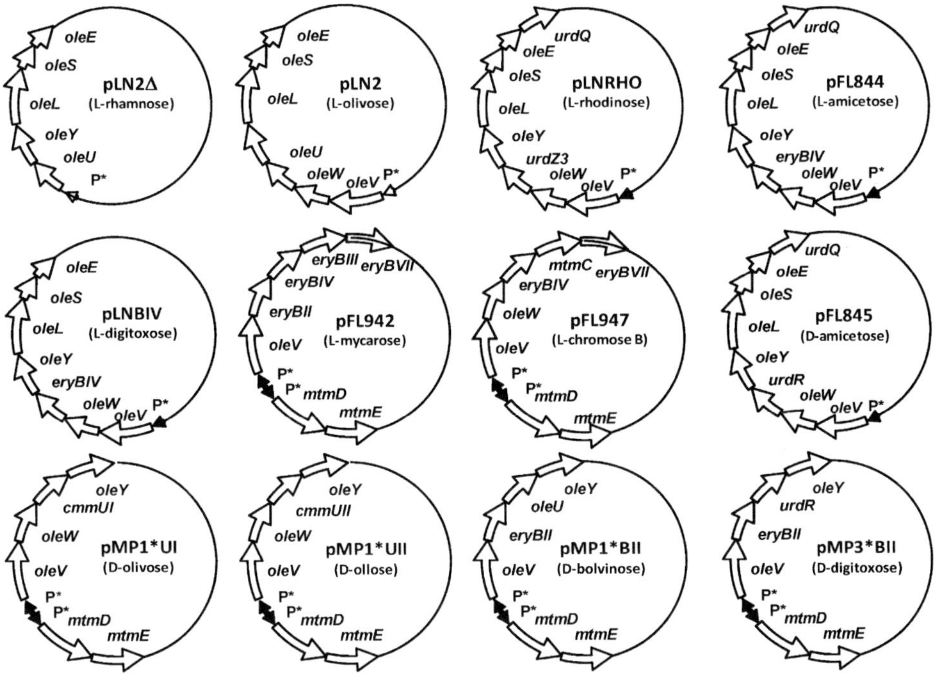

Figure 11 *Examples of deoxysugar plasmids encoding neutral D- or L-NDP-deoxysugar pathways. For the plasmids shown here, genes from five different pathways were used (oleandomycin, ole; urdamycin, urd; erythromycin, ery; mithramycin, mtm; chromomycin, cmm).*

Furthermore, the acceptor substrate specificity of various glycosyltransferases was assessed, in some cases exploited and even modified, particularly after some protein crystal structures of secondary metabolite pathway glycosyltransferases became available, e.g. through the design of glycosyltransferase chimera.[51-62] In addition to successfully exploiting the substrate flexibility of glycosyltransferases, other glycodiversification methods were described, such as glycosyltransfer-reversibility (for the generation of activated NDP-sugar substrates), glycorandomization and neoglycorandomization (enzymatic and chemico-enzymatic approaches, respectively),[63-70] but are so far only rarely applied in the praxis of combinatorial biosynthesis.[66,71,72]

Applying combinatorial biosynthesis involving the use of deoxysugar plasmids for the generation of new mithramycin analogues with altered glycoside patterns imposed a novel challenge, since the flexibility of five glycosyltransfer steps catalyzed by four glycosyltransferases needed to be interrogated. In addition, each newly expressed deoxysugar pathway encoded by a single deoxysugar-plasmid had to compete with the biosynthesis of three native deoxysugar pathways, those of D-olivose, D-oliose and D-mycarose, the building blocks found in the two saccharide chains of mithramycin. Initially, the wild-type mithramycin producer *S. argillaceus* was used as host (receptor) strain for the deoxysugar plasmids (leading to second generation mithramycin analogues).[73,74] Only recently, the approach was varied in that selected mutant strains were also considered as receptor strains.[74] The requirement for these mutant strains was that these were compromised in the biosynthesis of one of the natural sugar building blocks. Suitable mutant strains were the *mtmU*-minus mutant and the *mtmC*-minus mutant. MtmU is a 4-ketoreductase specific for the generation of the axial 4-OH group of the D-oliose building block (sugar D), and MtmC is the *C*-methyltransferase involved in the biosynthesis of the D-mycarose building block (sugar E). Thus, it was intended to replace only these natural sugars by those encoded by one of the deoxysugar plasmids, without compromising the glycosyltransferase activities. So far, only the experiments involving the *mtmU*-minus mutant were finished, leading to the first third generation mithramycin analogues.[74]

Overall, the use of 16 deoxysugar-plasmids (encoding the 12 neutral sugars shown in Figure 11) and of 4 plasmids (encoding the amino sugars NDP-L-ristosamine, NDP-D-desosamine, NDP-L-daunosamine, and NDP-D-demethylangolosamine) and two different host strains (the wild-type strain and the *mtmU*-minus mutant) yielded 11 new second and third generation mithramycin analogues, many with novel, unprecedented saccharide patterns (compounds **27-37**, Figure 12). Checking their biological activities now allowed the first SAR conclusions, e.g. regarding the (upper) disaccharide chain. As expected, the biologically most active compounds were those with a complete set of five sugars, the most active so far is demycarosyl-digitoxosylmithramycin **27**,[73] a compound that differs from mithramycin only in that it lacks the 3E-methyl group. Demycarosyl-digitoxosylmithramycin **27** was tested against various cancer cell lines and showed to be superior to mithramycin (5-10 times better activity) against almost all tested human cancer cell lines. Most intriguing is that the compound shows equally good inhibition of an estrogen-receptor (ER)-positive and ER-negative human breast cancer cell line, while mithramycin could only inhibit the ER-positive MCF-7 cell line (Figure 13).

The results show that subtle changes in the sugar patterns can clearly improve the biological activity of mithramycin, and that the idea to modify the sugar residues of mithramycin by combinatorial biosynthesis is promising, however the total number of found analogues remained far below the expectations. For instance, a single to multiple replacement of the natural mithramycin sugar moieties by one newly introduced sugar should yield already 27 new analogues with a complete sugar pattern of five sugars, plus various incompletely-decorated analogues plus analogues in which biosynthetic intermediates of the plasmid-encoded sugar pathway would be utilized. From that viewpoint, a result of only 11 mithramycin analogues utilizing two different host strains and the 12 deoxysugar plasmids shown in Figure 11 is rather low. In addition, none of the four used aminosugar encoding plasmids (not shown in Figure 11) yielded any new compounds, although the introduction of an amino sugar moiety would generate advantageous properties for drug delivery (e.g. by nano-delivery systems). Glycosyltransferases MtmGIV and MtmGII appeared to be very specific, since neither the B- nor the C-position could be affected by the approach. In many cases, only premithramycin-type compounds (with a tetracyclic aglycone) were produced, indicating that oxygenase MtmOIV was unable to convert some compounds with drastically changed sugar moieties.

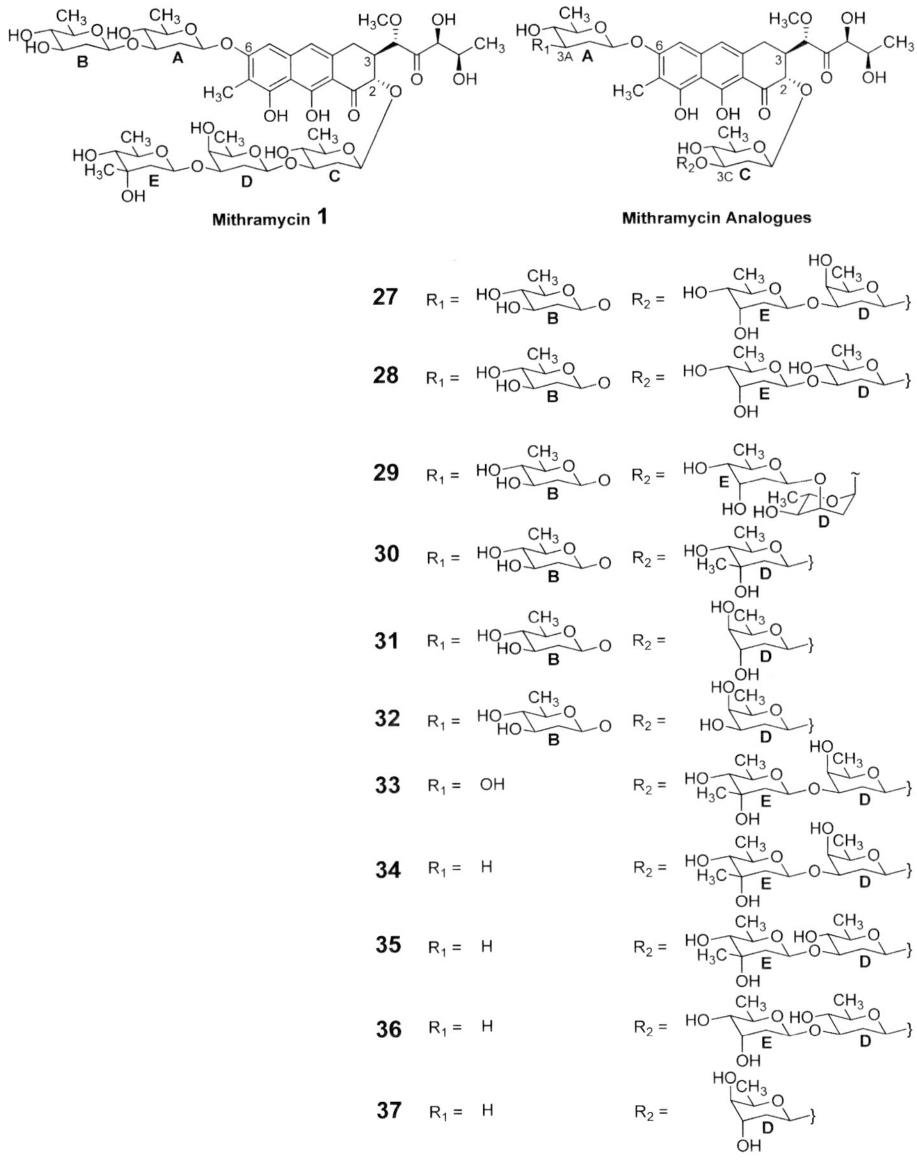

Figure 12 *New second and third generation mithramycin analogues from combinatorial biosynthesis using the deoxysugar-plasmid gene recombination approach.*

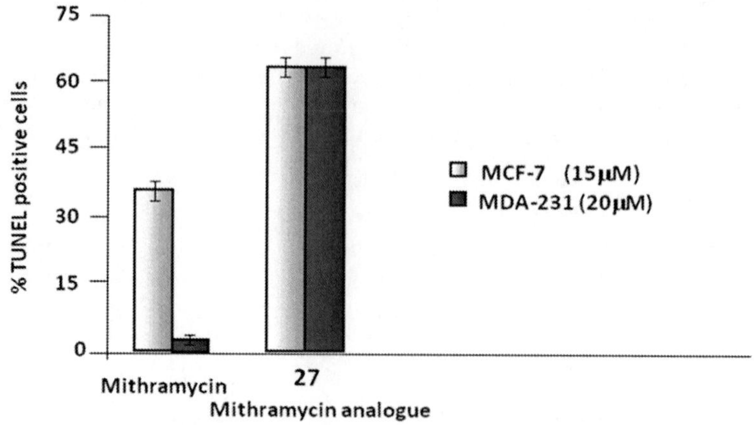

Figure 13 *Anticancer TUNEL assay showing mithramycin analogue **27** (demycarosyl-digitoxosylmithramycin) to be more active than mithramycin against the estrogen-receptor (ER)-positive human breast cancer cell line MCF-7 and showing a similar activity against the problematic ER-negative breast cancer line MDA-231, which is unaffected by mithramycin itself.*

5 MTMOIV, THE KEY OXYGENASE OF THE MITHRAMYCIN BIOSYNTHETIC PATHWAY

The results from the combinatorial biosynthetic gene recombination experiments involving deoxysugar plasmids (see previous section) indicated that MtmOIV, the oxygenase that catalyzes the oxidative cleavage of the fourth ring of premithramycin B, possesses only moderate substrate flexibility. This restricted all approaches to biosynthetically modify the sugar decoration pattern of mithramycin, since the MtmOIV step occurs *after* the completion of the sugar decoration. Consequently, many premithramycin-type compounds accumulated in deoxysugar-treated mithramycin producers. These were identified in HPLC plots from their UV spectrum, since the tetracyclic premithramycin chromophore is significantly different from the tricyclic mithramycin-chromophore.

From the viewpoint of a nature-optimized enzyme, the restricted substrate specificity of MtmOIV is not surprising, since the producer organism needs to avoid the co-production of many, presumably biologically less active, mithramycin congeners with incomplete sugar patterns. Indeed, none of the premithramycinones and premithramycins en route to premithramycin B is used as a substrate of MtmOIV, and not even trace amounts of mithramycin derivatives with incomplete sugar patterns can be detected in the wild-type producer strain.

Taking these facts into consideration, it was necessary for a broader applicability of the above described combinatorial biosynthetic approaches to study MtmOIV and its protein structure, with the long term goal of redesigning this enzyme towards broader substrate specificity.

As mentioned before, MtmOIV catalyzes the penultimate step of mithramycin biosynthesis, and oxidatively cleaves the fourth ring of premithramycin B, thereby introducing the oxygen atom, which ends up in the 3'-position of mithramycin. As another consequence of the MtmOIV reaction, carbon-1 of premithramycin B is lost, through oxidative decarboxylation. It was reasoned (and proven later, see below) that mithramycin DK, a compound with a labile β-diketo side chain, is the immediate product of the MtmOIV reaction. Finally, as the last step of mithramycin biosynthesis, the 4'-keto group of mithramycin DK is reduced by ketoreductase MtmW to a secondary alcohol function (Figure 14). Evidence for this sequence of events came from labelling studies with ^{13}C- and ^{18}O-labelled precursors[11] and the fact that premithramycin B accumulated upon inactivation of MtmOIV.[75] Furthermore, mithramycins SK **21**, SDK **22** and SA **23**, rearrangement and decomposition products of the unstable mithramycin DK **26**, respectively (see Figure 9), accumulate upon MtmW-inactivation.[39, 40]

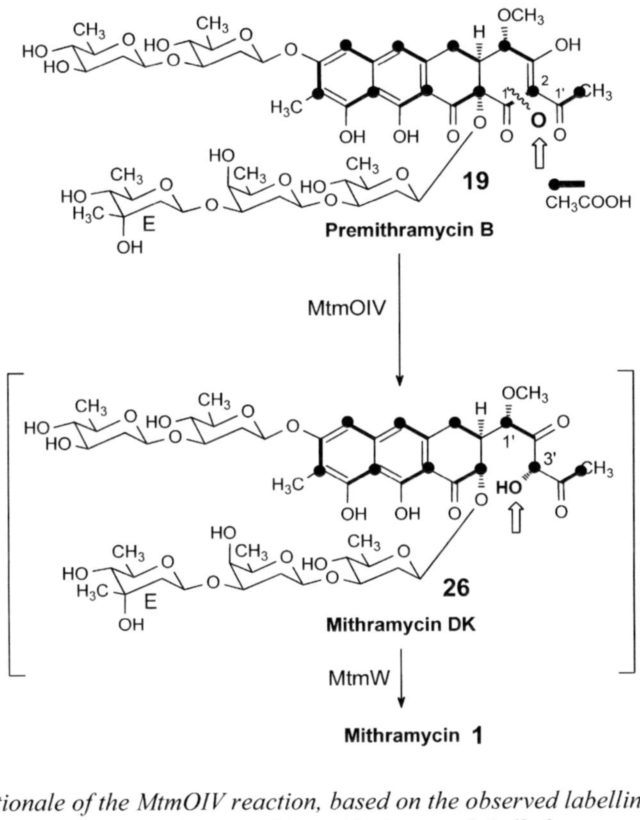

Figure 14 *Rationale of the MtmOIV reaction, based on the observed labelling pattern from incorporation experiments with stable-isotope labelled precursors, and gene inactivation experiments (see also text).*

It is assumed that MtmOIV (possibly along with MtmW) serves multiple roles for the producing organism *Streptomyces argillaceus*: (i) activation of an inactive (premithramycin B does not bind to DNA) into a biologically active molecule late in the pathway, presumably while or after channelling the molecules out of the bacterial cell, thereby (ii) protecting the DNA of the bacterial host organism, and (iii) controlling the biosynthetic flux and outcome (gatekeeper) to avoid the formation of unwanted, less active shunt products (Figure 15).

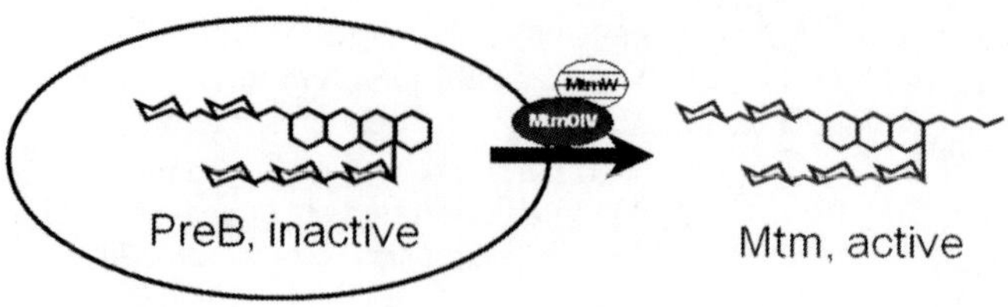

Figure 15 *The role of MtmOIV (possibly along with MtmW) as a gatekeeper enzyme that controls the biosynthetic pathway and helps to protect the host organism's DNA (PreB = premithramycin B **19**, Mtm = mithramycin **1**).*

First and foremost, kinetic data of the MtmOIV reaction, along with the identification of possible reaction intermediates, were needed, since the labelling patterns found after incorporation experiments with 1,2-$^{13}C_2$-acetate and a fermentation under an $^{18}O_2$-enriched atmosphere cannot distinguish between a Baeyer-Villiger (initiated through attack of the ketone C-1 in premithramycin B **19**) and a retro-aldol mechanism (initiated through hydroxylation of the 2-position in premithramycin B **19**, Figure 16).

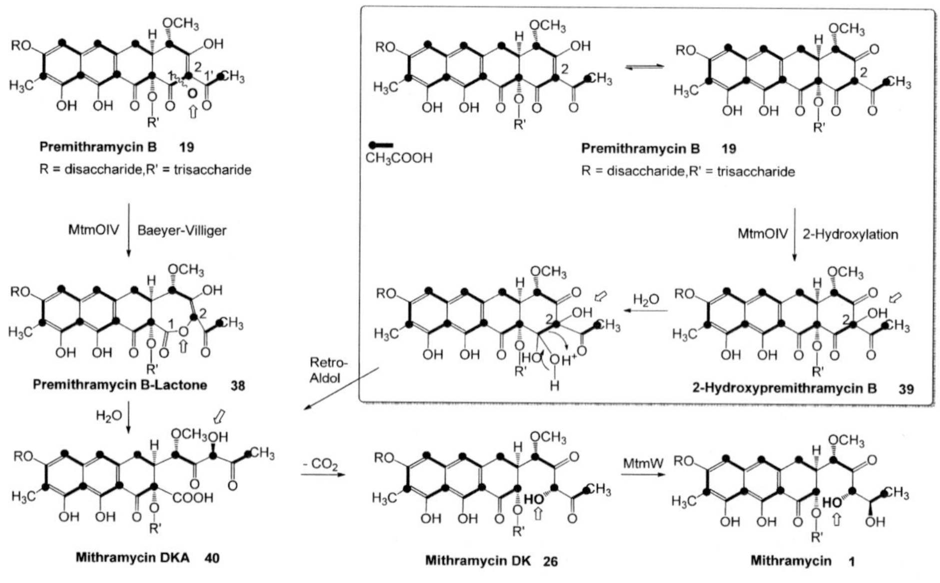

Figure 16 *Two alternative mechanisms leading from premithramycin B **19** to mithramycin **1** congruent with the isotope pattern found from incorporation experiments, (i) either via Baeyer-Villiger oxidation (outside the box), or (ii) via 2-hydroxylation followed by retro-aldol cleavage (box).*

Given these two alternatives, it was necessary to find and characterize a reaction intermediate unique for one or the other mechanism, here either premithramycin B-lactone

38, expected only for the Bayer-Villiger path, or 2-hydroxypremithramycin B **39**, expected only if the initial MtmOIV reaction were a 2-hydroxylation reaction.

After a suitable *E. coli*-overexpression system was found using plasmid pRSETB, and sufficient and purified His_6-tagged MtmOIV was available in soluble form, the stability co-factor dependence, kinetics and pH-dependence of the enzyme was established (Figure 17). MtmOIV is stable for a week when stored at -80 °C, uses FAD and NADPH as co-factors, has a K_M of 34 (± 12) μM and a V_{max} of 147 (± 22) nmol min^{-1} mg^{-1}, and operates best at alkaline pH >8.[76, 77]

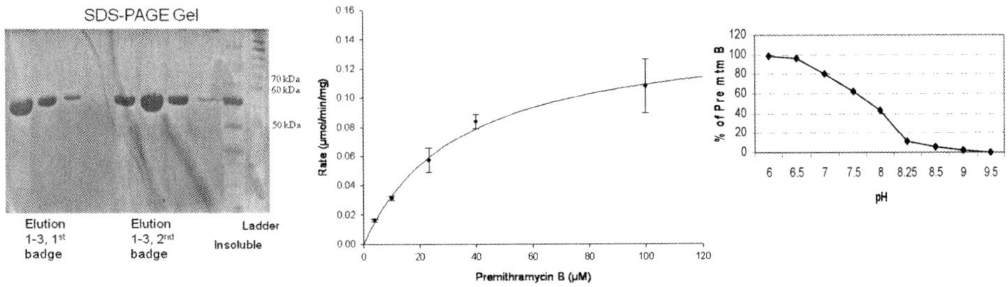

Figure 17 *Purification of MtmOIV (left), kinetics (middle) and pH-dependence (right) of the MtmOIV reaction (the conversion of premithramycin B was measured; the His_6-tagged protein was used for the generation of these data.*

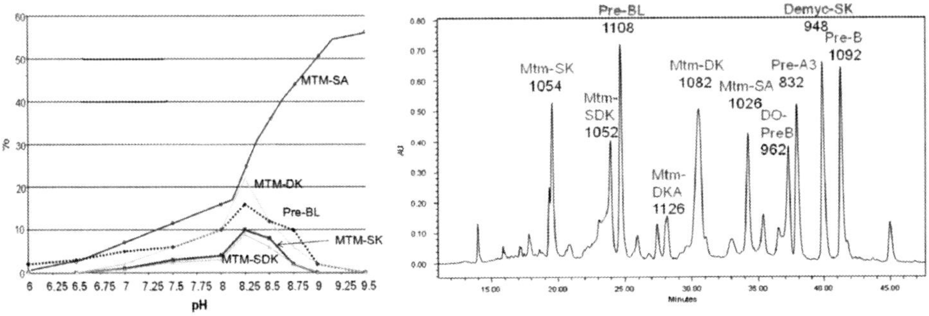

Figure 18 *Intermediates and shunt products of the MtmOIV-reaction appeared to be best accumulated when the reaction was run at pH 8.25 for 4 h. Left: observed intermediates, right: HPLC/MS diagram at 4 h and pH 8.25 (Pre-BL = premithramycin B-lactone, MTM-SK = mithramycin SK, MTM-SDK = mithramycin SDK, MTM-DK = mithramycin DK, MTM-SA = mithramycin SA, MTM-DKA = mithramycin DKA, DO-PreB = 3A-deolivosyl-premithramycinB, PreA3 = premithramycin A3, Demyc-SK = demycarosylmithramycin SK, PreB = premithramycin B; for structures see Figs 7, 8 and 16).*

Although the fastest conversion of substrate premithramycin B was observed at pH 9.5, a significant acceleration of the MtmOIV reaction compared to those at lower pHs could already be observed at pH 8.25. In addition, when the reaction was stopped at 4 h at pH 8.25, the largest number of apparent intermediates was detected in an HPLC plot (Figure 18).

Isolation of the major peaks visible in the HPLC yielded the proposed intermediates mithramycins DKA and DK, and – most important - premithramycin B-lactone **38**, but no trace of 2-hydroxy-premithramycin B **39** (see Figure 16). Mithramycin DKA **40** is extremely unstable and decarboxylates immediately into mithramycin DK **26**. Thus, the reaction sequence shown in Figure 16 (outside the box) is irreversible and there is no chance (particularly not in the alkaline milieu) that premithramycin B-lactone **38** could be formed from mithramycin DKA **40**.

Obviously, the decarboxylation of mithramycin DKA **40** is facilitated because of the β–keto group. All intermediates and shunt products were fully characterized, including by ^{1}H and ^{13}C NMR, except for mithramycin DKA **40**, for which only MS data could be obtained. In summary, the Baeyer-Villiger path was confirmed for the MtmOIV reaction, establishing MtmOIV as the first Baeyer-Villiger monooxygenase (BVMO) that was fully characterized with its proven natural substrate.[77] Note that numerous other BVMO steps were proposed (but not proven) for other biosynthetic pathways of natural products,[78] and most other investigations of BVMO were carried out with relatively simple small chemical molecules, usually mono-functional ketones (e.g. cyclohexanone, phenylmethylketone), which may or may not be (one of) their natural substrates.[79-81] MtmOIV seems to be involved in all steps leading to mithramycin DK (**26**), but its involvement in the formation of the short side chain mithramycins SK **21**, SDK **22** and SA **23** remains unclear (Figure 19).[77]

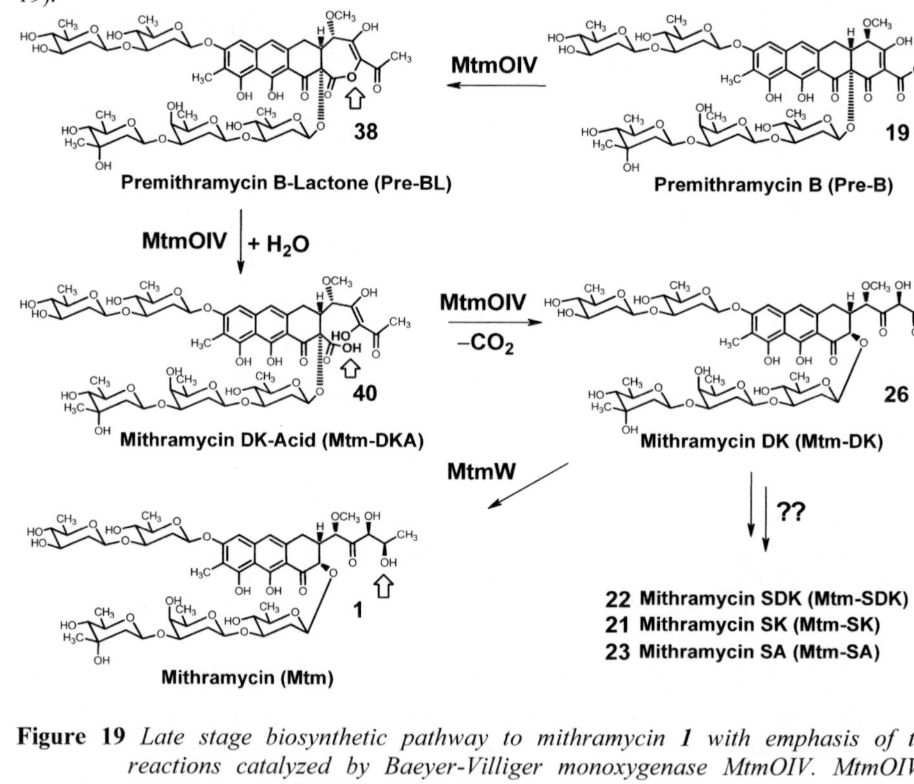

Figure 19 *Late stage biosynthetic pathway to mithramycin **1** with emphasis of the reactions catalyzed by Baeyer-Villiger monoxygenase MtmOIV. MtmOIV's involvement in the formation of the short chain shunt products mithramycins SDK **22**, SK **21** and SA **23** in the absence of ketoreductase MtmW remains obscure.*

6 THE MTMOIV CRYSTAL STRUCTURE AND TARGETED ENZYME DESIGN

After the kinetic parameters and the principle reaction mechanisms of MtmOIV were solved (see previous section), it was necessary to solve MtmOIV's protein structure, to gain insight into its co-factor (FAD) and substrate binding. From this, it was hoped that further insight into the exact reaction mechanism as well as guidance to modify the enzyme for broadening its substrate specificity could be acquired.

While X-ray structures of protein crystals (with and without Se-methionine incorporation), both with 2.9 Å resolution diffraction, were already acquired by the end of 2005,[82] it took another four years as well as the publication of the high-resolution structure of PgaE by J. Niemi et al.[83, 84] to solve MtmOIV's structure by molecular replacement.[78] MtmOIV is a 56 kDa homodimeric FAD and NADPH-dependent enzyme. Its protein structure is the second BVMO recorded in the protein data bank (after PAMO), and the first structure of a BMVO whose natural substrate was proven unambiguously. Compared to other so far described BVMOs, MtmOIV shows the two Rossman (GxGxx(G/A)) folds, residues G19-G24 and G162-G167, but lacks a so-called Baeyer-Villiger motif (FXGXXXHXXXW(P/D)),[85] all being proposed as typical for BVMOs (Figure 20). In contrast to MtmOIV, which was clearly established as BVMO catalyzing the oxidative C-C bond cleavage of the fourth ring of the complex substrate premithramycin B, PgaE is no BVMO, but proposed to be a hydroxylase catalyzing the 12-hydroxylation of UWM6 within the biosynthesis of angucyclinone gaudimycin C, however, no proof exists for PgaE's suggested substrate and reaction.

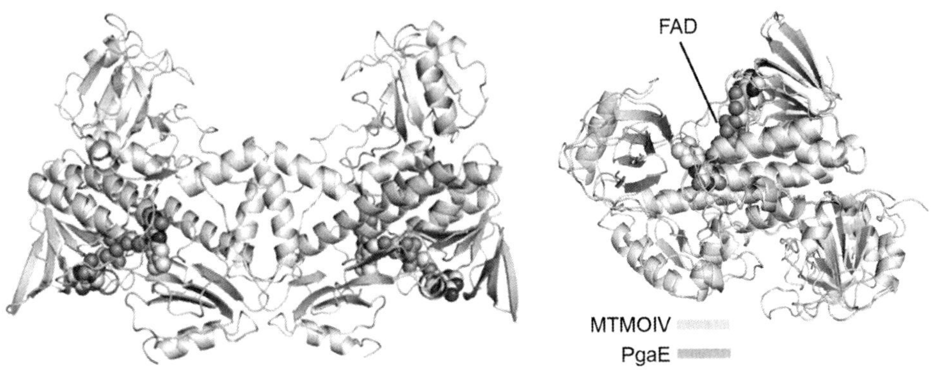

Figure 20 *MtmOIV crystal structure (left) and comparison (overlay) with PgaE (right).*

Interestingly, MtmOIV's arginine-52 residue in its active site, which is required to stabilize both the negative-charged peroxyflavin- and Criegee-intermediates, is located on the *si*-side of the FAD flavin ring, in contrast to the BVMO PAMO, whose stabilizing Arg residue was found on the *re*-side. Figure 21 shows this aspect and the proposed sequence of events of substrate binding and initial Baeyer-Villiger reaction resulting in the formation of premithramycin B-lactone.[78]

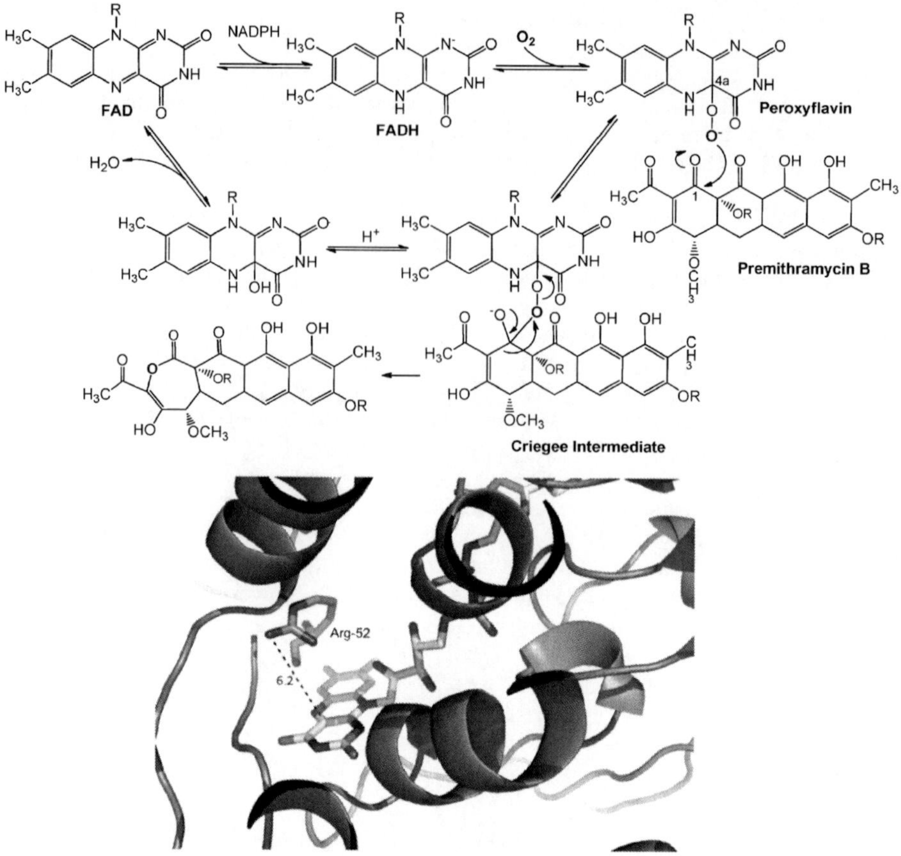

Figure 21 *MtmOIV's active site showing the Arg-52 residue on the si-side of the FAD flavin ring. On the right, the proposed sequence of events leading to premithramycin B-lactone is shown, including the two negatively-charged peroxyflavin and Criegee intermediates that need to be stabilized by the Arg-52 residue.*

While the FAD-binding could be clearly established from the MtmOIV protein crystal structure, the resolution of a premithramycin B/MtmOIV co-crystal was not good enough (3.6 Å) to determine the exact substrate binding site. However, Autodock 4 computational modelling allowed the preliminary determination of the premithramycin B substrate binding site. Important for these calculations was that the substrate could be placed in suitable 'striking distance' of the FAD flavin ring 4a-position, which carries the peroxy residue that initiates the Baeyer-Villiger reaction by attacking the C-1-keto group of premithramycin B. C-1 of premithramycin B and C-4a of the FAD flavin ring were 5.30 Å apart, typical for the required reaction (Figure 22). To obtain further evidence for the substrate binding, site-directed mutagenesis experiments were carried out targeting residues that were either (remotely) involved in FAD binding or substrate binding (Figure 22, Table 1).[78]

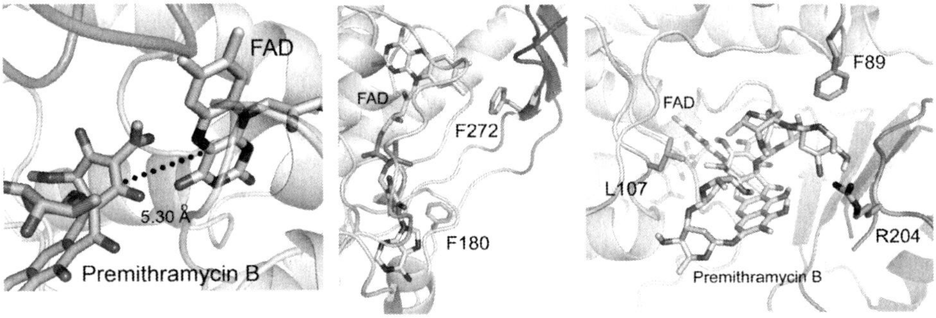

Figure 22 *The MtmOIV active site, left: substrate premithramycin B was modelled into the active side so that the two reacting centres C-4a of the FAD flavin ring and C-1of the substrate were in suitable distance; middle and right: green-labelled amino acid residues presumably involved in FAD (middle) and premithramycin B (right) binding that were targeted for site-directed mutagenesis to interrogate found FAD and substrate binding properties.*

Table 1 clearly shows that the enzyme activity (k_{cat}/K_m) is lost or significantly diminished when bulky natural residues needed for the FAD binding (F272, F180, L107) and function (R52, see Figure 21, left) were replaced by alanine residues. The total loss of activity was somewhat unexpected for the L107A mutant, since the leucine-107 residue was expected to interact more with the substrate than with the co-factor, but is located somewhere near where substrate and co-factor interact. However, replacing two bulky amino acid residues (phenylalanine-89 and arginine-204) deduced to be involved in substrate binding (interaction with the trisaccharide chain of premithramycin) by alanine improved the enzyme activity. The results provide further evidence that the correct substrate binding pocket was identified by the modelling studies. It should also be emphasized that the electron density found for the diffraction of the MtmOIV/premithramycin B co-crystal served as the starting point of the Autodock modelling studies.[78]

	K_m (μM) PreB	k_{cat} (s^{-1})	**k_{cat}/K_m (s^{-1} mM^{-1})**	**FAD mole ratio** (mol_{MtmOIV}/mol_{FAD})
WT	31 ± 10	1.1 ± 0.11	35	0.56
F89A	27 ± 9	1.27 ± 0.11	47	0.42
L107A	---	---	---	<0.01
R204A	18 ± 3	0.73 ±0.02	41	0.24
R52A	79 ± 10	2.21 ± 0.01	2.7	0.07
F180A	---	---	---	< 0.01
F272A	---	---	---	< 0.01

Table 1 *Kinetic data of selected active-site mutants*

Overall, the crystal structure of MtmOIV provided some deeper insight into further elucidation of the MtmOIV reaction mechanism and as a guide to modify MtmOIV towards broader substrate specificity, particularly to accommodate and process future premithramycin analogues with drastically altered saccharide patterns. A rationally designed MtmOIV enzyme may be a way out of the dilemmas the combinatorial biosynthetic approaches on mithramycin analogues seem to be faced with. Additional insight is expected from the crystal structure of CmmOIV, MtmOIV's counterpart in the chromomycin pathway, since chromomycin **7** has a significantly different sugar pattern than mithramycin **1**, notably its L-sugar in the E-position, the unusual α-D-connection of sugar B and its modified (O-acetyl, O-methyl derivatizations) sugar residues in A-, B- and E-positions (see Figure 6). The work on CmmOIV is in progress.

7 SUMMARY AND OUTLOOK

The work on the mithramycin biosynthetic pathway proves that combinatorial biosynthesis is a promising tool for anticancer drug improvement. More than in most other studied and reengineered pathways, the engineering of the mithramycin pathway by combinatorial biosynthesis was crowned by success in that several analogues with improved biological activities were generated, particularly mithramycins SK and SDK. The attempts to modify saccharide patterns were also successful (see demycarosyl-digitoxosylmithramycin), but were more challenging than expected, because of the specificity of MtmOIV, a biosynthetic bottleneck-enzyme acting after the glycosylation steps. Its specificity is logical from the viewpoint of the producing organisms (see introduction of section 5), which led to an evolutionary-optimized, cave-like binding pocket, designed for the accommodation of premithramycin B (Figure 23). Rational enzyme design may be the way of overcoming such late-acting bottleneck enzymes, and the mithramycin pathway is a good pioneer system to further explore this.

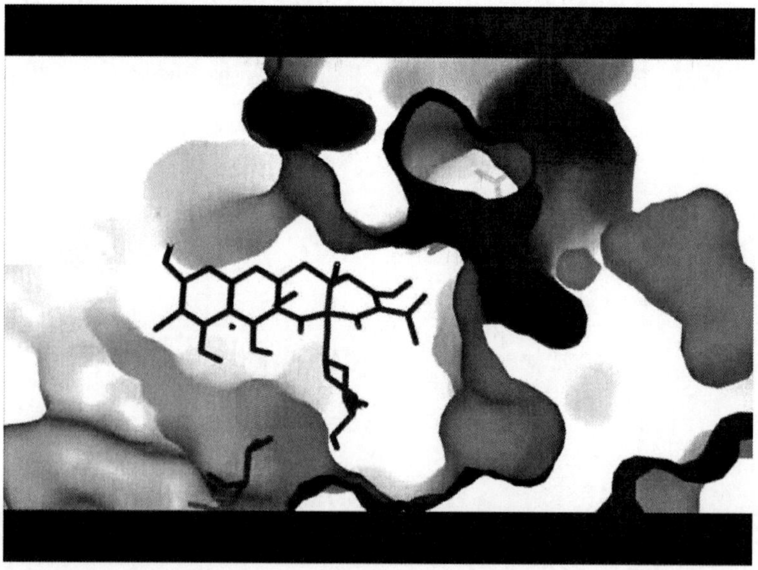

Figure 23 *Artistic rendering of the premithramycin B binding pocket of BVMO MtmOIV.*

References

1 H.G. Floss, *J. Biotechnol.*, 2006, **124**, 242.
2 J.K. Borchardt, *Modern Drug Discovery*, 1999, 22.
3 C.R. Hutchinson and R. McDaniel, *Curr Opin Investig Drugs*, 2001, **2**, 1681.
4 H.G. Menzella and C.D. Reeves, *Curr. Opin. Microbiol.*, 2007, **10**, 238.
5 J. Rohr and C. Hertweck, in *Comprehensive Natural Products II - Chemistry and Biology*, eds. L. Mander and H.-w. Liu, Elsevier, Amsterdam, 2010.
6 D. O'Hagan, *The Polyketide Metabolites*, Ellis Horwood, New York London, 1991.
7 B.J. Rawlings, *Nat. Prod. Rep.*, 1999, **16**, 425.
8 B.J. Rawlings, *Nat Prod Rep*, 2001, **18**, 231.
9 J. Staunton and K.J. Weissman, *Nat. Prod. Rep.*, 2001, **18**, 380.
10 J.D. Skarbek and M.K. Speedie, in *Antitumor compounds of natural origin: chemistry and biochemistry*, ed. A. Aszalos, CRC Press, Boca Raton, Florida, 1981, pp. 191.
11 J. Rohr, C. Méndez and J.A. Salas, *Bioorg. Chem.*, 1999, **27**, 41.
12 F. Lombo, N. Menendez, J.A. Salas and C. Mendez, *Appl. Microbiol. Biotechnol.*, 2006, **73**, 1.
13 S.E. O'Connor, *Chem Biol*, 2004, **11**, 8-10.
14 H. Jayasuriya, R.B. Lingham, P. Graham, D. Quamina, L. Herranz, O. Genilloud, M. Gagliardi, R. Danzeisen, J.E. Tomassini, D.L. Zink, Z.Q. Guan and S.B. Singh, *J. Nat. Prod.*, 2002, **65**, 1091.
15 R. Katahira, Y. Uosaki, H. Ogawa, Y. Yamashita, H. Nakano and M. Yoshida, *J. Antibiot.*, 1998, **51**, 267.
16 M. Tagashira, T. Kitagawa, N. Nozato, S. Isonishi, A. Okamoto, K. Ochiai and Y. Ohtake, *Chem. Pharm. Bull.*, 2000, **48**, 575.
17 W.R. Roush, K. Briner, B.S. Kesler, M. Murphy and D.J. Gustin, *J Org Chem*, 1996, **61**, 6098.
18 T.J. Lee, E.M. Jung, J.T. Lee, S. Kim, J.W. Park, K.S. Choi and T.K. Kwon, *Mol. Cancer Ther.*, 2006, **5**, 2737.
19 S.V. Reddy, N. Kurihara, C. Menaa and G.D. Roodman, *Rev Endocr Metab Disord*, 2001, **2**, 195.
20 G. Koutsodontis and D. Kardassis, *Oncogene*, 2004, **23**, 9190.
21 S. Chatterjee, K. Zaman, H. Ryu, A. Conforto and R.R. Ratan, *Annals of Neurology*, 2001, **49**, 345.
22 R.J. Ferrante, H. Ryu, J.K. Kubilus, S. D'Mello, K.L. Sugars, J. Lee, P. Lu, K. Smith, S. Browne, M.F. Beal, B.S. Kristal, I.G. Stavrovskaya, S. Hewett, D.C. Rubinsztein, B. Langley and R.R. Ratan, *J. Neurosci.*, 2004, **24**, 10335.
23 H. Ryu, J. Lee, K. Zaman, J. Kubilis, R.J. Ferrante, B.D. Ross, R. Neve and R.R. Ratan, *J Neurosci*, 2003, **23**, 3597.
24 L.L. Remsing, H.R. Bahadori, G.M. Carbone, E.M. McGuffie, C.V. Catapano and J. Rohr, *Biochemistry*, 2003, **42**, 8313.
25 M. Sastry, R. Fiala and D.J. Patel, *Journal of Molecular Biology*, 1995, **251**, 674.
26 M. Sastry and D.J. Patel, *Biochemistry*, 1993, **32**, 6588.
27 D.J. Silva, R. Goodnow, Jr. and D. Kahne, *Biochem istry*, 1993, **32**, 463.
28 D.J. Silva and D.E. Kahne, *J. Am. Chem. Soc.*, 1993, **115**, 7962.
29 G. Blanco, E. Fernandez, M. J. Fernandez, A. F. Brana, U. Weissbach, E. Kunzel, J. Rohr, C. Mendez and J.A. Salas, *Mol. Gen. Genet.*, 2000, **262**, 991.
30 E. Fernandez, U. Weissbach, C. Sanchez Reillo, A.F. Braña, C. Méndez, J. Rohr and J.A. Salas, *J. Bacteriol.*, 1998, **180**, 4929.

31 J. Garcia-Bernardo, A.F. Braña, C. Méndez and J.A. Salas, *FEMS Microbiology Letters*, 2000, **186**, 61.
32 F. Lombó, A.F. Braña, C. Méndez and J.A. Salas, *J. Bacteriol.*, 1999, **181**, 642.
33 M. Nur-e-Alam, C. Méndez, J.A. Salas and J. Rohr, *ChemBioChem*, 2005, **6**, 632.
34 L. Prado, F. Lombó, A.F. Braña, C. Méndez, J. Rohr and J.A. Salas, *Molecular & General Genetics*, 1999, **261**, 216.
35 D. Rodriguez, L.M. Quiros and J.A. Salas, *J. Biol. Chem.*, 2004, **279**, 8149.
36 H. Decker, S. Haag, G. Udvarnoki and J. Rohr, *Angew. Chem. Int. Ed. Engl.*, 1995, **34**, 1107.
37 L.L. Remsing, J. Garcia-Bernardo, A.M. Gonzalez, E. Künzel, U. Rix, A.F. Braña, D. W. Bearden, C. Méndez, J.A. Salas and J. Rohr, *J. Am. Chem. Soc.*, 2002, **124**, 1606.
38 F. Barcelo, C. Scotta, M. Ortiz-Lombardia, C. Mendez, J.A. Salas and J. Portugal, *Nucleic Acids Res*, 2007, **35**, 2215.
39 L.L. Remsing, A.M. Gonzalez, M. Nur-e-Alam, M.J. Fernandez-Lozano, A. F. Brana, U. Rix, M. A. Oliveira, C. Mendez, J.A. Salas and J. Rohr, *J Am Chem Soc*, 2003, **125**, 5745.
40 V. Albertini, A. Jain, S. Vignati, S. Napoli, A. Rinaldi, I. Kwee, M. Nur-e-Alam, J. Bergant, F. Bertoni, G.M. Carbone, J. Rohr and C.V. Catapano, *Nucl. Acids Res.*, 2006, **34**, 1721.
41 A. Gonzalez, L.L. Remsing, F. Lombó, M.J. Fernandez, L. Prado, A.F. Braña, E. Künzel, J. Rohr, C. Méndez and J.A. Salas, *Mol. Gen. Genet.*, 2001, **264**, 827.
42 J. Rohr, U. Weissbach, C. Beninga, E. Künzel, K. Siems, K. Bindseil, L. Prado, F. Lombó, A.F. Braña, C. Méndez and J.A. Salas, *Chem. Commun.*, 1998, 437.
43 S.-E. Wohlert, G. Blanco, F. Lombo, E. Fernandez, A.F. Brana, S. Reich, G. Udvarnoki, C. Mendez, H. Decker, J. Frevert, J.A. Salas and J. Rohr, *J. Am. Chem. Soc.*, 1998, **120**, 10596.
44 C. Fischer, L. Rodríguez, E.P. Patallo, F. Lipata, A.F. Braña, C. Méndez, J.A. Salas and J. Rohr, *J. Nat. Prod.*, 2002, 1685.
45 A. Ramos, F. Lombo, A.F. Brana, J. Rohr, C. Mendez and J.A. Salas, *Microbiology*, 2008, **154**, 781.
46 F. Lombó, M. Gibson, L. Greenwell, A.F. Braña, J. Rohr, J.A. Salas and C. Méndez, *Chem. Biol.*, 2004, **11**, 1709.
47 G. Blanco, E.P. Patallo, A.F. Braña, A. Trefzer, A. Bechthold, J. Rohr, C. Méndez and J.A. Salas, *Chem. Biol.*, 2001, **8**, 253.
48 C. Méndez and J.A. Salas, *Trends in Biotechnology*, 2001, **19**, 449.
49 J.A. Salas and C. Mendez, *Trends Microbiol*, 2007, **15**, 219.
50 C. Sanchez, L. Zhu, A.F. Braña, A.P. Salas, J. Rohr, C. Méndez and J.A. Salas, *Proc. Natl. Acad. Sci. (USA)*, 2005, **102**, 461.
51 C. Duerr, D. Hoffmeister, S.-E. Wohlert, K. Ichinose, M. Weber, U. von Mulert, J.S. Thorson and A. Bechthold, *Angew. Chem. Int. Ed.*, 2004, **43**, 2962.
52 S. A. Borisova, L. Zhao, I.C. Melancon, C.L. Kao and H.W. Liu, *J Am Chem Soc*, 2004, **126**, 6534.
53 S. Ha, D. Walker, Y. Shi and S. Walker, *Protein Sci.*, 2000, **9**, 1045.
54 D. Hoffmeister, G. Drager, K. Ichinose, J. Rohr and A. Bechthold, *J. Am. Chem. Soc.*, 2003, **125**, 4678.
55 D. Hoffmeister, K. Ichinose and A. Bechthold, *Chem. Biol.*, 2001, **8**, 557.
56 D. Hoffmeister, B. Wilkinson, G. Foster, P.J. Sidebottom, K. Ichinose and A. Bechthold, *Chem. Biol.*, 2002, **9**, 287.

57 A. Luzhetskyy, M. Fedoryshyn, C. Dürr, T. Taguchi, V. Novikov and A. Bechthold, *Chem. Biol.*, 2005, **12**, 725.

58 A. Luzhetskyy, C. Mendez, J.A. Salas and A. Bechthold, *Current Topics in Medicinal Chemistry*, 2008, **8**, 680.

59 M. Mittler, A. Bechthold and G.E. Schulz, *J. Mol. Biol.*, 2007, **372**, 67.

60 A.M. Mulichak, H.C. Losey, C.T. Walsh and R.M. Garavito, *Structure*, 2001, **9**, 547.

61 G.J. Williams, R.D. Goff, C. Zhang and J.S. Thorson, *Chem. Biol.*, 2008, **15**, 393.

62 A.W. Truman, M.V. Dias, S. Wu, T.L. Blundell, F. Huang and J.B. Spencer, *Chem Biol*, 2009, **16**, 676.

63 C. Zhang, B.R. Griffith, Q. Fu, C. Albermann, X. Fu, I.-K. Lee, L. Li and J. S. Thorson, *Science*, 2006, **313**, 1291.

64 D.L. Jakeman, *Chem. Biol.*, 2008, **15**, 307.

65 S. Blanchard and J.S. Thorson, *Curr. Opinion Chem. Biol.*, 2006, **10**, 263.

66 A. Ahmed, N.R. Peters, M.K. Fitzgerald, J.A. Watson, Jr., F.M. Hoffmann and J.S. Thorson, *J. Am. Chem. Soc.*, 2006, **128**, 14224.

67 W.A. Barton, J. Lesniak, J.B. Biggins, P.D. Jeffrey, J.Q. Jiang, K.R. Rajashankar, J.S. Thorson and D.B. Nikolov, *Nature Struct. Biol.*, 2001, **8**, 545.

68 B.R. Griffith, J.M. Langenhan and J.S. Thorson, *Curr. Opin. Biotechnol.*, 2005, **16**, 622.

69 J.M. Langenhan, B.R. Griffith and J.S. Thorson, *J. Nat. Prod.*, 2005, **68**, 1696.

70 J. Yang, D. Hoffmeister, L. Liu, X. Fu and J.S. Thorson, *Bioorg. Med. Chem.*, 2004, **12**, 1577.

71 X. Fu, C. Albermann, C. Zhang and J.S. Thorson, *Org Lett*, 2005, **7**, 1513.

72 J.M. Langenhan, N.R. Peters, I.A. Guzei, F.M. Hoffmann and J.S. Thorson, *Proc. Natl. Acad. Sci. USA*, 2005, **102**, 12305.

73 I. Baig, M. Pérez, A. F. Braña, R. Gomathinayagam, C. Damodaran, J.A. Salas, C. Méndez and J. Rohr, *J. Nat. Prod.*, 2008, **71**, 199.

74 M. Pérez, I. Baig, A.F. Braña, J.A. Salas, J. Rohr and C. Méndez, *ChemBioChem*, 2008, **9**, 2295.

75 L. Prado, E. Fernandez, U. Weissbach, G. Blanco, L.M. Quiros, A.F. Brana, C. Méndez, J. Rohr and J.A. Salas, *Chem. Biol.*, 1999, **6**, 19.

76 D. Rodriguez, L.M. Quiros, A.F. Braña and J.A. Salas, *J. Bacteriol.*, 2003, **185**, 3962.

77 M. Gibson, M. Nur-e-Alam, F. Lipata, M.A. Oliveira and J. Rohr, *J Am Chem Soc*, 2005, **127**, 17594.

78 M.P. Beam, M.A. Bosserman, N. Noinaj, M. Wehenkel and J. Rohr, *Biochemistry*, 2009, **48**, 4476.

79 D.E. Torres Pazmino, B.-J. Baas, D.B. Janssen and M.W. Fraaije, *Biochemistry*, 2008, **47**, 4082.

80 A. Willetts, *Trends in Biotechnology*, 1997, **15**, 55.

81 N.M. Kamerbeek, D.B. Janssen, W.J.H. van Berkel and M.W. Fraaije, *Adv. Synth. Catal.*, 2003, **345**, 667.

82 C. Wang, M. Gibson, J. Rohr and M.A. Oliveira, *Acta Cryst.* , 2005, **F61**, 1023.

83 H. Koskiniemi, M. Metsa-Ketela, D. Dobritzsch, P. Kallio, H. Korhonen, P. Mäntsälä, G. Schneider and J. Niemi, *J. Mol. Biol.*, 2007, **372**, 633.

84 P. Kallio, Z. Liu, P. Mantsala, J. Niemi and M. Metsa-Ketela, *Chem. Biol.*, 2008, **15**, 157.

85 M.W. Fraaije, N.M. Kamerbeek, W.J. van Berkel and D.B. Janssen, *FEBS Lett*, 2002, **518**, 43.

STUDIES ON THE BIOSYNTHESIS OF THE ANTHRAX SIDEROPHORE PETROBACTIN

Daniel Oves-Costales and Gregory L. Challis

Department of Chemistry, University of Warwick, Coventry, CV4 7AL, United Kingdom

1 INTRODUCTION

1.1 Importance of Iron for Living Organisms

Iron is an element that, due to its redox chemistry and its involvement in numerous metabolic functions,[1] is essential for almost all living organisms. Bacteria require a cytoplasmic iron concentration of 10^{-6} M for their growth.[2] However, despite being the fourth most abundant element in the earth's crust, the bioavailability of iron is very low. In aerobic, neutral pH environments, iron is present as insoluble ferric hydroxy/oxyhydroxy complexes.[3] Thus, the concentration of free Fe(III) in water is 10^{-18} M.[4] In human serum, the action of different iron-binding proteins reduces the free iron concentration to 10^{-24} M.[5] Therefore, both pathogenic and non-pathogenic bacteria need to overcome the scarcity of uncomplexed iron to survive.

The production, excretion and uptake of high-affinity Fe(III) chelators, termed siderophores, is arguably the most important mechanism for the bacterial iron uptake.[6-8] Several hundred siderophores, exhibiting high structural diversity, have been isolated from microbes and higher plants.[9] The biosynthesis of these compounds in bacteria is usually transcriptionally-regulated by intracellular iron levels.[10] Once a siderophore is biosynthesised, it is excreted to the environment or the host, where it can chelate iron. The ferric-siderophore complex thus formed is recognized by a specific protein-receptor associated with a transport system, which is responsible for its uptake back into the cytoplasm of the producing organism.[11,12] Reductive or hydrolytic mechanisms will then break apart the ferric-siderophore complex,[13,14] thus making available the iron required by the bacteria.

Pathogenic microbes need to scavenge iron from their host, and in many cases the siderophore production is essential for pathogens to reach full virulence levels;[15] examples of pathogenicity-conferring siderophores include mycobactin in *Mycobacterium tuberculosis*,[16] yersiniabactin in *Yersinia pestis*,[17] and aerobactin in *Escherichia coli*.[18]

There are two main pathways for siderophore biosynthesis. One relies on the nonribosomal peptide synthetase (NRPS) multienzyme superfamily, which has been extensively studied over the last two decades.[7] The other, termed the NRPS-independent siderophore (NIS) pathway,[19] utilises a family of synthetases that has only very recently been biochemically and structurally characterised.[20]

1.2 Siderophores in *Bacillus* Species: the Importance of Petrobactin

Three siderophores are known to be produced by *Bacillus* species; bacillibactin, petrobactin and 3,4-dihydroxybenzoic acid (Figure 1). Bacillibactin is a catechol-based siderophore which is NRPS-produced.[21] Its structure resembles that of enterobactin from *E. coli*, and its ferric complex formation constant is very high (~10^{48} M^{-1}).[22] Petrobactin, on the other hand, was first isolated from the marine bacterium *Marinobacter hydrocarbonoclasticus*,[23] and it is one of the only three known siderophores that contains the unusual 3,4-dihydroxybenzoyl moiety, the other two being petrobactin sulfonate and petrobactin disulfonate.[24,25] Its ferric-affinity constant is ~10^{23} M^{-1}, which is surprisingly low for a catechol-based siderophore.[26] Finally, 3,4-dihydroxybenzoic acid (protocatechuic acid) is an intermediate in the biosynthesis of petrobactin which has been observed to be present in culture supernatants of both *Bacillus anthracis* and *Bacillus cereus*.[27] Very recently, a protein receptor for the ferric complex of this compound has been identified and characterised in *Bacillus cereus*, thus indicating its likely siderophore function.[28]

Given the extraordinarily high affinity of bacillibactin for iron, one might wonder why redundant iron-uptake mechanisms exist in *Bacillus anthracis*. Indeed, the iron-affinity constant displayed by petrobactin is several orders of magnitude lower than that of bacillibactin, and protocatechuic acid has a relatively low affinity for iron.[29] Interestingly, Hanna and coworkers reported that a *B. anthracis* mutant unable to biosynthesise petrobactin showed impaired ability to grow in depleted-iron media and in a murine infection model.[30] The virulence levels displayed by this mutant in mice were also attenuated by ~1000 fold.[30] However, a *B. anthracis* mutant unable to biosynthesise bacillibactin exhibited similar levels of growth and virulence to those of the wild-type.[30] Taken together, these data indicate that the production of petrobactin, but not bacillibactin, is essential in *Bacillus anthracis* to reach full levels of growth and virulence in mice.

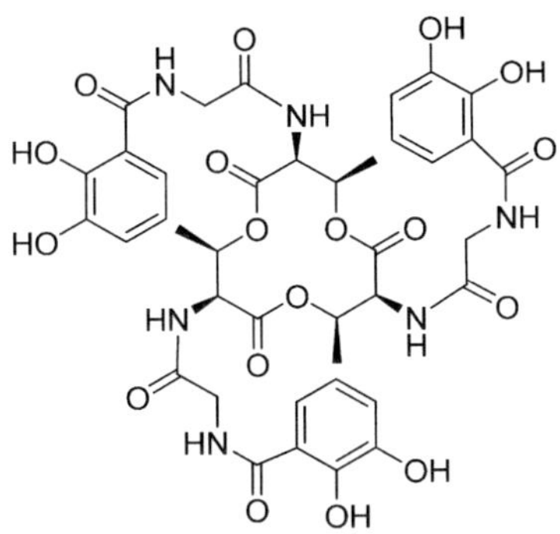

Bacillibactin

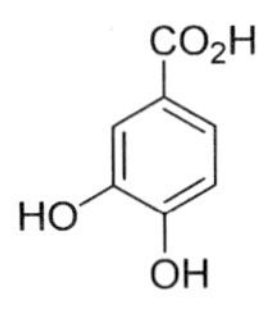

3,4-dihydroxybenzoic acid

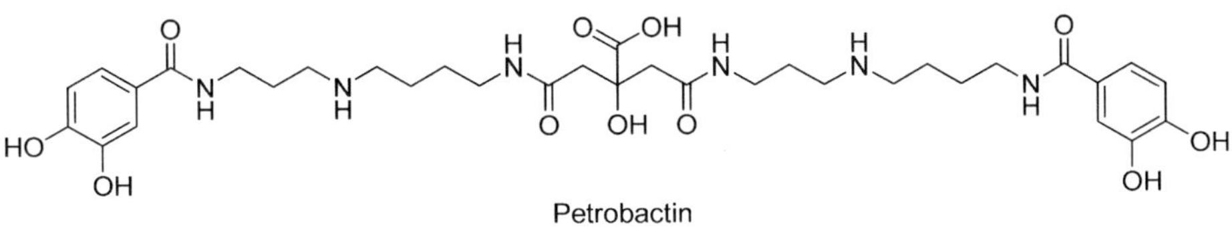

Petrobactin

Figure 1 *Siderophores produced by* Bacillus *species*

Raymond and coworkers have suggested that a plausible explanation for these observations is the ability of petrobactin to evade the action of siderocalin.[31] Siderocalin (also known as lipocalin-2, neutrophil gelatinase-associated lipocalin or NGAL) is a 25 kDa protein from the mammalian immune system which can recognize and chelate iron-bound catecholic siderophores such as ferric-enterobactin and ferric-bacillibactin.[31,32] In doing so, this component of the immune system disrupts the bacterial iron-uptake process, thus impairing bacterial growth. In response to this mammalian defence system some bacteria have evolved the ability to produce "stealth" siderophores, such as petrobactin, that cannot be recognized by siderocalin.[31] Another example of a "stealth" siderophore is salmochelin, a C-glycosylated derivative of enterobactin produced by *Salmonella typhimurium* and some pathogenic strains of *Escherichia coli*,[33] which cannot be bound by siderocalin in either its ferri- or desferri- forms.[34] Consequently, the production of salmochelin by *Salmonella typhimurium* or *Escherichia coli*, and petrobactin by *Bacillus anthracis* provides a secure route to acquire host iron for these pathogens, thus making the siderophores virulence determinants.

The dependence of *Bacillus anthracis* on petrobactin to acquire iron within the host suggests that all the molecular machinery involved in the biosynthesis, excretion and uptake of this siderophore could offer new targets for the development of antibiotics.

2 BIOSYNTHESIS OF PETROBACTIN

Two gene clusters encoding siderophore biosynthesis enzymes in *Bacillus anthracis*, *asbABCDEF* (*a*nthrax *s*iderophore *b*iosynthesis) and *bacACEBF* (*b*acillus *a*nthracis *c*atechol), were identified by Hanna and coworkers.[30] Shortly after, Byers and coworkers showed that *asbABCDEF* is responsible for the biosynthesis of petrobactin, whereas *bacACEBF* directs the production of bacilibactin.[35]

The petrobactin biosynthetic gene cluster contains 6 genes arranged in a single operon (Figure 2). *asbA* and *asbB* encode type A and type C NIS synthetases, respectively; *asbC* and *asbD* encode NRPS-like enzymes; *asbE* encodes an acyl-transferase; and *asbF* encodes a 3,4-dihydroxybenzoic acid (3,4-DHBA) synthase. Upstream of *asbABCDEF* there is a putative Fur box, which suggest that this gene cluster may be regulated by a Fur-like protein.[30]

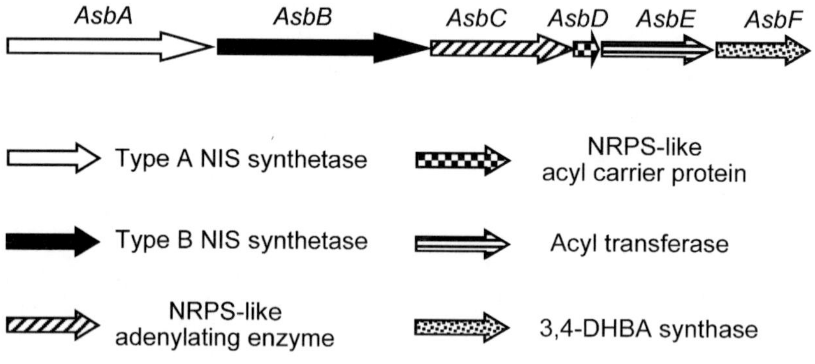

Figure 2 *Organization of the* asbABCDEF *gene cluster that directs petrobactin biosynthesis*

2.1 Genetic Studies on *asbABCDEF*

The first studies on petrobactin biosynthesis were carried out by Sherman and coworkers, and consisted of in-frame deletions of each of the genes in the *asb* locus, followed by identification of accumulated petrobactin-related metabolites.[36]

HPLC analysis of culture supernatants of Δ*asbA*, Δ*asbB* and the double mutant Δ*asbAB* showed accumulation of a common metabolite, not present in the wild type, with *m*/*z* 282.1, consistent with the presence of (3,4-dihydroxybenzoyl)-spermidine (3,4-DHB-SP).[36] In addition, a new compound in the Δ*asbB* mutant was found with *m*/*z* 456.1, which could correspond to (3,4-dihydroxybenzoyl)-citryl-spermidine (3,4-DHB-CT-SP).[36] Database comparisons of AsbA and AsbB revealed that these enzymes have sequence similarity with IucA and IucC, which are gene-products involved in the biosynthesis of aerobactin, a siderophore from *Escherichia coli*.[37] IucA was proposed to catalyze the condensation of the α-amino group in N^6-acetyl-N^6-hydroxy-L-lysine (ahLys) with one of the prochiral carboxyl groups of citric acid,[38,39] and IucC was proposed to be responsible for the linkage of a second molecule of ahLys with the product of the IucA-catalysed reaction.[38,39]

Taken together, these data led Sherman and coworkers to propose the biosynthetic pathway outlined in Scheme 1.[36] Thus, AsbCDEF would be responsible for the biosynthesis of N^1-(3,4-dihydroxybenzoyl)-spermidine **3**. AsbA would then catalyze the condensation of the primary amine of **3** with one of the prochiral carboxyl groups of citric acid, to afford N^1-(3,4-dihydroxybenzoyl)-N^8-citryl-spermidine **4**. Finally, AsbB could catalyze the linkage of a second molecule of (3,4-DHB-SP) **3** with AsbA-generated **4**, affording petrobactin.[36]

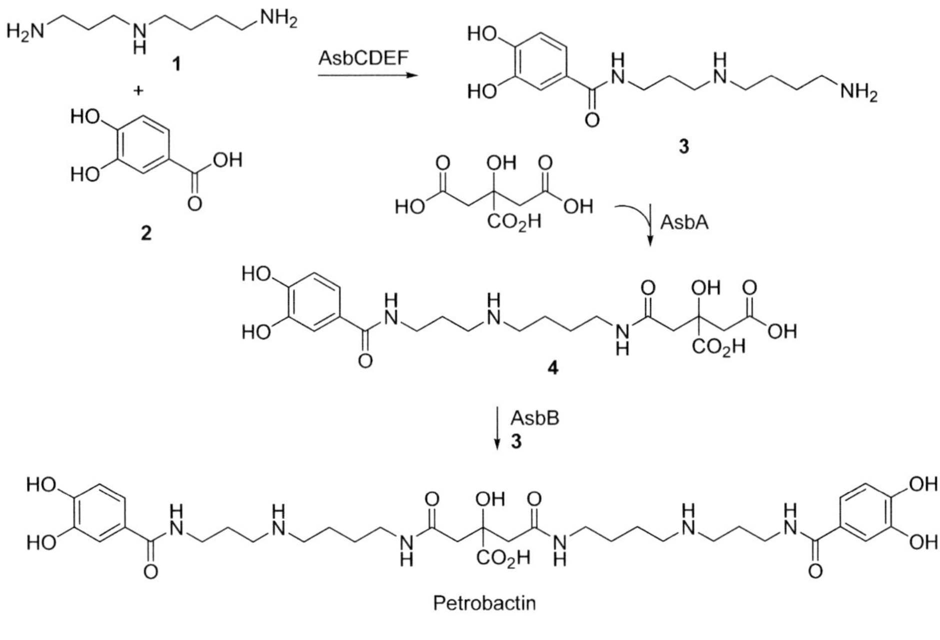

Scheme 1 *Pathway for petrobactin biosynthesis proposed by Sherman and coworkers*

Surprisingly, Sherman and coworkers also reported that SORI-CAD MS/MS data of the isolated 3,4-DHB-SP and 3,4-DHB-CT-SP showed the compounds to be a mixture of regioisomers. Thus, 3,4-DHB-SP from Δ*asbA*, Δ*asbB* and Δ*asbAB* was found to be a mixture of N^1-(3,4-dihydroxybenzoyl)-spermidine **3** and its regioisomer N^8-(3,4-dihydroxybenzoyl)-spermidine.[36] In a similar way, 3,4-DHB-CT-SP from Δ*asbB* consisted of a mixture of N^1-(3,4-dihydroxybenzoyl)-N^8-citryl-spermidine **4** and its regioisomer N^8-(3,4-dihydroxybenzoyl)-N^1-citryl-spermidine.[36] However, SORI-CAD MS/MS data of petrobactin showed the exclusive presence of the fragments **3** and **4**, suggesting that their corresponding regioisomers were not used in the assembly of a regioisomeric form of petrobactin.[36]

2.2 Biochemical characterization of AsbC, AsbD and AsbE

Shortly after the publication of these genetic studies, the same authors published the biochemical characterization of AsbC, AsbD and AsbE.[40] Thus, *asbC*, *asbD* and *asbE* were cloned into an overexpression vector which was used to obtain the corresponding recombinant proteins. AsbC was found to be a NRPS-like adenylating enzyme that was able to activate the carboxyl group of 3,4-dihydroxybenzoic acid (3,4-DHBA) **2**. The AMP-activated 3,4-DHBA is then loaded onto AsbD, which is a NRPS-like acyl carrier protein. Finally, AsbE was shown to be an acyl-transferase which can catalyze the transfer of the 3,4-dihydroxybenzoyl group from AsbD to N^1 and N^8 of spermidine, to afford N^1-(3,4-dihydroxybenzoyl)-spermidine **3** and its regioisomer N^8-(3,4-dihydroxybenzoyl)-spermidine **5**, respectively (Scheme 2).[40]

AsbE was shown to have relaxed substrate specificity. Indeed, N^1-(3,4-dihydroxybenzoyl)-spermidine **3**, which was the intermediate proposed to be required for petrobactin biosynthesis, was the minor compound obtained from this enzymatic reaction, whereas its regioisomer, N^8-(3,4-dihydroxybenzoyl)-spermidine **5**, was the major product obtained.[40] This observation, although a bit surprising, offered an explanation for the presence of compound **5** in culture supernatants of the Δ*asbA*, Δ*asbB* and Δ*asbAB* mutants.

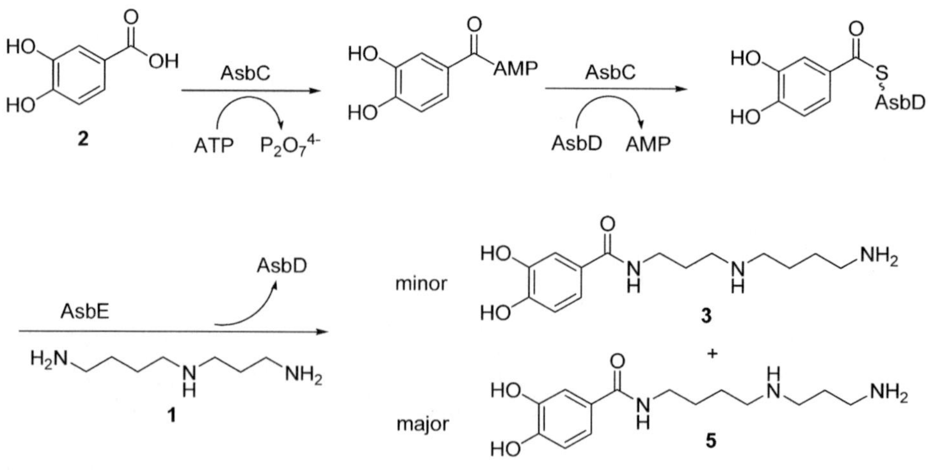

Scheme 2 *Reactions catalysed by AsbC, AsbD and AsbE*

2.3 Biochemical Characterisation of AsbA

Sequence comparison had indicated that AsbA is a type A NIS synthetase.[19] Thus, it was proposed that this enzyme would catalyze the condensation of citric acid with either spermidine **1** or N^1-(3,4-dihydroxybenzoyl)-spermidine **3**.[41] Purified recombinant AsbA was therefore incubated with mixtures containing citric acid, ATP, Mg^{2+} and either **1** or **3**. LC-ESI-MS analysis of the incubations showed that the enzyme was able to catalyze efficient regioselective acylation of citric acid with N^8 of spermidine (but not its N^1-3,4-dihydroxybenzoyl derivative) to afford N^8-citryl-spermidine (Scheme 3).[41] Through the use of enantiomerically pure [1,2-$^{13}C_2$]-labelled citric acid, it was shown that the AsbA-catalysed reaction is highly enantioselective, generating (3′S)-N^8-citryl-spermidine **6** (Scheme 3).[42]

The role of ATP in the reaction was investigated using continuous coupled assays for AMP and ADP formation. These assays showed time-dependent formation of the former but not the latter.[41] Taken together, these data suggests that AsbA selectively activates the *pro-R* carboxyl group of citric acid to generate an acyl-adenylate intermediate, which subsequently reacts with N^8 of spermidine, thus affording (3'S)-N^8-citryl-spermidine **6** (Scheme 3).[41,42]

Finally, investigations of the susbtrate specificity of AsbA were carried out. The enzyme can catalyze the ATP- and Mg^{2+}-dependent condensation of citric acid with several di- and triamine analogues of spermidine, allowing the access to a small library of optically-actived citric acid derivatives.[42] On the other hand, the enzyme is much more specific for citric acid, and only tricarballylic acid, a citric acid derivative lacking the hydroxyl group, could be employed in place of citric acid in AsbA-catalysed condensations with spermidine.[42]

The inability of AsbA to catalyse efficient condensation of citric acid with N^1-(3,4-dihydroxybenzoyl)-spermidine **3** indicated that the biosynthetic pathway proposed by Sherman and coworkers (Scheme 1) is unlikely to be correct.[41] The AsbE-catalysed transfer of the 3,4-dihydroxybenzoyl moiety onto N^1 of spermidine must occur after the AsbA-catalysed condensation of citric acid with N^8 of spermidine.

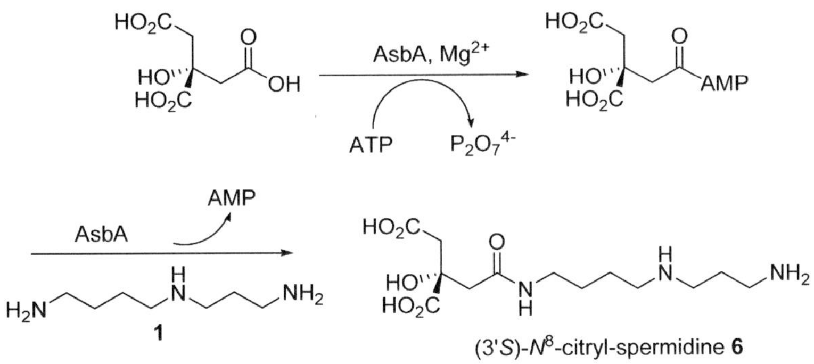

Scheme 3 *Reactions catalysed by AsbA*

2.4 Biochemical Characterisation of AsbB

Sequence comparison had shown that AsbB is a type C NIS synthetase,[19] suggesting that this enzyme could catalyse the condensation of a derivative of citric acid with an amine or alcohol. We therefore envisioned four possible different roles for AsbB. It could catalyse the condensation of AsbA-generated (3'S)-N^8-citryl-spermidine **6** with either spermidine **1** or N^1-(3,4-dihydroxybenzoyl)-spermidine **3**; alternatively, N^1-(3,4-dihydroxybenzoyl)-N^8-citryl-spermidine **4** could undergo AsbB-catalysed condensation with either **1** or **3** (Scheme 4).[43] To investigate these hypotheses, separate incubations using chemically synthesised racemic **4** and **6** with either **1** or **3**, ATP, Mg^{2+} and purified recombinant AsbB were carried out. LC-ESI-MS analysis of these mixtures showed products for all four incubations with the *m/z* values expected for **7**, **8** and petrobactin, suggesting that all potential AsbB roles envisioned were possible.[43]

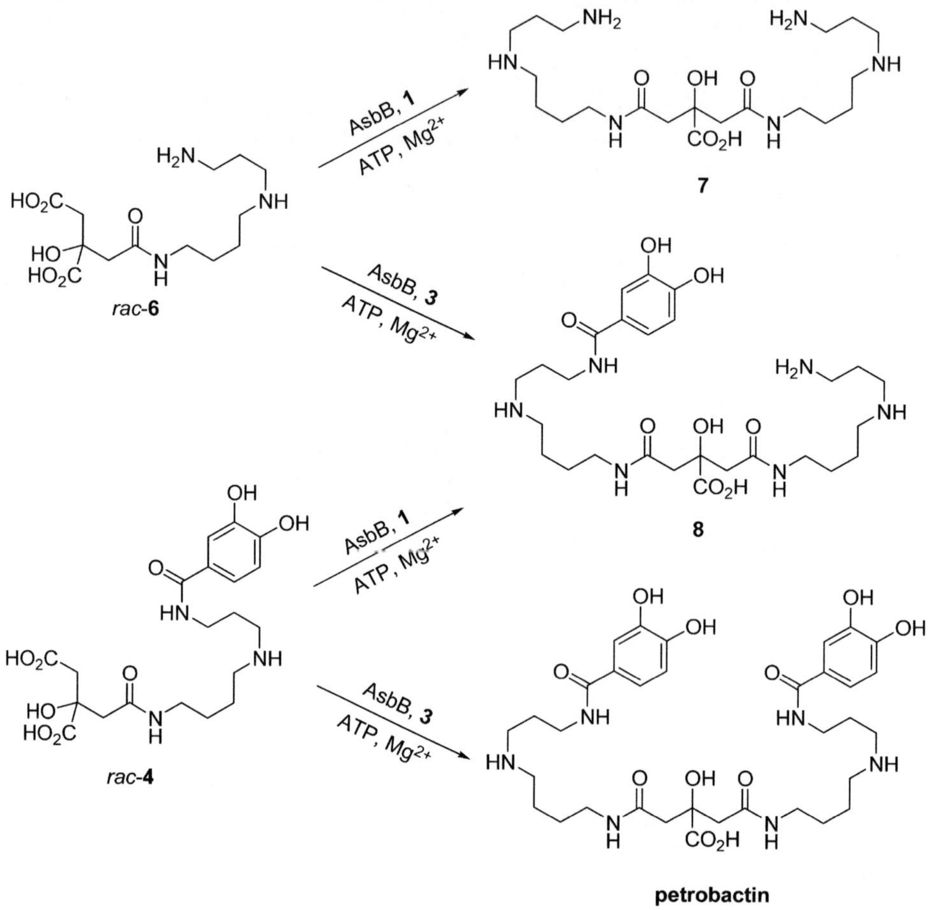

Scheme 4 *Reactions catalysed by AsbB*

Continuous coupled assays for AMP and ADP formation were used with the double purpose of gaining more insight into the mechanism and the relative rate of the different reactions. Time-dependent production of AMP, but not ADP, was observed for the four different AsbB-catalysed condensations.[43] These data are consistent with the reaction of

ATP with enzyme-bound **4** or enzyme-bound **6** to generate an acyl-adenylate intermediate, which subsequently reacts with either **1** or **3**, thus generating the new amide bond. In order to establish which of the AsbB-catalysed reactions were most important for petrobactin biosynthesis, the relative rates of AMP formation were measured using saturating concentrations of ATP, **1** or **3** and *rac*-**4** or *rac*-**6**.[43] The rates of AMP formation from condensation of **1** with *rac*-**4** or *rac*-**6** were ~ 30 to 55 times higher than the rates of AMP formation from the reaction of **3** with *rac*-**4** or *rac*-**6**. These data suggested that AsbB has a strong preference for spermidine over N^1-(3,4-dihydroxybenzoyl)-spermidine as a substrate.[43]

2.5 Biochemical Characterisation of AsbF

The last enzyme from the *asbABCDEF* gene cluster to be biochemically characterized was AsbF. Sherman's and Koppisch's groups independently showed that the enzyme has dehydratase activity, that catalyses the generation of 3,4-diydroxybenzoic acid from 3-dehydroshikimate, an intermediate of the shikimate pathway (Scheme 5).[44,45]

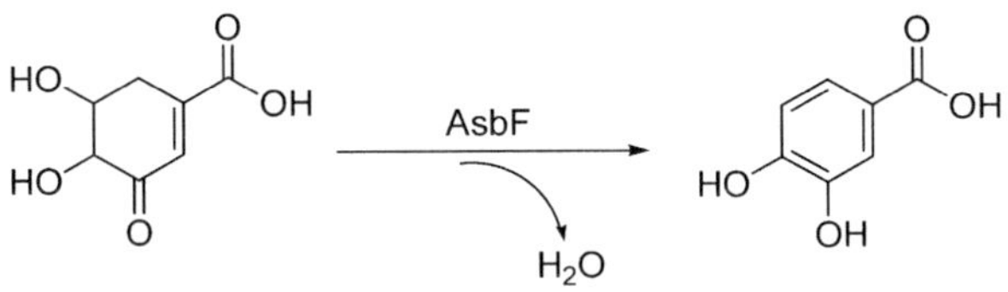

Scheme 5 *Reaction catalysed by AsbF*

2.6 Revised Pathway for Petrobactin Biosynthesis

The data obtained during the biochemical characterization of the NIS synthetases AsbA and AsbB indicate that N^1-(3,4-dihydroxybenzoyl)-spermidine **3** can not be used efficiently as a substrate by either enzyme.[41,43] These findings, together with the observation that compound **3** is the minor product obtained in the AsbE-catalysed reaction (Scheme 2) indicate that N^1-(3,4-dihydroxybenzoyl)-spermidine **3** is very unlikely to be a significant intermediate in petrobactin biosynthesis. The presence of compounds **3** and **5** in culture supernatants of the Δ*asbA*, Δ*asbB* and Δ*asbAB* mutants can be explained by the relaxed substrate specificity of AsbE, suggesting that these two compounds are shunt metabolites, rather than real intermediates in petrobactin biosynthesis.[41]

The biochemical characterization data of AsbABCDEF reported by us and others has led us to propose a revised pathway for petrobactin biosynthesis (Scheme 6) in which (*S*)-**4**, (*S*)-**6**, **7** and **8** are all intermediates.[43] AsbE is proposed to transfer the 3,4-dihydroxybenzoyl moiety from the phosphopantetheine group in AsbD to the primary amines in (*S*)-**6**, **7**, and **8**.[43]

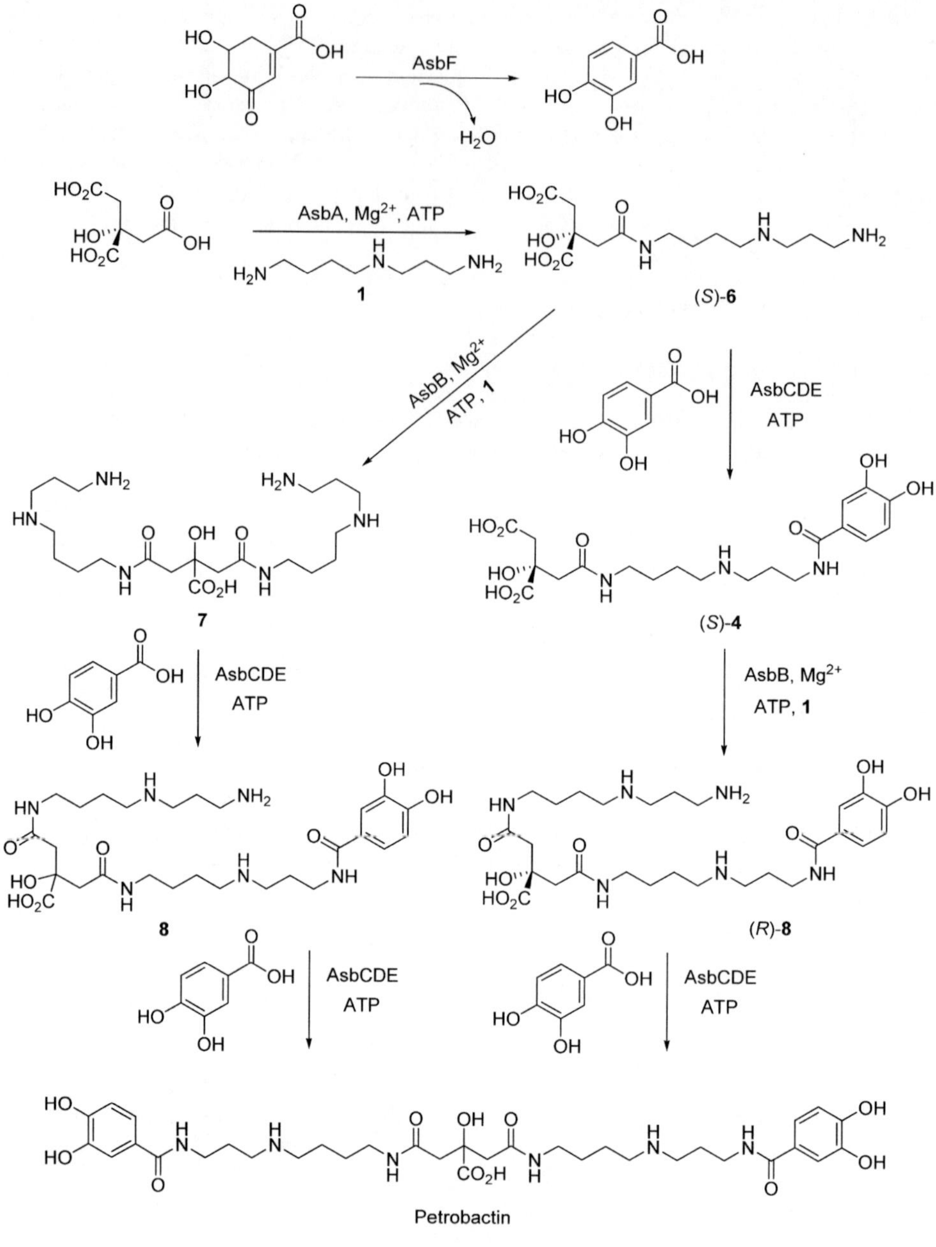

Scheme 6 *Revised pathway for petrobactin biosynthesis*

3 CONCLUSION

Petrobactin is a siderophore produced by several *Bacillus* species. It has been recently discovered that the production of this siderophore is essential for *B. anthracis* to reach full levels of growth and virulence. These observations have sparked a considerable interest in the study of the biosynthetic pathway of petrobactin, because an understanding of the biochemistry involved in its biosynthesis is the first step required to develop inhibitors as potential new antibiotics against anthrax.

References

1 J. J. R. Fraústo da Silva and R. J. P. Williams, *The Biological Chemistry of the Elements*, Oxford University Press, New York, USA, 2001, pp. 340-399.
2 S. C. Andrews, A. K. Robinson and F. Rodriguez-Quinones, *FEMS Microbiol. Rev.*, 2003, **27**, 215.
3 W. L. Lindsay, in *Iron in Soils and Clay Minerals*, ed. J. W. Stucki, B. A. Goodman and U. Schwertmann, NATO ASI Series C, Vol 127, 1988, 37-62.
4 K. N. Raymond and C. J. Carrano, *Acc. Chem. Res.*, 1979, **12**, 183.
5 K. N. Raymond, E. A. Dertz and S. S. Kim, *Proc. Nat. Acad. Sci. U. S. A.*, 2003, **100**, 3584.
6 J. R. Telford and K. N. Raymond, *Compr. Supramol. Chem.*, 1996, **1**, 245.
7 J. H. Crosa and C. T. Walsh, *Microbiol. Mol. Biol. Rev.*, 2002, **66**, 223.
8 M. Miethke and M. A. Marahiel, *Microbiol. Mol. Biol. Rev.*, 2007, **71**, 413.
9 G. Winkelmann and H. Drechsel, in *Biotechnology*, ed. H.-J. Rehm, G. Reed, A. Pühler and P. Stadler, Weinheim, Germany, 1997, vol. 7, ch. 5, pp. 200-246.
10 K. Hantke, *Curr. Opin. Microbiol.*, 2001, **4**, 172.
11 W. Koster, *Res. Microbiol.*, 2001, **152**, 291.
12 K. D. Krewulak and H. J. Vogel, *Biochim. Biophys. Acta*, 2008, **1778**, 1781.
13 I. Schroder, E. Johnson and S. de Vries, *FEMS Microbiol. Rev.*, 2003, **27**, 427.
14 T. J. Brickman and M. A. Mcintosh, *J. Biol. Chem.*, 1992, **267**, 12350.
15 U. E. Schaible and S. H. E. Kaufmann, *Nat. Rev. Microbiol.*, 2004, **2**, 946.
16 G. A. Snow, *Bacteriol. Rev.*, 1970, **34**, 99.
17 S. W. Bearden, J. D. Fetherston and R. D. Perry, *Infect. Immun.*, 1997, **65**, 1659.
18 P. J. Warner, P. H. Williams, A. Bindereif and J. B. Neilands, *Infect. Immun.*, 1981, **33**, 540.
19 G. L. Challis, *ChemBioChem*, 2005, **6**, 601.
20 D. Oves-Costales, N. Kadi and G. L. Challis, *Chem. Commun.*, 2009, 6530.
21 J. J. May, T. M. Wendrich, and M. A. Marahiel, *J. Biol. Chem.*, 2001, **276**, 7209.
22 E. A. Dertz, J. Xu, A. Stintzi and K. N. Raymond, *J. Am. Chem. Soc.*, 2006, **128**, 22.
23 K. Barbeau, G. Zhang, D. H. Live and A. Butler, *J. Am. Chem. Soc.*, 2002, **124**, 378.
24 S. J. H. Hickford, F. C. Küpper, G. Zhang, C. J. Carrano, J. W. Blunt, and A. Butler, *J. Nat. Prod.*, 2004, **67**, 1897.
25 V. V. Homann, K. J. Edwards, E. A. Webb and A. Butler, *Biometals*, 2009, **22**, 565.
26 R. J. Abergel, A. M. Zawadzka and K. N. Raymond, *J. Am. Chem. Soc.*, 2008, **130**, 2124.
27 B. L. Garner, J. E. L. Arceneaux and B. R. Byers, *Curr. Microbiol.*, 2004, **49**, 89.
28 A. M. Zawadzka, R. J. Abergel, R. Nichiporuk, U. N. Andersen and K. N. Raymond, *Biochemistry*, 2009, **48**, 3645-3657.
29 P. K. Migal and V. A. Ivanov, *Zh. Neorg. Khim+.*, 1973, **18**, 1019.
30 S. Cendrowski, W. MacArthur and P. Hanna, *Mol. Microbiol.*, 2004, **51**, 407.

31 R. J. Abergel, M. K. Wilson, J. E. L. Arceneaux, T. M. Hoette, R. K. Strong, B. R. Byers and K. N. Raymond, *Proc. Nat. Acad. Sci. U. S. A.*, 2006, **103**, 18499.

32 D. H. Goetz, M A. Holmes, N. Borregaard, M. E. Bluhm, K. N. Raymond and R. K. Strong, *Mol. Cell*, 2002, **10**, 1033.

33 K. Hantke, G. Nicholson, W. Rabsch and G. Winkelmann, *Proc. Nat. Acad. Sci. U. S. A.*, 2003, **100**, 3677.

34 M. A. Fischbach, H. Lin and L. Zhou, *Proc. Nat. Acad. Sci. U.S.A.*, 2006, **103**, 16502.

35 M. K. Wilson, R. J. Abergel, K. N. Raymond, J. E. Arceneaux and B. R. Byers, *Biochem. Biophys. Res. Commun.* 2006, **348**, 320.

36 J. Y. Lee, B. K. Janes, K. D. Passalacqua, B. F. Pfleger, N. H. Bergman, H. Liu, K. Håkansson, V. S. Ravindranadh, C. A. Courtney, S. Cendrowski, P. C. Hanna, D. H. Sherman, *J. Bacteriol.*, 2007, **189**, 1698.

37 F. Gibson and D. I. Magrath, *Biochim. Biophys. Acta*, 1969, **192**, 175.

38 V. De Lorenzo, A. Bindereif, B. H. Paw and J. B. Neilands, *J. Bacteriol.*, 1986, **165**, 570.

39 V. De Lorenzo, and J. B. Neilands, *J. Bacteriol.*, 1986, **167**, 350.

40 B. F. Pfleger, J. Y. Lee, R. V. Somu, C. C. Aldrich, P. C. Hanna and D. H. Sherman, *Biochemistry*, 2007, **46**, 4147.

41 D. Oves-Costales, N. Kadi, M. J. Fogg, L. Song, K. S. Wilson and G. L. Challis, *J. Am. Chem. Soc.*, 2007, **129**, 8416.

42 D. Oves-Costales, L. Song and G. L. Challis, *Chem. Commun.*, 2009, 1389.

43 D. Oves-Costales, N. Kadi, M. J. Fogg, L. Song, K. S. Wilson and G. L. Challis, *Chem. Commun.*, 2008, 4034.

44 B. F. Pfleger, Y. Kim, T. D. Nusca, N. Maltseva, J. Y. Lee, C. M. Rath, J. B. Scaglione, B. K. Janes, E. C. Anderson, N. H. Bergman, P. C. Hanna, A. Joachimiak and D. H. Sherman, *Proc. Natl. Acad. Sci. U. S. A.,* 2008, **105**, 17133.

45 D. T. Fox, K. Hotta, C. Y. Kim and A. T. Koppisch, *Biochemistry*, 2008, **47**, 12251.

PROPOSED QUATERNARY STRUCTURE OF TYPE I (MODULAR) POLYKETIDE SYNTHASES

A. M. Hill[1] and J. Staunton[2]

[1]School of Biosciences, University of Exeter, Hatherly Laboratories, Prince of Wales Road, Exeter EX4 4PS, UK
[2]The University Chemical Laboratory, University of Cambridge, Lensfield Road, Cambridge CB2 1EW, UK

1 INTRODUCTION

The enzymes responsible for the chain extension process in polyketide assembly are structurally and mechanistically homologous across nature, however, the arrangement of the enzymes does differ. Enzymes can be discrete (type II) or covalently linked (type I) and can operate iteratively using of a single set of enzymes, or there can be a separate enzyme for every step of the chain extension process (modular assembly line operation).[1-3] This review will concentrate on type I modular polyketide synthases (PKS), specifically the erythromycin DEBS genes and proteins which are the best studied.

Erythromycin A is a macrolide antibiotic produced by *Saccharapolyspora erythraea*. The *ery* PKS is responsible for the biosynthesis of the first enzyme free intermediate 6dEB from one propionyl CoA and six methylmalonyl CoAs (Figure 1).[1,4-6] Three giant proteins, DEBS1,2,3 carry out the six condensation reactions with each protein responsible for two chain extensions. There is a total of 28 catalytic domains spread across the three proteins, arranged into six chain extension clusters (or modules) with a loading di-domain at the beginning of DEBS1, and a thioesterase domain (TE) domain to catalyse lactonisation at the terminus of DEBS 3. 6dEB is converted to the final product by a series of conventional elaboration reactions using a series of independent enzymes.

The primary structure of the three giant proteins can be inferred with a high degree of confidence from the gene sequences. Analysis of the primary sequence, and comparison with other closely related PKS and fatty acid synthases, allows a succession of folded globular regions (domains) to be identified, whose predicted catalytic activities are appropriate to the predicted chemical steps.[1,2] For example, the keto-synthase (KS) domains carry out the condensation reactions that add successive C-3 residues to the growing chain of carbon atoms. In the interior of the tertiary structure, there is an active thiol residue (cysteine) to carry the acyl chain as a thio ester ready, for extension.[7] The partner in this process, the acyl carrier proteins (ACP), also has an active thiol, in this case at the terminus of a long, flexible phosphopantetheine chain (denoted by a wavy line), attached to the hydroxyl group of a serine residue on the interior of the globular residue.[8] The chain extension unit, methyl malonate, is attached to this thiol. The flexible arm allows insertion of the chain extender unit into the loaded KS domain[9] so that chain extension can take place by a Claisen condensation reaction. In all, there are six KS domains, one for each chain extension cycle. They are strategically distributed along the three giant proteins with a partner ACP domain conveniently positioned downstream. This regularity of this pattern of distribution led to the concept of modules, each containing the appropriate set of catalytic domains for the chemical steps in each successive chain extension cycle.

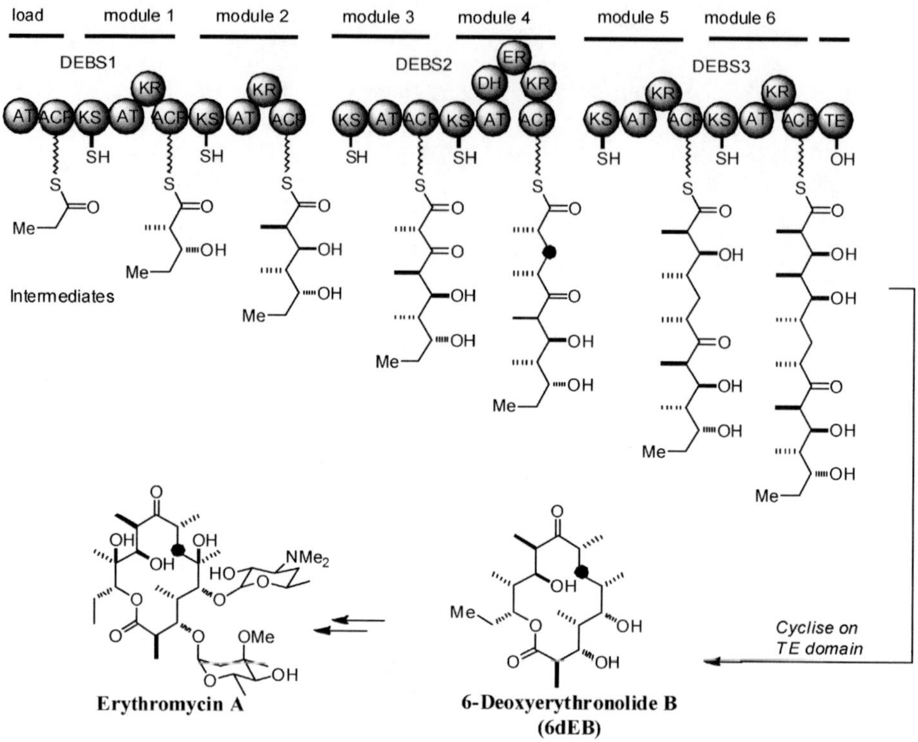

Figure 1 *The erythromycin PKS. Each cycle of chain assembly is carried out by a dedicated set of enzymes so that there is a separate enzyme for every step. The enzymes are organised in modules, one for each cycle with each module containing the correct number of enzymes for keto group modification (keto, hydroxyl, methylene). At the terminus there is a TE domain that carries out cyclisation and release of the first enzyme-free product, 6dEB, as a macrocyclic lactone.*

The ketoester product of each condensation step is then appropriately modified prior to transfer to the next downstream KS for the next chain extension step. The maximum extent of modification results in reduction of the keto group to a methylene, as in fatty acid biosynthesis. Three enzymes are responsible for these operations. First, a ketoreductase (KR) converts the keto group to an alcohol group. Next, a dehydratase (DH), converts the hydroxy acyl to an enoyl residue. Finally, an enoyl reductase (ER) reduces the double bond to generate a saturated methylene residue.

Only module 4 sees the complete conversion of the keto group to a methylene. As expected, this contains the full set of keto group modification enzymes. Other modules are depleted in keto group modification domains to an extent that exactly corresponds to the required extent of keto group modification prior to transfer downstream to the KS of the next module.

There are two other types of domain forming parts of the synthase, AT and TE. The acyl transferase (AT) domains in the six modules load methylmalonate from the external pool of the CoA derivative, onto the thiol of the ACP, ready for the condensation step. In addition, there is a seventh AT domain located at the start of DEBS 1 that loads the starter acyl propionate onto the thiol residue of the KS of module 1. The TE domain is responsible for cyclisation of the completed linear chain to the macrolide ring, resulting in the release of 6-dEB.

The growing chain remains bound to the synthase throughout the 28 synthetic steps required to generate 6dEB. In effect this modular PKS acts as an assembly line that is programmed to produce the required product by the substrate selectivity of the various AT domains, the number of modules and their individual complement of reductive domains, and the ability of the TE to form a macrolactone as its mechanism for releasing the chain from the assembly line.

2 QUATERNARY STRUCTURE OF TYPE I FAS

Polyketide synthases are often compared with fatty acid synthases (FAS), and polyketides can be regarded as oxidised fatty acids, or, alternatively, fatty acids as fully reduced polyketides. There are strong homologies in structure between all systems at the level of tertiary structure of individual enzymes or domains. It is also evident that there are strong homologies in terms of quaternary structure between the type I FAS and PKS systems.[1,2] In the following account, the focus is on the erythromycin modular PKS and various Type I FAS systems found in animals.

Both the PKS and FAS enzyme systems are homodimers and the reactions catalysed use equivalent enzyme activities. Structural information for PKS systems are limited and restricted to X-ray[10,11] and NMR[12,13] structures of individual domains or small multidomain fragments, rather than whole proteins. In contrast, structural information for FAS is much more advanced with a high-resolution X-ray structure of a complete porcine fatty acid synthase reported in 2008[14] and a cryoelectron microscopy (EM) study in 2009.[15] Unfortunately, neither technique revealed the location of the key ACP and TE domains, presumably because they are relatively mobile or disordered even in the crystal. Their positioning in the observed structure can be inferred, however, from their location in the primary sequence.

The structure is divided into two clusters of domains, one containing the KS and AT domains, the other the keto-processing domains (KR, DH, ER) together with two extra globular domains designated S (spacer; these are called ψ-Me and ψ-KR, respectively, in the original paper) which help maintain the quaternary structure (Figure 2).[14] Linker regions join the two clusters via a narrow waist region in the middle of the overall structure. This is where the ACP and TE are proposed to be located. It is envisaged that the ACP shuffles up and down the complementary clefts on opposite side of the axis in carrying out the synthetic operations.

The EM study[15] showed images that could reasonably be interpreted in terms of the same clustering of the two sets of domains. However, in contrast to the single orientation of the two clusters seen by the X-ray,[14] the EM showed many different orientations of the two clusters, caused by rotation of the clusters relative to each other about the central axis of the image together with lateral rocking motions. It would appear that the AT and KS domains of rat FAS can make movements from side to side, and swivel (± 80-100°) about the axis relative to the cluster of reductive domains.[15] Smith has shown that a heterodimer comprising a wild-type subunit partnered with a mutant lacking all seven functionalities is still functionally active which would require a full 180° swivel during each catalytic cycle.[2]

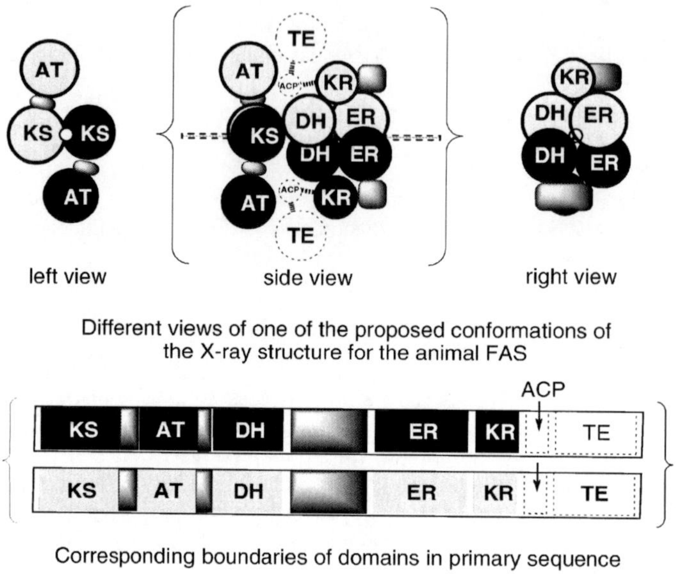

Figure 2 *The X-ray structure of the animal FAS with the position of the axis shown as a dashed horizontal line. The spacer region is shown as a grey box.*

It can be seen from the X-ray structure that each ACP operates in a separate cleft lined with the full set of domains required for the chain extension cycle which requires the ACP to deliver its payload several times around the active-site cleft.[14] Chemical cross-linking studies carried out by Smith and coworkers[2,16] on mammalian FAS (Figure 3a) showed that the ACP can interact with the KS and AT on the partner chain which is not possible from the Ban structure.[14] While highly informative, the solid-state picture is not truly representative of what has to be a highly mobile system as it can only provide a single snap-shot. The EM study[15] can address these shortcomings to some extent providing many images of the workings of the mFAS. However, EM also has its drawbacks and limitation as the high energy radiation to which the sample is subjected may distort the shape of the individual molecular clusters.[17] Consequently, caution must be exercised in drawing conclusions from observations of multi-domain proteins in which the relative positions of domains are varied.

Clearly the FAS structure is highly mobile and may adopt radically different conformations during its cycle of synthetic operations. X-ray studies on the human FAS KS-MAT[18] di-domain show there is a strong similarity in the orientations of the domains with those found in the Ban structure,[14] which is reassuring. They also reveal that there are significant structural variations which may have implications for catalysis. Rudolph and co-workers[18] suggest that some catalytic sites may be formed at the interface between globular domains and linkers and they raise the possibility that the KS-AT linker may be more flexible than it appears in the X-ray.

To account for these discrepancies between the biochemical and X-ray studies Smith has suggested that the pattern of cooperation between the two ACP domains and the catalytic domains in these quaternary conformations can be explained by differing degrees of twist of the linker region between the AT and DH domains (Figure 3b, 3c).[2] Considering the structure depicted in Figure 3b first, it can be seen that all the cooperating domains reside in the same chain, including the KS and AT domains. This is consistent with the Smith complementation experiments for the reductive domains (Figure 3a) but it does not explain the observation that the ACP can reach both pairs of the KS and AT domains, one on its own chain and the other on the partner chain. If we unwind the linker region we arrive at the second, less twisted conformation shown in Figure 3c which now allows the ACP to interact with the AT on the partner chain.

An interesting observation from the human FAS study[18] provides an important warning for assessments of X-ray studies of large multifunctional assemblies. The structure revealed by the X-ray image was not the same as the structure submitted to the crystallisation process because a sizable section of C-terminal residues had been removed by proteolytic cleavage in the course of crystallisation. In general, therefore, if a part of a large multidomain structure is not revealed by an X-ray image, the cause may not be attributable to localised disordering, but to proteolytic cleavage of the missing residues in the crystallisation process.

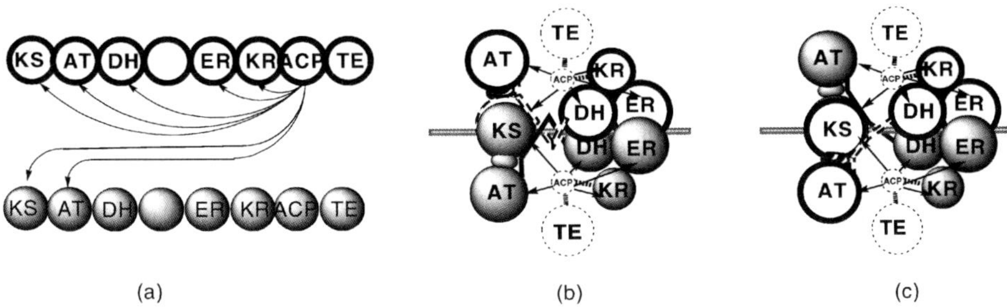

Figure 3 *(a) Summary of mutant complementation experiments with allowed interactions of an ACP domain in the FAS megasynthase during a chain-extension. The spacer is shown as a shaded circle. (b) Expanded view of the FAS showing the route followed by the AT-DH linkers and the pattern of cooperation of the two ACP domains; the extent of the twist in the waist region corresponds to a full turn and the ACP interacts only with the domains on its own chain. The arrows indicate which domains each ACP can reach. (c) Same expanded view of the FAS with only half a turn twist in the waist region which allows the ACP to interact with the KS and AT domains on the partner chain.*

3 QUATERNARY STRUCTURE OF THE DEBS MODULES

There is a very high level of confidence in the primary sequence of the erythromycin PKS based on the gene sequence, and a reasonable level of confidence in predictions of the secondary and tertiary folding of the protein chains in the regions of globular domains based on homology with closely related enzymes that exist as isolated entities in other PKS and FAS systems.[1,2] In contrast, making predictions concerning the quaternary organisation necessary to allow domains to interact in the course of synthetic operations is a much more challenging task. Despite their large size, these megasynthases are homodimeric, making them true giants with a massive increase in the range of possible quaternary structures.

There has been no report of an image of a complete module of a PKS equivalent to the FAS system mentioned above. There is an X-ray structure for the KS-AT portion of module 5[10] and for the reductive loop region of DEBS module 1,[11] both of which have been used to predict a structure for a type I PKS module which has ended up looking essentially identical to the Ban structure for the FAS.[14] However, Weissman has pointed out that the ACP domains in modules of a PKS have very short linkers to the KS homodimer in the downstream module, or to another tightly dimerised structural residue.[19,20] A structure in which the ACP resides in a cleft in a FAS type quaternary structure would therefore require breaking the dimerisation motif downstream, with massive disruption of any downstream module.

3.1 The Cambridge Topology for the PKS Module

It was recognition of this major constraint that led to the proposal of the 'Cambridge Double Helical Model'.[21] The inspiration came from early proteolysis studies on DEBS modules, in which individual domains could be isolated.[22,23] It came as a surprise at the time that the free KS domains existed as strong homodimers. That led to the idea that the KS domains might be in contact within the homodimeric module, and that the two chains might run side by side in the 'head-to-head' sense, rather than the then assumed 'head-to-tail'. The latter arrangement was originally proposed to account for evidence from cross-linking that the ACP of one chain lies in contact with the KS domain of the other. In the Cambridge model, this contact is achieved in a head to head arrangement simply by twisting the chains through 180 degrees about the axis of symmetry.

The other crucial feature of the Cambridge model is the looping out of the intervening domains away from the axis to allow the two ACP domains to make direct contact with the upstream KS group in the same module.[21] This looping out generates a double helical topology, although that description it is not to be taken as a precise description of the positioning of individual domains. This head to head arrangement with a twisted topology is consistent with all the available evidence that has emerged since, including studies by proteolysis,[24,25] X-ray,[10,11,14] and mutant complementation,[2,16] and NMR[12,13] studies of docking domains.

The starting point is the primary sequence and the proposed division into globular and linker sections. The KS and TE domains of the DEBS are strongly homodimeric in isolation from the multienzyme and so are placed in contact inside the homo-dimeric multienzyme leading to a head-to-head arrangement of the chains in **A** (Figure 4). The DH and ER domains are shown as potential *temporary* dimerisation sites based on the FAS X-ray studies.

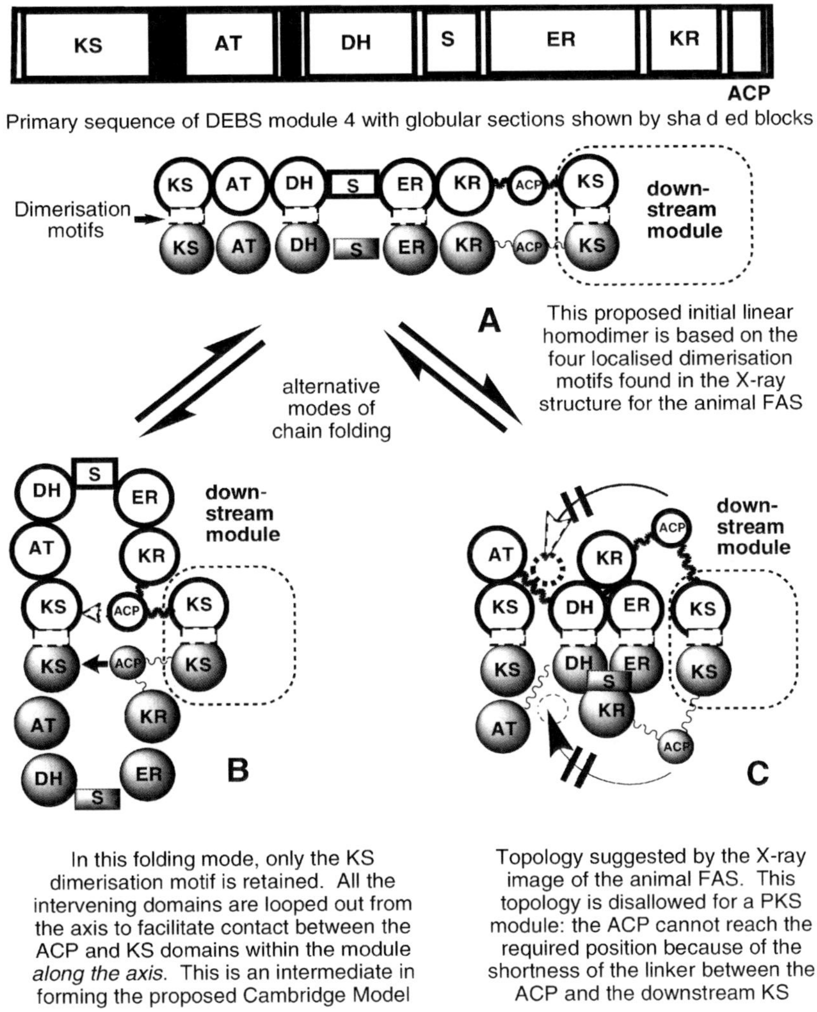

Figure 4 *Alternative modes of folding for DEBS module 4.*

The unfolded dimer, **A**, can then fold in two different ways to give either quaternary conformation, **B**, in which the domains between the ACP and upstream KS are looped out, or **C**, in which the DH and ER domains are left associated on the axis, and only the KR and spacer domains are looped out (Figure 4). Structure **C** corresponds to the latest FAS structure but in terms of linker lengths, there is a problem in a PKS: the shortness of the linker between the ACP and the downstream KS prevents the ACP from reaching its KS partner upstream. This logistical barrier is overcome in the alternative conformation, **B**, by looping out the intervening domains from the axis of symmetry. Structure **C** can also convert to structure **B** simply by peeling apart the ER and DH domains. The problem with conformation **B** is that it predicts the wrong pairing between ACP and KS domains for the condensation step. By twisting the downstream module through a half-turn to give

D, we can resolve this problem (Figure 5). Now when the ACP domains move to make contact with the KS domains in the more compact structure **E**, (the Cambridge Model) the correct pairing between KS and ACP domains in opposite chains is achieved.

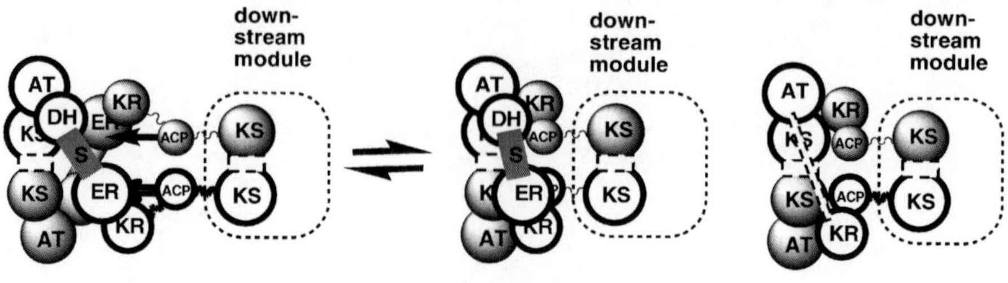

D

The two chains of quaternary structure B in Fig 4 are twisted through half a turn to give the double helical toplology of the Cambridge Model.

E

This is a shortened version of structure D in which the pair of ACP domains have moved to the left to make contact with the KS domains. A backwards and forwards motion exemplified by structures D and E probably allows the ACP to interact with other domains at different stages of the chain extension cycle.

F

A cut-away version of conformation **E** to show the ACP-KS face-to-face interaction

Figure 5 *Cambridge Double Helical model of DEBS module 4, a type I modular PKS module with a full set of reductive domains.*

All the views in Figures 4 and 5 are from the side and an alternative set of views of the Cambridge Model are shown in Figure 6. A full reductive loop such as that found in DEBS module 4 forms a complete circle around the ACP domains located in the core. Extra stability from vertical interactions between the AT domains and the KR domains of the opposite chain are possible. Once the loop has been cut down to three domains (AT, KT, and a spacer, Figure 6c) this potential connectivity is no longer possible unless the spacer adopts a more extended conformation as shown in Figure 6d. It is interesting to note that in modules that have no reductive activity, such as DEBS module 3, both the spacer and the dead KR are usually retained, consistent with the need to retain a sufficient size of reductive loop for effective operation of the module.[6]

3.2 Proteolysis Studies on the DEBS Multienzymes

The X-ray[10,11,14] and EM[15] results together with the mutant complementation studies[2,16] have shed light on the quaternary structure of the DEBS proteins, however, they do not provide any insight into how the assembly line actually works. Do we have one propionyl-CoA starter progressively extended and offloaded before another chain is built, or do we have a continuous stream of intermediates on one or both chains? Moreover, how is the fidelity of the assembly line maintained? Do the partner chains interact and/or cooperate in the assembly of a particular intermediate? For, while we know much about these giant enzymatic assemblies, we do not know how they are controlled and coordinated. It is known that modular PKS are capable of by-passing (or 'skipping') modules[26,27] or using a module more than once (iteration or 'stuttering').[28-30]

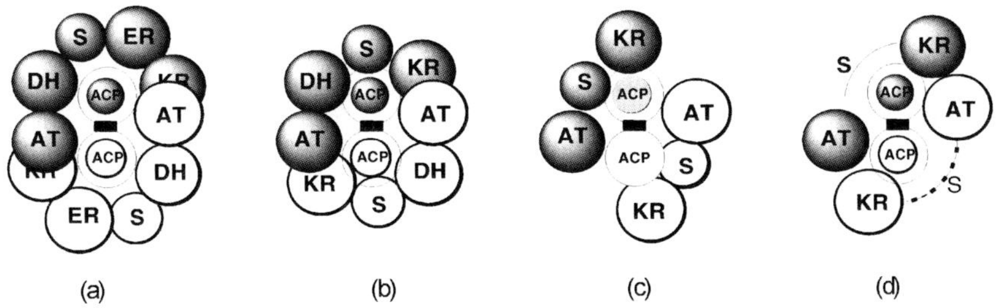

Figure 6 *A view looking down the axis of modules with different complements of reductive domains shows that the outer domains AT to KR, form a ring of protein structure around the pair of ACP domains at the core. The KS domains are shown as see-through circles to reveal the positions of the two ACP domains. The relative orientation of the KS pair of domains to the AT pair is based on the module 4 KS-AT X-ray structure.[10] (a) Full reductive loop as found in DEBS module 4; (b) KR, DH and spacer; (c) KR and spacer as found in DEBS modules 1, 2, 5 and 6; (d) same as (c) but spacer adopts a more extended conformation.*

Modular type I PKS are functional homodimers and their huge size precludes direct analysis using techniques such as mass spectrometry, which have provided invaluable information about acyl intermediates attached to dissociated (Type II) FAS and PKS.[31,32] One successful approach has been to cleave the multi-enzymes and to extract out the active-site peptides and identify the covalently attached acyl-groups using high-resolution electrospray ionization Fourier-Transform mass spectrometry (ESI/FTMS).[33,34]

Proteolysis of DEBS1 gives rise to a number of single and di-domains.[21-25] Using an engineered diketide synthase (DKS)[35] (Figure 7) comprising the DEBS loading module and first chain-extension module with a C-terminal thioesterase, Hong et al. have been able to study chain elongation directly.[24,25] The DKS is subjected to limited proteolysis to generate domain-sized fragments (Figure 8a) which are separated and analysed by liquid chromatography(LC)/MS.[24] If the DKS is incubated with the natural starter unit, propionyl CoA prior to proteolysis, both of the loading module domains and the KS of module 1 are found to be completely acylated (Figure 8b).[24] Alternative starter units such as acetyl CoA, butyryl CoA and valeryl CoA give similar results. When the experiment was repeated with the chain extender unit, methylmalonyl CoA, both AT1 and ACP1 were acylated but the loading module domains and the KS were completely unloaded (Figure 8c).[25] Incubation with malonyl CoA under the same conditions resulted in no acylated domains demonstrating that the AT1 domain is highly substrate specific in contrast with the loading domain AT.

The next step was to look at the synthesis of the diketide in terms of the occupancy levels of the catalytic domains and the nature of the acyl-intermediates.[25] Incubation of the DKS with the starter and chain-extender units and NADPH, required for the KR-catalysed reduction step, gave the results shown in Figure 8d.[25] The KS domain was found to be fully loaded with propionate which shows that the transfer of the starter unit from the loading domain is not rate-limiting under the experimental conditions. Interestingly, the TE domain in both the ACP1-TE di-domain and the TE did not have a covalently attached acyl group. Examination of the ACP-TE di-domain showed that the ACP moiety was partially acylated with the diketide (58%) or unloaded (42%), with no methylmalonyl-ACP detected.[25] Hence, it would appear that under these conditions, both chains of the DKS are involved in the chain-extension and that the rate limiting step is the transfer of the diketide intermediate from the chain-extension ACP to the TE.

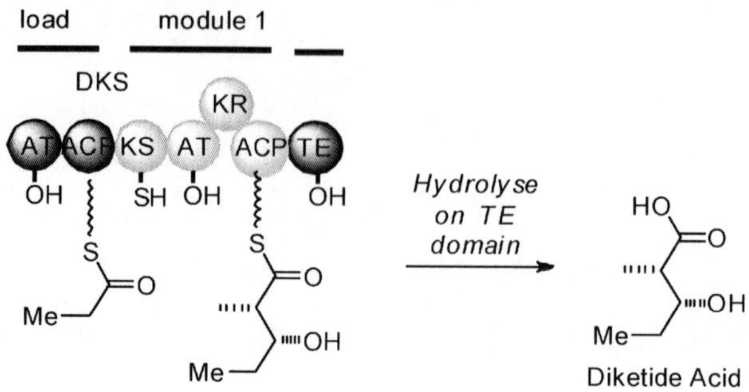

Figure 7 *The diketide synthase (DKS) carries out the operation of DEBS module 1 and then releases the diketide product as a fatty acid.*

Replacing the starter unit with butyryl CoA and valeryl CoA still resulted in fully loaded KS1 and AT1 domains, but striking differences were observed in ACP1 occupancy.[24] With butyrate, half of the ACP was loaded with diketide. The drop in occupancy can be accounted for by the slowing of the rate of the condensation step relative to the offloading step. For both butyrate and propionate no methylmalonyl-ACP was observed. However, this species accounted for more than a third of the occupancy of ACP1 when valerate was used as the starter unit, with the corresponding diketide only accounting for 28%. In this case it appears that an additional bottleneck has been introduced into the chain assembly process.

If we extrapolate from the diketide system to the full DEBS assembly line, the picture that emerges is one where both chains have high occupancy levels along the full length of the assembly line, with the ACP and KS domains all loaded with their intermediates and the AT domains all primed with their relevant building blocks. As a consequence, movement along the assembly line would be controlled by the release of the heptaketide intermediate from ACP6 in the last module (Figure 1) which is reminiscent of a drinks can dispenser.[1] As each KS domain becomes free, it would immediately be reloaded with a substrate from the upstream ACP thereby preventing back-transfer of biosynthetic intermediates and suppressing iterative use of the module. This mode of operation can be described as occupancy-level control[1] and implies that the overall rate of chain assembly is dependent on the release of the final product by the exit mechanism. The price that is paid for iteration suppression is a much slower throughput on the assembly line.

The retardation also enables sufficient time for all the reductive loop domains to operate before moving to the next chain extension step. We use the term residence-time control[1] to describe these effects. Clearly both occupancy-level control and residence-time control are achieved by slowing the rate of release of the final intermediate on the assembly line. There is a subtle but important difference between them though: occupancy-level control is a spatial effect and will reach a maximum once the KS domains are fully occupied whereas residence-time control is a temporal effect that could continue to increase after the occupancy levels of the KS and ACP domains are fully loaded.[1]

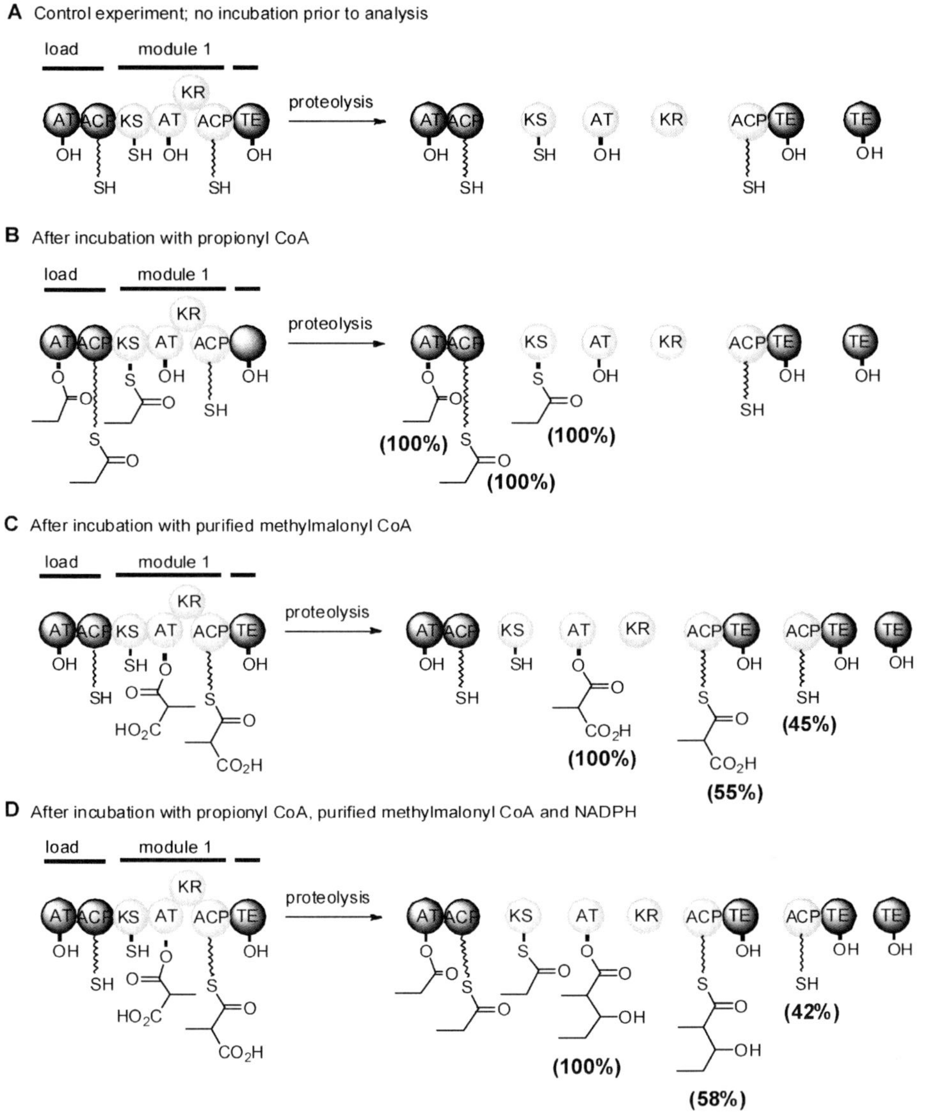

Figure 8 *Proteolysis of the diketide synthase followed by HPLC-ESMS. Levels of occupancy are shown in brackets.*[24,25]

When the NADPH cofactor was left out of the reaction mixture, no diketide intermediate was found on the ACP1 or the TE and only the methylmalonyl form of ACP1 was observed (55%) along with the free thiol (45%).[25] This result is in direct contrast with those obtained from an *in vitro* system consisting of individually expressed and purified domain (ACP) and di-domains (KS-AT) which cooperate *in trans* to carry out a chain extension cycle.[36] Incubation of a diketide thioester substrate, methylmalonyl CoA and

ACP afforded the β-keto triketide tethered to the ACP. However, using the diketide synthase system, no diketide intermediate was detected. To explain the inability of the diketide to accumulate on the DKS, Hong *et al.* suggested that an allosteric effect is in operation caused from a change in quaternary structure when NADPH binds to the KR domain.[25]

Based on the Cambridge Model, Figure 9 shows a cartoon version of the topology of a representative chain extension module in which the reductive loop is truncated by deletion of the DH and ER domains. The two KS and ACP domains need to come close enough to enable the Claisen condensation step to take place. The double-helical rope-like twist of the two chains in the homodimer enables the two ACP domains to move along the axis bringing them into contact with the upstream KS domains.

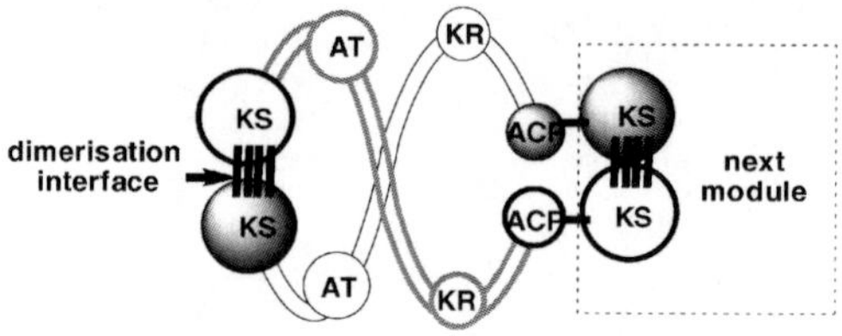

The relative positioning of the KS and AT domains conforms to that shown in the X-ray structure of an isolated KS-AT fragment with the AT domains in an outer position remote from the axis. However, the KS-AT linker undergoes facile proteolysis, so it must be in equilibrium with an unfolded form, not revealed in the X-ray image.

With the linker unfolded, the two AT domains would be free to move to inner positions closer to the axis. In the outer position, they would reload with methylmalonate. The subsequent move to an inner position (the original cartoon illustrating the structural principles of the Double Helical topology showed this quaternary conformation) would facilitate transfer of the methylmalonate to the ACP, and, in addition, to block premature access of the unloaded ACP thiol to the KS active site, a situation that would lead to skipping of the module.

Figure 9 *The Cambridge Topology for the truncated DEBS module 1. The two identical chains of the homodimeric structure are differentiated by red and grey colouring. The KS domains have strong homodimeric interfaces and are placed in contact at the 'head' of the structure. The strongly homodimeric KS domains are also placed in contact with each other at the 'tail'. The remaining three domains are not homodimeric and so can move away from the common axis running through the pairs of KS and TE domains. The pair of ACP domains, however, are held close to the KS domains by short linker regions, and so must remain in close proximity to each other and the axis. To aid visualization, the two domains, KR and AT, in the mid section of the homodimer are shown as small circles rather than spheres of appropriate size. These domains are sited away from the axis of the proposed structure to free up a central passage. The pair of ACP domains can now make contact with the pair of KS domains by moving parallel to the axis with the KS domains in tow. The structure is also given a helical twist of 180 degrees to accord with evidence that the ACP of one chain interacts with the KS of the opposite chain.*

Inspection of the model shows that the AT, KR and intervening space regions form a 'collar' around the ACP domains shielding the acyl intermediates from the surrounding aqueous medium. It is conceivable that binding of the NADPH cofactor to the KR domain induces a conformational change that permits the collar to open. In the resulting quaternary structure, each ACP domain can access the appropriate KS domain for the condensation step (and other domains in succession through the chain extension cycle) by moving backwards and forwards (dashed arrows) along the axis within the central core of the structure. Because of the restricting effect of the short ACP to KS linker, and the anticipated need for coordinated movements of the two looped out domains, it is likely that the pair of ACP domains move backwards and forwards in tandem, rather than independently. As a result, the successive reactions of the chain extension cycle on the two chains might also be constrained to operate in tandem. At the start of each step of the chain extension cycle, the pair of ACP domains would be loaded with identical intermediates and both would bind with the appropriate domain for the next step. The first intermediate to complete the reaction would be free to leave its catalytic domain, but its ACP would stay put until the second intermediate has completed the same operation. Both intermediates are now free of the catalytic sites and the two ACP domains would then move in tandem to cooperate with the next pair of catalytic domains. These predictions of coordinated movements of domains in the Cambridge Double Helical Model are consistent with the intriguing patterns of occupancy of ACP1 found in the Hong *et al.* paper.[25]

4 CONCLUSION

X-ray images of the mammalian FAS and fragments of PKS modules have greatly enhanced our understanding of the quaternary structure of type I modular FAS and PKS systems. The picture that is emerging from EM studies and solution work is that the quaternary structure must be highly mobile with the ACP domain making considerable changes in position as it delivers its payload to the reductive enzymes. It is also possible that other domains undergo movements between inner and outer locations. What is evident is that more images are required to capture the different conformations adopted by a module during the chain-extension process. Experiments on the DEBS diketide synthase suggest that the assembly of polyketides uses both chains of the homodimer and involves high occupancy levels of the catalytic domains to ensure the fidelity of the assembly line synthetic operations.

References

1 A.M. Hill and J. Staunton, In *Comprehensive Natural Products II Chemistry and Biology*, L. Mander and H.-W. Lui (Eds.), Elsevier, Oxford, 2010, **1**, 386.

2 S. Smith and S.-C. Tsai, *Nat. Prod. Rep.*, 2007, **24**, 1041.

3 C. Hertweck, *Angew. Chem. Int. Ed. Engl.*, 2009, **48**, 4688.

4 J. Cortés, S.F Haydock, G.A. Roberts, D.J. Bevitt and P.F. Leadlay, *Nature*, 1990, **348**, 176.

5 J.S. Tuan, J.M. Weber, M.J. Staver, J.O. Leung, S. Donadio and L. Katz, *Gene*, 1990, **90**, 21.

6 C. Khosla, Y. Tang, A.Y. Chen, N.A. Schnarr and D.E. Cane, *Ann. Rev. Biochem.*, 2007, **76**, 195.

7 Y. Tang, C.-Y. Kim, I.I. Matthews, D.E. Cane and C. Khosla, *Proc. Natl. Acad. Sci. USA*, 2006, **103**, 11124.

8 M.A. Reed, M. Schweizer, A.E. Szafranska, C. Arthur, T.P. Nicholson, R.J. Cox, J. Crosby, M.P. Crump and T.J. Simpson, *Org. Biol. Chem.*, 2003, **1**, 463.

9 Y.Y. Tang, A.Y. Chen, C.Y. Kim, D.E. Cane and C. Khosla, *Chem. Biol.*, 2007, **14**, 931.

10 Y. Tang, C.Y. Kim, I.I. Mathews, D.E. Cane and C. Khosla, *Proc. Natl. Acad. Sci. USA*, 2006, **103**, 11124.

11 A.T. Keatinge-Clay and R.M. Stroud, *Structure*, 2006, **14**, 737.
12 R.W. Broadhurst, D. Nietlispach, M.P. Wheatcroft, P.F. Leadlay and K.J. Weissman, *Chem. Biol.*, 2003, **10**, 723.
13 L. Tran, M. Tosin, J.B. Spencer, P.F. Leadlay and K.J. Weissman, *ChemBioChem*, 2008, **9**, 905.
14 T. Maier, N. Leibunggut and N. Ban, *Science*, 2008, **321**, 1315.
15 E.J. Brignole, S. Smith and F.J. Asturias, *Nat. Struct. Mol. Biol.*, 2009, **16**, 190.
16 A.K. Joshi, V.S. Rangan, A. Witkowski and S. Smith, *Chem. Biol.*, 2003, **10**, 169.
17 C.M.T. Spahn, P.A. Penczek, *Curr. Opin. Struct. Biol.*, 2009, **19**, 623.
18 G. Pappenburger, J. Benz, B. Gsell, M. Hennig, A. Ruf, M. Stihle, R. Thoma and M. G. Rudolph, *J. Mol. Biol.*, 2010, **397**, 508.
19 K.J. Weissman, *ChemBioChem*, 2008, **9**, 2929.
20 K.J. Weissman and R. Müller, *ChemBioChem*, 2008, **9**, 826.
21 J. Staunton, P. Caffrey, J.F. Aparicio, G.A. Roberts, S.S. Bethel and P.F. Leadlay, *Nat. Struct. Biol.*, 1996, **3**, 188.
22 A.F.A. Marsden, P. Caffrey, J.P. Aparicio, M.S. Loughran, J. Staunton and P.F. Leadlay, *Science*, 1994, **263**, 378.
23 J.F. Aparicio, P. Caffrey, A.F.A. Marsden, J. Staunton and P.F. Leadlay, *J. Biol. Chem.*, 1994, **269**, 8524.
24 H. Hong, A.N. Appleyard, A.P. Siskos, J. Carcia-Bernado, J. Staunton and P.F. Leadlay, *FEBS J.*, 2005, **272**, 2372.
25 H. Hong, P.F. Leadlay and J. Staunton, *FEBS J.*, 2009, **276**, 7057.
26 C.J. Rowe, I.U. Böhm, I.P. Thomas, B. Wilkinson, B.A.M. Rudd, G. Foster, A.P. Blackaby, P.J. Sidebottom, Y. Roddis, A.D. Buss, J. Staunton and P.F. Leadlay, *Chem. Biol.*, 2001, **8**, 475.
27 I. Thomas, C.J. Martin, C.J. Wilkinson, J. Staunton and P.F. Leadlay, *Chem. Biol.*, 2002, **9**, 781.
28 B. Wilkinson, G. Foster, B.A.M. Rudd, N.L. Taylor, A.P. Blackaby, P.J. Sidebottom, D.J. Cooper, M.J. Dawson, A.D. Buss, S. Gaisser, I.U. Böhm, C.J. Rowe, J. Cortés, P.F. Leadlay and J. Staunton, *Chem. Biol.*, 2000, **7**, 111.
29 B.J. Beck, C.C. Aldrich, R.A. Fecik, K.A. Reynolds and D.H. Sherman, *J. Am. Chem. Soc.*, 2003, **125**, 4683.
30 S.J. Moss, C.J. Martin and B. Wilkinson, *Nat. Prod. Rep.*, 2004, **21**, 575.
31 A.M. Bridges, P.F. Leadlay, W.P. Revill and J. Staunton, *Chem. Commun.*, 1991, **11**, 778.
32 T.S. Hitchman, J. Crosby, K.J. Byrom, R.J. Cox and T.J. Simpson, *Chem. Biol.*, 1998, **5**, 35.
33 M.T. Mazur, C.T. Walsh and N.L. Kelleher, *Biochemistry*, 2003, **42**, 13393.
34 L.M. Hicks, S.E. O'Connor, M.T. Mazur, C.T. Walsh and N.L. Kelleher, *Chem. Biol.*, 2004, **11**, 327.
35 L.H. Østergaard, L. Kellenberger, J. Cortés, M.P. Roddis, M. Deacon, J. Staunton and P.F. Leadlay, *Biochemistry*, 2002, **41**, 2719.
36 R. Catonguay, W. He, A.Y. Chen, C. Khosla and D.E. Cane, *J. Am. Chem. Soc.*, 2007, **129**, 13758.

BIOSYNTHESIS OF POLYENES IN *AUXARTHRON UMBRINUM*

B.R. Clark, D. Fox and C.D. Murphy

School of Biomolecular and Biomedical Science, University College Dublin, Belfield, Dublin 4, Ireland. E-mail: cormac.d.murphy@ucd.ie

1 INTRODUCTION

Rumbrin **1** (Figure 1) is a fungal polyketide containing a highly unusual chloropyrrole moiety, which possesses potent anticancer[1] and cytoprotective properties.[2] First isolated in 1993 from the fungus *Auxarthron umbrinum*, it has since been isolated several times from related fungi along with the closely related auxarconjugatins.[3] Very little is known about the biosynthesis of rumbrin, though it is presumably formed through the action of a polyketide synthase (PKS) enzyme complex. In this study we describe an investigation into the biosynthesis of **1** using feeding studies with isotopically labelled precursors and the production of unnatural analogues *via* precursor directed biosynthesis with substituted pyrrole carboxylic acids.

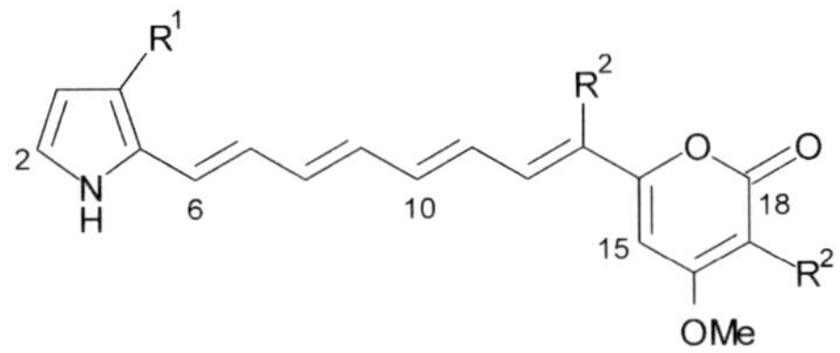

rumbrin (**1**) R^1 = Cl, R^2 = Me
dechlororumbrin (**2**) R^1 = H, R^2 = Me
auxarconjugatin D (**3**) R^1 = H, R^2 = H

Figure 1 *Structures of naturally-occurring pyrrolylpolyenes produced by* A. umbrinum *DSM-3193.*

2 THE BIOSYNTHESIS OF RUMBRIN

2.1 The Incorporation of Primary Building Blocks

The commercially available strain *A. umbrinum* DSM-3193 was used for the studies. The major polyene produced was dechlororumbrin **2**, though rumbrin **1** and the new metabolite

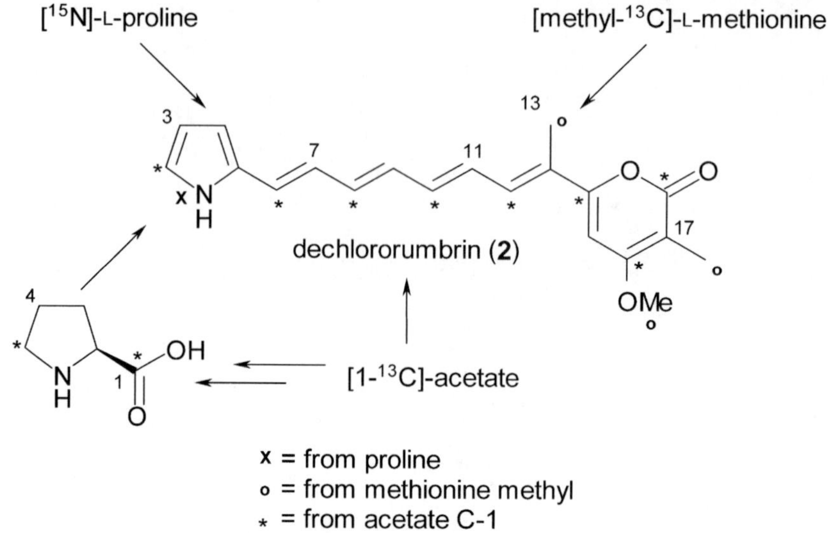

Figure 2 *Origins of the atoms in dechlororumbrin were determined by feeding studies with labelled precursors.*

auxarconjugatin D **3** were also detectable (Figure 3a). Given the high titre of **2**, the biosynthetic studies focused on this as a model for production of all polyenes.

[1-^{13}C]-acetate, [^{15}N]-L-proline, and [methyl-^{13}C]-L-methionine were added to separate, growing cultures after three days to determine whether they were incorporated into the polyene. The mycelium was extracted and the extracts analysed by mass spectrometry and HPLC. Percentage incorporations of the labelled substrates are given in Table 1.

Precursor added	Conc.	Number of units; % Incorporation			
	(mM)	1	2	3	4+
[^{15}N]-L-proline	2	22			
	5	43			
[^{15}N]-glycine	2	5.1			
	5	8.1			
[Me-^{13}C]-L-methionine	2	14	5.5		
[1-^{13}C]-acetate	25	18	12	12	13.7

Table 1 *Percentage incorporation of labeled precursors into dechlororumbrin* ***2****. The percentages were assessed by comparison of the peak heights at m/z of (M+1) and (M+2) in the mass spectrum of* ***2****.*

^{13}C NMR was used to determine the location of the labelled carbons. For the methionine feeding study, resonances corresponding to the 13-methyl (8.7 ppm, 8.5%), 17-methyl (12.8 ppm, 6.1%), and 16-O-methyl (55.5 ppm, 10.1%) groups of dechlororumbrin **2** were enhanced. For the acetate study ^{13}C NMR revealed incorporation of C-1 of acetate as shown in Figure 2. The unexpected labels at C-2 and C-6 were incorporated *via* proline, which is derived from acetate *via* the TCA cycle. While the incorporation of acetate and methionine was expected, this is the first example of proline being incorporated into a fungal polyketide. The origins of the atoms in **2** are illustrated in Figure 2.

2.1 Competitive Incorporation of Pyrrole-2-carboxylic Acid and Proline

While little is known about the biosynthesis of fungal pyrroles, several bacterial pyrrole-containing antibiotics are synthesised *via* pyrrolyl-2-carboxylate intermediates.[4] To determine whether pyrrole-2-carboxylic acid **4** was incorporated into dechlororumbrin **2**, a label dilution experiment was conducted in which both [^{15}N]-L-proline and **4** (2 mM each) were added to an *A. umbrinum* culture. The extracted **2** showed reduced incorporation of ^{15}N (2.6%) relative to the [^{15}N]-proline-only control (22%), indicating that **4** is a direct precursor to rumbrin. Addition of **4** also lead to an increase in the production of non-halogenated polyenes (Figure 3b). A proposed biosynthesis for **2** is given in Scheme 1. In contrast to bacterial pyrrole biosynthesis, we suggest that proline is oxidised to **4** as a free acid, which is then used as the starter unit for polyketide biosynthesis without the involvement of additional carrier proteins. Timing of the chlorination step for rumbrin **1** is not known.

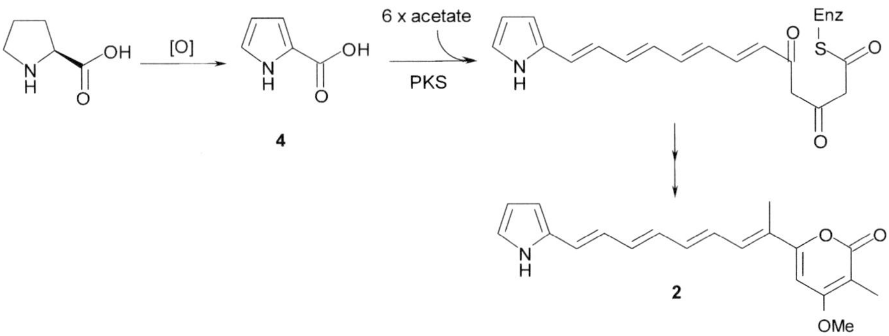

Scheme 1 *Proposed biosynthetic route for dechlororumbrin* **2**.

3 INCORPORATION OF UNNATURAL PYRROLE CARBOXYLATES

It has been shown that substitution of the pyrrole ring has a drastic effect on the anticancer activity of rumbrin derivatives.[5] We therefore attempted to produce un-natural analogues through precursor-directed biosynthesis. Pyrrole-2-carboxylic acid derivatives (Scheme 3)

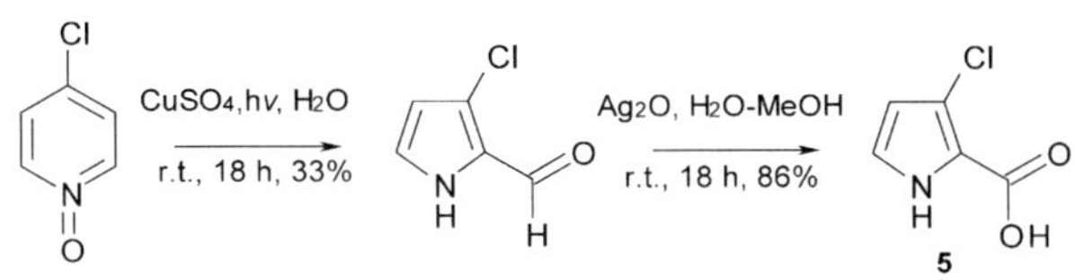

Scheme 2 *Synthesis of 3-Chloropyrrole-2-carboxylate* **5**.

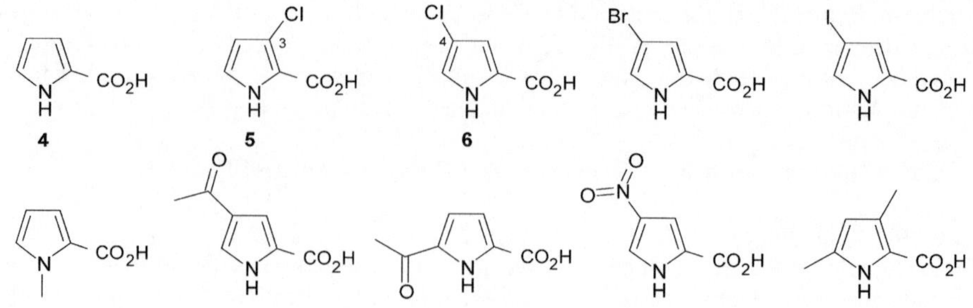

Scheme 3 *Pyrrole carboxylates used in precursor-directed biosynthesis studies.*

were purchased or were synthesised using literature methods,[6] with the exception of **5** which was synthesized in two steps as shown in scheme 2.[7,8]

A. umbrinum fermentations were fed with the selected compounds and the extracts analysed by HPLC and MS. Acids **5** and **6** were incorporated into the polyenes and the HPLC chromatograms in Figures 3c and 3d show the change in the eluted u/v profile

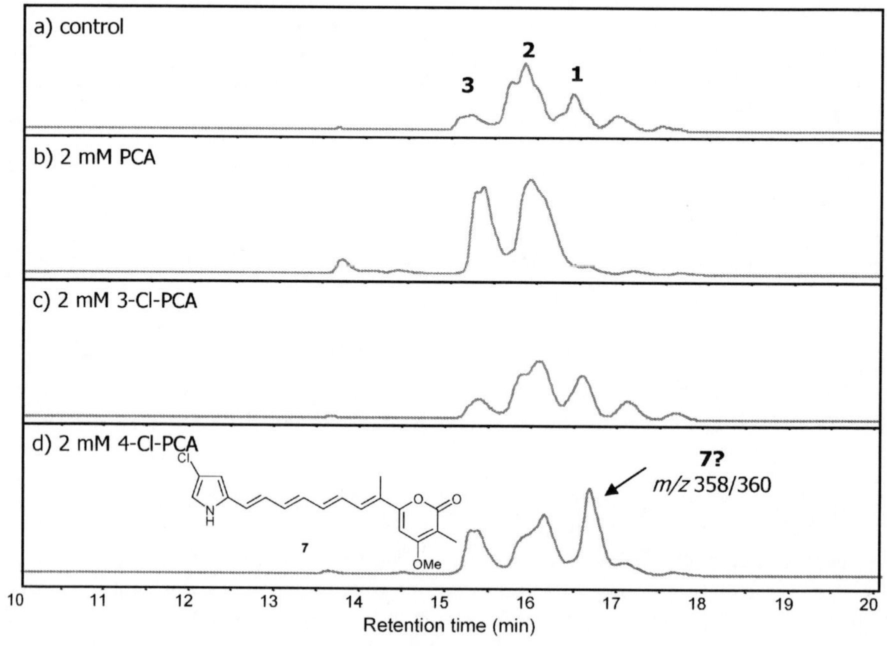

Figure 3 *HPLC analysis of organic extracts from* A. umbrinum *cultures showing polyene production. Addition of the acids* ***4-6*** *led to significant changes in the polyene profile. Analysis: Thermo Hypersil* C_{18} *4.6 × 150 mm column, 0.8 mL/min, 10% MeCN-*H_2O *to 100% MeCN over 20 minutes, 440 nm.*

caused by the feeding of the unnatural precursors relative to that of an unfed control (Figure 3a). The incorporation of **5** into **1** was low and as such it is not believed to be the natural precursor. Addition of **6** led to production of a new polyene for which we suggest the structure **7**. Large-scale studies are in progress to isolate the new polyene and determine its biological properties.

References

1 B.R. Clark, R.J. Capon, E. Lacey, S. Tennant and J.H. Gill, *Org. Lett.*, 2006, **8**, 701.
2 Y. Yamagishi, K. Shindo and H. Kawai, *J. Antibiot.*, 1993, **46**, 888.
3 T. Hosoe, K. Fukushima, K. Takizawa, M. Miyaji and K.-I. Kawai, *Phytochemistry*, 1999, **52**, 459.
4 S. Garneau, P.C. Dorrestein, N.L. Kelleher and C.T. Walsh, *Biochemistry*, 2005, **44**, 2770.
5 B.R. Clark, E. Lacey, J.H. Gill and R.J. Capon, *J. Nat. Prod.*, 2007, **70**, 665.
6 P. Bélanger, *Tetrahedron Lett.*, 1979, 2504.
7 F. Bellamy, P. Martz and J. Streith, *Heterocycles*, 1975, **3**, 395.
8 P. Hodge and R.W. Rickards, *J. Chem. Soc.*, 1963, 2543.

Acknowledgements

Parts of this work have been have been published: B. R. Clark and C. D. Murphy, *Org. Biomol. Chem.* 2009, **7**, 111-116.
BRC would like to acknowledge funding from IRCSET in the form of an Embark Fellowship.

5 Summary of Other Lectures

FUNCTIONAL MOLECULES FROM NATURAL SOURCES: BRIEF SUMMARY OF OTHER CONFERENCE LECTURES

Stephen K. Wrigley

Hypha Discovery Ltd,. Russell Building No.1, Brunel Science Park, Kingston Lane, Uxbridge, Middlesex UB8 3PQ, UK

1 INTRODUCTION

Some lecturers at the conference did not provide manuscripts based on their contributions. The principal reason for this is that the work they described has already been published piecemeal in a variety of publications. In order to make this volume a more complete account of the conference proceedings for the delegates, and an accurate record for those who were unable to attend, this chapter summarises these contributions with a brief synopsis of the lectures and key references to the relevant literature.

2 FUNCTIONAL MOLECULES AND THEIR ENHANCEMENT

2.1 Biosynthetic Engineering of Nonribosomal Lipopeptides

Jason Micklefield (University of Manchester) described the elucidation of the key steps in the biosynthesis of the calcium-dependent antibiotic (CDA) from *Streptomyces coelicolor* and the biosynthetic engineering of new lipopeptides with improved biological properties. CDA **1** (Figure 1) belongs to the family of non-ribosomal acidic lipopeptides, which includes the structurally-related daptomycin, produced by fermentation of *S. roseosporus*, which in 2003 became the first natural antibiotic representing a new structural class to be approved for clinical use in more than thirty years.[1] Daptomycin is used for the treatment of infections caused by Gram positive pathogens and has become one of the most successful intravenous antibiotics launched in US history. Studies of the biosynthetic origins of CDA were described in detail including the elucidation at the chemical and genetic levels of the pathways leading to the non-proteinogenic amino acids in CDA including 3-phosphohydroxyasparagine,[2] *Z*-dehydrotryptophan,[3] hydroxyphenylglycine,[3] and 3-methylglutamic acid,[4,5] as well as the 2,3-epoxyhexanoyl fatty acid side chain.[6] Information garnered from these investigations has been used to develop a range of strategies for engineering the combinatorial biosynthesis of lipopeptides based on CDA.

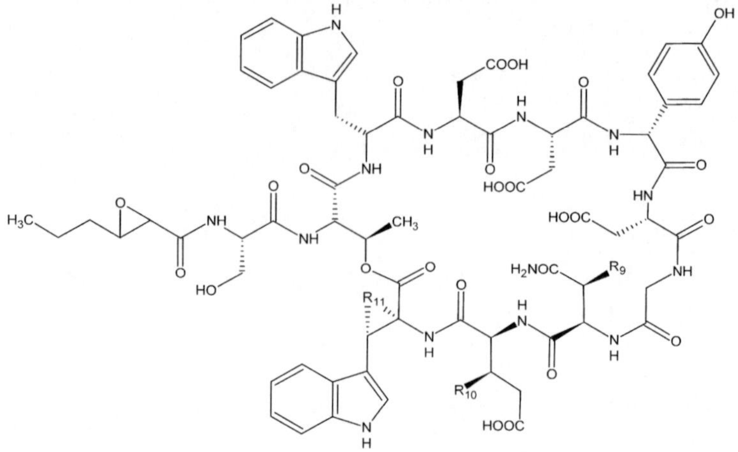

1

Figure 1 *Structural variation in native members of the CDA complex: R_9 = OPO_3H_2 or OH; R_{10} = H or CH_3; R_{11} = H,H or π-bond.*

These included: mutasynthesis,[3,7] auxotrophic precursor-directed biosynthesis[8], active site modifications of adenylation domains,[9] module or domain swaps of the non-ribosomal peptide synthetase, and modification of the fatty acid synthetase.[6]

2.2 Development of Lantibiotics for Treatment of Nosocomial Infections

Mike Dawson (Novacta) described the company's research into the optimisation of lantibiotics as antibacterial lead compounds. Lantibiotics are antimicrobial peptides that are attractive as antibacterial molecules with development potential both because of their mode of action and the presence of extensive post-translational modifications that confer increased stability under physiological conditions.

They are divided into Types A and B lantibiotics, both of which bind the bacterial cell wall intermediate lipid II; while the type A lantibiotics exert antibacterial activity mainly through pore formation, members of the type B series are cell wall biosynthesis inhibitors. These compounds are produced by Gram positive bacteria as ribosomally-synthesised peptides that undergo dehydration of serine and/or threonine residues followed by electrophilic addition of the sulphur or cysteine to generate multiple cyclic structures. Type B lantibiotics are highly stable and have drug-like properties. The biosynthesis of lantibiotics is amenable to the generation of modified structures by mutagenesis of the prepropeptide. Novacta has developed a platform for investigation and manipulation of lantibiotic structure and biological activity using a combination of molecular biology and chemistry, as exemplified by the tetracyclic lantibiotic mersacidin **2** (Figure 2), through mutagenesis of the peptide residues not involved in cycle formation.[10,11] A new lantibiotic entity had also been discovered and a derivative of this has been selected as a drug candidate for treatment of *Clostridium difficile* infections. This compound is now undergoing formal preclinical development.

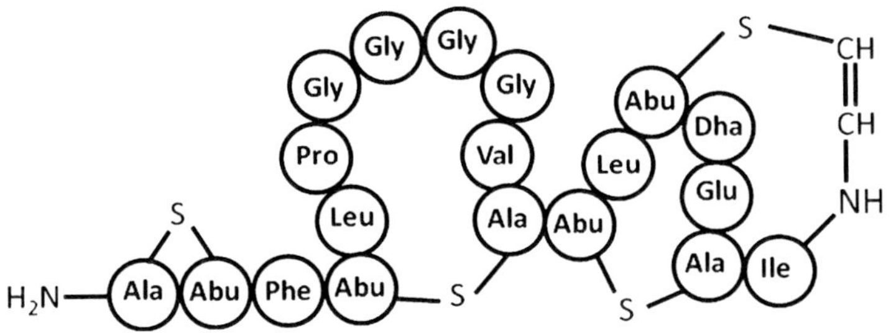

Figure 2 *The structure of mersacidin* **2***; Dha = didehydroalanine; Abu = aminobutyric acid.*

3 IMPROVED STRATEGIES FOR NATURAL PRODUCT EXPLOITATION

3.1 Acetogenic Anthraquinones and Alkaloids – Online Structural Elucidation, Biosynthesis, Bioactivities and Total Synthesis

Gerhard Bringmann (University of Würzburg) described the composition of an analytical triad platform comprising LC-MS/MS-NMR-CD, used in combination with quantum mechanical CD calculations, for the directed search for novel acetogenic metabolites present in crude extracts.[12] This platform has resulted in the early recognition and online-structure elucidation of novel natural products, including absolute configurations, of natural products such joziknipholone A **3**,[13] an unprecedented dimeric phenyl anthraquinone,[14] and ancistrocladinium A **4**, the first *N,C*-coupled naphthyldihydroisoquinoline alkaloid.[15] He drew attention to the remarkable contrast in biosynthetic pathways leading to the anthraquinone chrysophanol **5** by different modes of folding of a common octaketide precursor, namely F in eukaryotes and S and S' in prokaryotes.[16,17] Chrysophanol is the anthraquinone portion of joziknipholone A and its monomeric half, knipholone.[18] The structures of these compounds are shown in Figure 3.

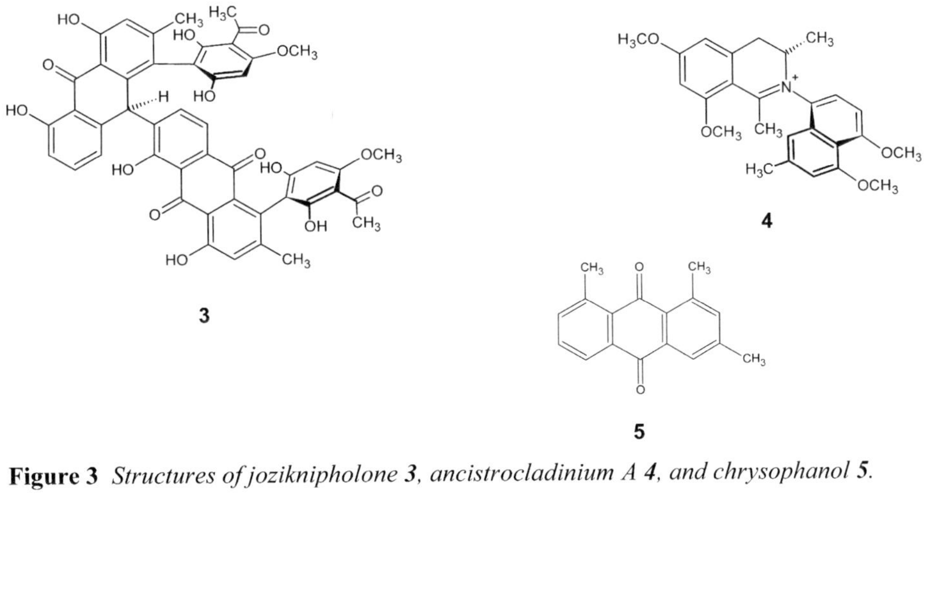

Figure 3 *Structures of joziknipholone* **3***, ancistrocladinium A* **4***, and chrysophanol* **5***.*

4 BIOSYNTHESIS AND GENETICS

4.1 Chemical Diversity by Other means: the Biosynthesis of Polyketide Drugs

Peter Leadlay (University of Cambridge) highlighted the large numbers of cryptic or silent biosynthetic gene clusters that are being identified by genome-scale sequencing of *Streptomyces* and other polyketide-producing organisms. Studies of *Streptomyces violaceusniger* DSM4137 had resulted in the characterisation genes for multiple products including the diolide elaiophylin **6**,[19] the macrocyclic neuroprotectant meridamycin **7**,[20] and the polyether nigericin **8**,[21] the structures of which are shown in Figure 4. There was now further genetic and chemical evidence for the presence in this strain of multiple additional gene clusters encoding modular polyketide synthase enzymes. Cross talk between different gene clusters was evident as knockout of one cluster often affected the products of others. Borrelidin **9** (Figure 4) from *Streptomyces parvulus* Tü4055 was discussed as another polyketide macrolide with potent *in vitro* and *in vivo* anti-angiogenic properties. Precursor-directed biosynthesis by a mutasynthetic approach using a mutant disrupted in the gene *borG* had generated a series of borrelidin analogues and SAR analysis indicated that modification at the C17 side chain provided increased selectivity for *in vitro* angiogenesis inhibition over cytotoxicity.[22,23]

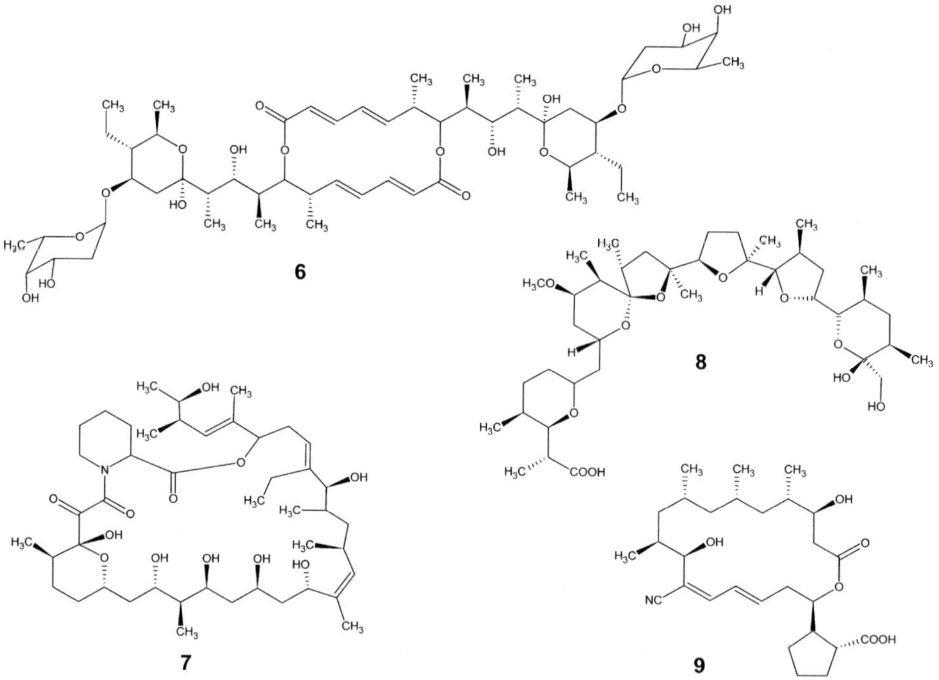

Figure 4 *Structures of elaiophylin **6**, meridamycin 7, nigericin **8** and borrelidin **9**.*

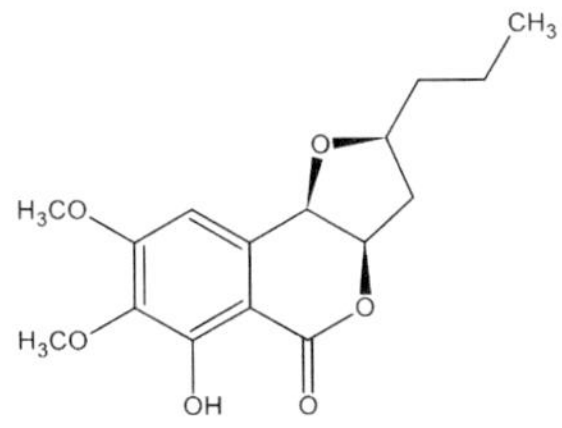

10

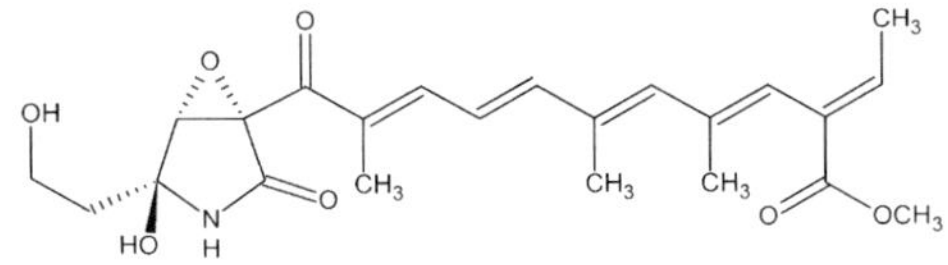

12

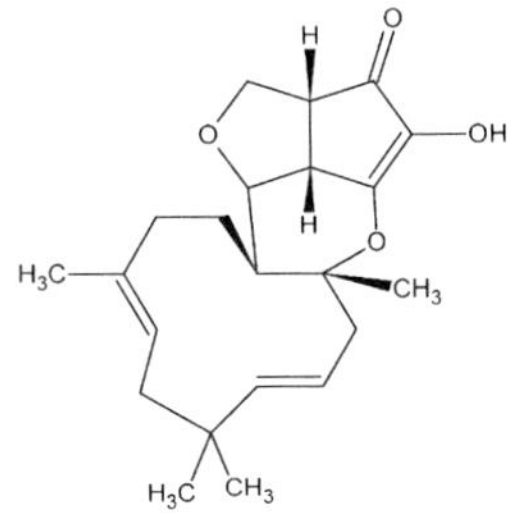

11

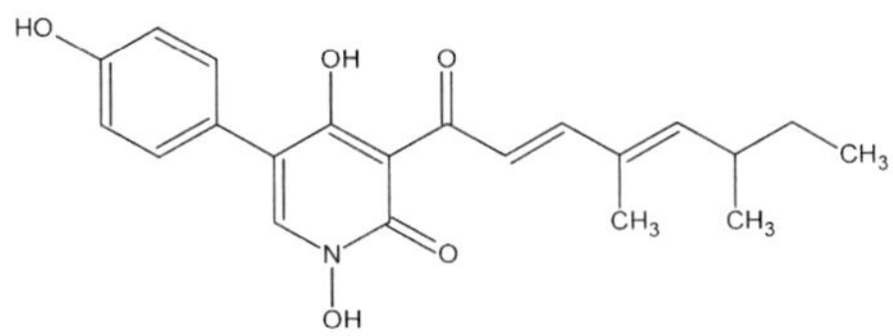

13

Figure 5 *Structures of monocerin* ***10****, xenovulene A* ***11****, fusarin C* ***12****, and tenellin A* ***13****.*

4.2 New Paradigms in Fungal Polyketide Biosynthesis

In contrast to bacterial biosynthetic pathways, fungal pathways are more challenging in terms of genetic manipulation. Tom Simpson (University of Bristol) described results of the complementation of classical biosynthetic studies with molecular genetic approaches, heterologous gene expression, gene disruption and silencing, and analysis of metabolite patterns in mutant strains for a number of fungal metabolites. These included monocerin **10** from *Drechslera ravenelii*;[24] xenovulene A **11** from *Acremonium strictum*, where heterologous expression studies of a PKS gene transferred to *Aspergillus oryzae* identified 3-methylorcinaldehyde as a key intermediate;[25] fusarin C **12** from *Fusarium moniliforme*;[26] and tenellin A **13** from the pathogenic fungus *Beauveria bassiana*, where genes encoding cytochrome P450 oxidase and oxygenase enzymes had been identified as being responsible for oxidative ring expansion of pretenellin A to form the 2-pyridone core and selective *N*-hydroxylation of the 2-pyridone moiety, respectively.[27] The structures of these compounds are shown in Figure 5.

5 NEW AGROCHEMICAL AGENTS AND STRATEGIES

5.1 Switching on Defence Genes with Plant Secondary Metabolites

John Pickett (Rothamsted Research) outlined the use of genetic modification strategies for reducing losses by insect and weed pests, focusing on the manipulation of the biosynthesis of small lipophilic molecules (SLMs) rather than the heterologous expression of toxin genes encoding bioactive proteins. These SLMs may be able to provide effective crop and

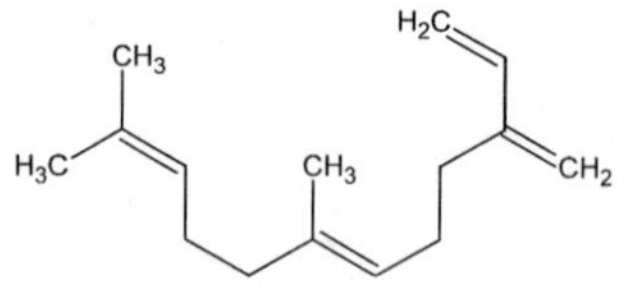

14

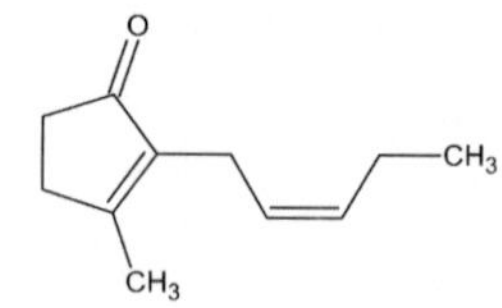

15

Figure 6 *Structures SLMs (*E*)-β-farnesene **14** and* cis*-jasmone **15**.*

livestock protection through intrinsically more benign mechanisms. Natural products have previously provided leads for synthetic pesticides such as the pyrethroids and more recently developed insecticides such as spinosad but these compounds have modes of action relating to single gene products, and the development of resistance by pests is a major issue. As neurotoxins they are also, unfortunately, perceived in some quarters as damaging to the environment and to human health. A new generation of genetically modified organisms is therefore being generated to target natural SLMs that act by non-toxic mechanisms of action, affecting behavioural and developmental processes in pest organisms in sophisticated ways. These strategies were exemplified by studies in which the biosynthesis of SLMs such as the aphid alarm pheromone, (*E*)-β-farnesene **14**,[28] or the semiochemical and plant activator *cis*-jasmone **15**,[29,30] is manipulated to provide crop protection. The structures of these compounds are shown in Figure 6. The "push-pull" or "stimulo-deterrant diversionary" strategy, in which semiochemicals are used to repel insect pests from one crop ("push") and attract them to a second, trap crop ("pull") was also described.[31,32] These companion crops can have added advantages in providing forage for farm animals, helping with water retention and reduced land erosion and, in some cases, providing control of the major plant pest, African witchweed (striga).

5.2 Natural Products as Leads for New and Innovative Crop Protection Chemicals

John Clough outlined the value of natural products to Syngenta as leads for the development of new synthetic herbicides, insecticides and fungicides through a number of case studies. The invention of new crop protection chemicals is challenging: products need to be potent and selective, have good chemical stability but not be too persistent in the environment, possess physico-chemical properties that allowed transport to the target site and, most importantly, be cost effective for treatment of large areas. Bioactive natural products are not generally suitable for use in agriculture in their own right – they usually have insufficient potency, or may be too unstable in light to be used in foliar sprays. Those with suitable chemical structures can, however, form starting points for programmes of synthesis leading to fully synthetic compounds for use in agriculture. The case studies covered included pseudomonic acid A **16**, initially discovered as an antibacterial agent but found to have herbicidal properties as a result of inhibiting protein biosynthesis *via* isoleucyl tRNA synthetase. Semi-synthetic analogues (monic acids) had potent and selective herbicidal activity, but they were not robust enough and there were gaps in the weed spectrum so they were not developed further.[33] In contrast, the β-methoxyacrylate basidiomycete metabolites, strobilurins and oudemansins, have led to highly successful commercial fungicides. These compounds inhibit mitochondrial respiration by binding to cytochrome b. The commercial fungicides such as azoxystrobin have much improved systemic availability and light stability compared to the parent natural products such as

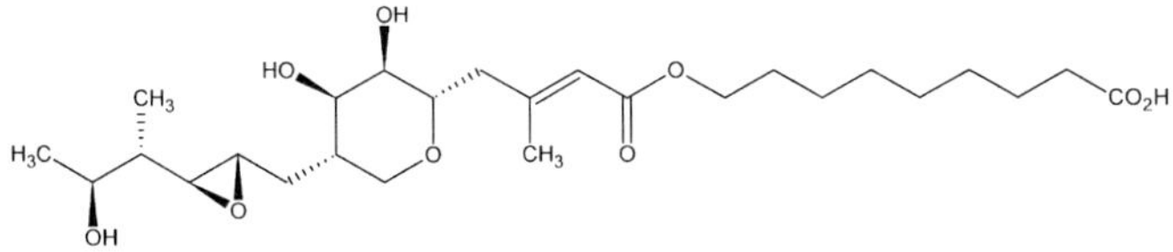

15

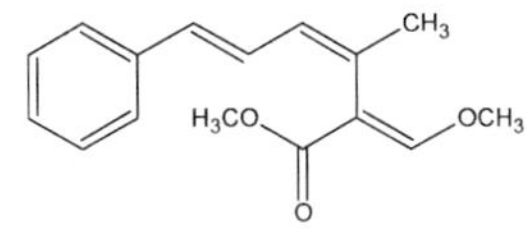

16

Figure 7 *Structures of pseudomonic acid A* ***16*** *and strobilurin A* ***17****.*

strobilurin A **17**.[34] The structures of strobilurin A and pseudomonic acid A are shown in Figure 7.

References

1. R.H. Baltz, V. Miao and S.K. Wrigley, *Nat. Prod. Rep.*, 2005, **22**, 717.
2. J.M. Neary, A. Powell, L. Gordon, C. Milne, F. Flett, B. Wilkinson, C.P. Smith and J. Micklefield, *Microbiology*, 2007, **153**, 768.
3. Z. Hojati, C. Milne, B. Harvey, L. Gordon, M. Borg, F. Flett, B.Wilkinson, P. J. Sidebottom, B. A. M. Rudd, M. A. Hayes, C. P. Smith and J. Micklefield, *Chem. & Biol.*, 2002, **9**, 1175.
4. C. Milne, A. Powell, J. Jim, M. Al Nakeeb, C. P. Smith and J. Micklefield. *J. Am. Chem. Soc.*, 2006, **128**, 11250.
5. C. Mahlert, F.Kopp, J. Thirlway, J. Micklefield, and M. A. Marahiel *J. Am. Chem. Soc* **2007,** *129*, 12011.
6. A. Powell, M. Borg, B. Amir-Heidari, J. M. Neary, J. Thirlway, B. Wilkinson, C.P. Smith and J. Micklefield. *J. Am. Chem. Soc.*, 2007, **129**, 15182.
7. A. Powell, M. Al Nakeeb, B. Wilkinson and J. Micklefield *Chem. Commun.* **2007**, 2683.
8. B. Amir-Heidari, J. Thirlway and J. Micklefield, *Org. Biomol. Chem.*, 2008, **6**, 975.
9. G. C. Uguru, C. Milne, M. Borg, F. Flett, C. P. Smith and J. Micklefield *J. Am. Chem. Soc.* **2004**, *126*, 5032.
10. J. Cortés, A.N. Appleyard and M.J. Dawson, *Meth. Enzymol.*, 2009, **458**, 559.
11. A.N. Appleyard, S. Choi, D.M. Read, A. Lightfoot, S. Boakes, A. Hoffmann, I. Chopra, G. Bierbaum, B.A.M. Rudd, M.J. Dawson and J. Cortés, *Chem. & Biol.*, 2009, **16**, 490.
12. G. Bringmann, T.A.M. Gulder, M. Reichert and T. Gulder, *Chirality*, 2008, **20**, 628.
13. G. Bringmann, J. Mutanyetta-Comar, K. Maksimenka, J. M. Wanjohi, M. Heydenreich, R.Brun, W.E.G. Müller, M.G. Peter, J.O. Midiwo and A. Yenesew, *Chem. Eur. J.*, 2008, **14**, 1420.
14. G. Bringmann, B. Abegaz, M. Knauer and J. Mutanyatta, *Nat. Prod. Rep.*, 2008, **25**, 696.
15. G. Bringmann, I. Kajahn, M. Reichert, S.E.H. Pedersen, J.H. Faber, T. Gulder, R. Brun, S.B. Christensen, A. Ponte-Sucre, H. Moll, G. Heubl and V. Mudogo, 2006, **71**, 9348.
16. G. Bringmann, T.F. Noll, T.A.M. Gulder, M. Grüne, M. Dreyer, C. Wilde, F. Pankewitz, M. Hilker, G.D. Payne, A.L. Jones, M. Goodfellow and H.P. Fiedler H-P, *Nature Chem. Biol.*,2006, **2**, 429.
17. R. Thomas, *ChemBioChem.* 2001, **2**, 612.
18. G. Bringmann, T.F.Noll, T. Gulder, M. Grüne and D. Moskau, *J. Org. Chem.*, 2007, **72**, 3247.

19. S.F. Haydock, T. Mironenko, H.I. Ghoorahoo and P.F. Leadlay, *J. Biotechnol.*, 2004, **113**, 55.
20. Y. Sun, H. Hong, M. Samborskyy, T. Mironenko, P.F. Leadlay and S.F. Haydock, *Microbiology*, 2006, **152**, 3507.
21. B.M. Harvey, T. Mironenko, Y. Sun, H. Hong, Z. Deng, P.F. Leadlay, K.J. Weissmann and S.F. Haydock, *Chem. & Biol.*, 2007, **14**, 703.
22. S.J. Moss, I. Carletti, C. Olano, R.M. Sheridan, M. Ward, V. Math, M. Nur-E-Alam, A.F. Braña, M.Q. Zhang, P.F. Leadlay, C. Méndez, J.A. Salas and B. Wilkinson, *Chem. Commun.*, 2006, 2341.
23. B. Wilkinson, M.A. Gregory, S.J. Moss, I. Carletti, R.M. Sheridan, A. Kaja, M. Ward, C. Olano, C. Méndez, J.A. Salas, P.F. Leadlay, R. vanGinckel and M.Q. Zhang, *Biorg. Med. Chem. Letts.*, 2006, **16**, 5814.
24. L.C. Axford, T.J.Simpson and C.J.Willis, *Angew. Chem. Int. Ed.,* 2004, **43**, 727.
25. A.M. Bailey, R.J. Cox, K. Harley, C.M. Lazarus, T.J. Simpson and E. Skellam, *Chem. Commun.*, 2007, 4053.
26. D.O. Rees, N. Bushby, R.J. Cox, J.R. Harding, T.J.Simpson and C.J. Willis, *ChemBioChem*, 2007, **8**, 46.
27. L.M. Halo, M.N. Heneghan, A.A. Yakasai, Z. Song, K.Williams, A.M. Bailey, R.J. Cox, C.M. Lazarus and T.J. Simpson, *J. Am. Chem. Soc.*, 2008, **130**, 17988.
28. M.H. Beale, M.A. Birkett, T.J. Bruce, K. Chamberlain, L.M. Field, A.K. Huttley, J.L. Martin, R. Parker, A.L. Phillips, J.A. Pickett, I.M. Prosser, P.R. Shewry, L.E. Smart, L.J. Wadhams, C.M. Woodcock and Y. Zhang, *Proc. Natl. Acad, Sci. USA*, 2006, **103**, 10509.
29. M.A. Birkett, C.A.M. Campbell, K. Chamberlain, CE. Guerrieri, A.J. Hick, J.L. Martin, M. Matthes, J.A. Napier, J. Pettersson, J.A. Pickett, G.M. Poppy, E.M. Pow, B.J. Pye, L.E. Smart, G.H. Wadhams, L.J. Wadhams and C.M. Woodcock, *Proc. Natl. Acad, Sci. USA*, 2000, **97**, 9329.
30. T.J.A. Bruce, M.C. Matthes, K. Chamberlain, C.M. Woodcock, A. Mohib, B. Webster, L.E. Smart, M.A. Birkett, J.A. Pickett and J.A. Napier, *Proc. Natl. Acad, Sci. USA*, 2008, **105**, 4555.
31. S.M. Cook, Z.R. Khan and J.A. Pickett, *Ann. Rev. Entemology*, 2007, **52**, 375-400
32. A. Hassanali, H. Herren, Z.R. Khan, J.A. Pickett and C.M. Woodcock, *Phil. Trans. Roy. Soc. London B*, 2008, **363**, 611.
33. P. Bellini, J.M. Clough and G. Hatter, *Pesticide Sci.*, 1999, **55**, 491.
34. D.W. Bartlett, J.M. Clough, J.R. Godwin, A.A. Hall, M. Hamer and B. Parr-Dobrzanski, *Pest Man. Sci.*, 2002, **58**, 649.

Subject Index